POULTRY PRODUCTION IN HOT CLIMATES

UNIVERSITY OF MAINE

RAYMOND H. FOGLER LIBRARY

Poultry Production in Hot Climates

Edited by

N.J. Daghir
Faculty of Agricultural Sciences United Arab Emirates University
Al-Ain, UAE

CAB INTERNATIONAL

CAB INTERNATIONAL   Tel: Wallingford +44(0) 1491 832111
Wallingford   Telex: 847964 (COMAGG G)
Oxon OX10 8DE   E-mail: cabi@cabi.org
UK   Fax: +44(0) 1491 833508

© CAB INTERNATIONAL 1995. All rights reserved. No part of this publication may be reproduced in any form or by any means, electronically, mechanically, by photocopying, recording or otherwise, wihout the prior permission of the copyright owners

A catalogue entry for this book is available from the British Library.

ISBN 0 85198 907 1

Typeset by Colset Private Limited, Singapore
Printed and bound in the UK at the University Press, Cambridge

Contents

Contributors	vii
Preface	xi
1 Present Status and Future of the Poultry Industry in Hot Regions *N.J. Daghir*	1
2 Breeding for Resistance to Heat Stress *R.S. Gowe and R.W. Fairfull*	11
3 Behavioural, Physiological, Neuroendocrine and Molecular Responses to Heat Stress *R.J. Etches, T.M. John and A.M. Verrinder Gibbins*	31
4 Housing for Improved Performance in Hot Climates *R.A. Ernst*	67
5 Nutrient Requirements of Poultry at High Temperatures *N.J. Daghir*	101
6 Feedstuffs Used in Hot Regions *N.J. Daghir*	125
7 Mycotoxins in Poultry Feeds *N.J. Daghir*	157

8 Broiler Feeding and Management in Hot Climates 185
  N.J. Daghir

9 Replacement Pullet and Layer Feeding and Management in
  Hot Climates 219
  N.J. Daghir

10 Breeder and Hatchery Management in Hot Climates 255
   N.J. Daghir and R. Jones

Index 293

Contributors

About the Editor

Dr N.J. Daghir was born and raised in Lebanon where he received his primary and secondary education at the International College in Beirut. In 1954, he lived and worked on poultry farms in the states of Indiana and Arkansas as the first Lebanese participant in the International Farm Youth Exchange Programme sponsored by the Ford Foundation. He received his BSc from the American University of Beirut (AUB) in 1957 and was immediately appointed by AUB to provide agricultural extension services to the central and northern Beqa'a region in Lebanon where he introduced commercial poultry production to that area. In 1958, he left for the US for graduate work where he earned both his MSc and PhD degrees from the Iowa State University in 1959 and 1962 respectively. In 1962, he helped establish a Lebanese branch of the World Poultry Science Association and became president of that branch until 1984. During the same year, he started his teaching and research career at the AUB as Assistant Professor of Poultry Science and Nutrition. In 1967, he was promoted to Associate Professor and in 1975 to full professor. His teaching covered regular undergraduate and graduate courses to students from all over the Middle East area. He has served as adviser for over 30 MSc graduate students, many of whom have later received PhD degrees from US universities and are now occupying key positions all over the world.

Dr Daghir spent two sabbatical years in the USA, one at the University of California, Davis, in 1969 as visiting associate professor and one at the Iowa State University in 1979 as visiting professor. He is a member of several professional and honorary organizations such as the American Institute of Nutrition, the American Poultry Science Association and the

World Poultry Science Association. He has travelled widely in over 60 different countries of Asia, Europe, Africa and America. He has served as consultant to poultry companies in Lebanon, Jordan, Syria, Iraq, Iran, Egypt, Kuwait, Tunisia, Saudi Arabia and Yemen, and participated in lecture tours on poultry production in these countries sponsored by organizations such as ASA, USFGC, WPSA, etc. Dr Daghir has also served on special assignments for the Food and Agriculture Organization of the United Nations (FAO) and Aramco and participated in preparing feasibility and pre-tender studies for poultry projects in several Middle East countries.

Dr Daghir has had over 100 articles published in scientific journals and the proceedings of international meetings as well as several chapters in books and compendia. His research has covered a wide range of subjects, such as factors affecting vitamin requirements of poultry, utilization of agricultural by-products in poultry feeds, nutrient requirements of poultry at high temperature conditions, seeds of desert plants as potential sources of feed and food, single-cell protein for poultry and plant protein supplements of importance to hot regions. His research has received funding from the US National Institutes of Health, the International Development Research Centre and LNCSR as well as from AUB. He has served in many administrative positions at the university, such as Chairman of the Animal Science Department, Associate Dean and Acting Dean of his Faculty and, for two years (1984–1986), served as team leader of the American University of Beirut technical mission to Saudi Arabia. From September 1986 to June 1992, he served as Director of Technical Services at the Shaver Poultry Company in Cambridge, Ontaria, Canada, and nutrition consultant to ISA breeders in North America. Since September 1992, he has been serving as Dean of the Faculty of Agricultural Sciences and Professor of Poultry Science at the United Arab Emirates University, Al-Ain, UAE.

Other Contributors

R.A. Ernst, Extension Poultry Specialist, Department of Avian Science, University of California, Davis, California 95616-8532, USA.

R.J. Etches, Department of Animal and Poultry Science, University of Guelph, Guelph, Ontaria, Canada, N1G 2W1.

R.W. Fairfull, Centre for Food and Animal Research, Research Branch, Agriculture Canada, Ottawa, Ontario, Canada K1A 0C6.

R.S. Gowe, Centre for Genetic Improvement of Livestock, Department of Animal and Poultry Science, University of Guelph, Guelph, Ontario, Canada N1G 2W1.

T.M. John, Department of Zoology, University of Guelph, Guelph, Ontario, Canada N1G 2W1.

R. Jones, Shaver Poultry Breeding Farms, Box 400, 16 Branchton Road, Cambridge, Ontario, Canada N1R 5V9.

A.M. Verrinder Gibbins, Department of Animal and Poultry Science, University of Guelph, Guelph, Ontario, Canada N1G 2W1.

Preface

This book was first envisioned in the early 1970s when the editor was still teaching at the American University of Beirut, Lebanon. It was felt at the time that a reference was needed in the teaching of a senior-level course on Poultry Production in Hot Climates. When the literature was screened at that time it was found that the references on the subject did not exceed 100 and covered very few areas. Today, more than 20 years later, there has been extensive work in the areas of breeding for resistance to heat stress, heat stress physiology, housing for improved performance in hot climates, nutrient requirements at high temperatures, feedstuff composition and nutritional value, and management of broilers, layers and breeders in hot climates, and a book covering information on these various subjects that includes over a thousand references became possible.

Chapter 1 gives an overview of the poultry industry in the warm regions of the world (Africa, Latin America, Middle East, South Asia and Southeast Asia) and covers some of the constraints to future development of the industry in those regions. Chapter 2 covers research on breeding for heat resistance and concludes that it is feasible to select for resistance to heat stress, but the challenge of the breeder is to introduce heat stress tolerance while retaining and improving the economic traits needed in commercial chickens. Chapter 3 discusses several aspects of heat stress physiology and the behavioural, physiological, neuroendocrine and molecular responses to heat stress. It also includes a section on heat-shock proteins. Chapter 4 addresses several issues in housing for improved performance in hot climates. It deals with the principles related to housing design as well as factors affecting poultry house design for hot climates. Poultry house maintenance and monitoring of house performance are also covered. Chapter 5 highlights some of the findings on nutrient requirements of chickens and turkeys at high environmental temperatures and emphasizes the fact that nutritional manipulations can reduce the detrimental effects of high temperatures, but cannot fully correct them, for only part of the impairment in performance is due to poor nutrition. Because of the shortage and high cost of feed ingredients in many hot regions of the world, Chapter 6 was included to focus on the importance of research on the composition and nutritional value of available ingredients that can be used and those that are being used for poultry feeding in hot regions. It is hoped that data presented in this chapter will help countries in these regions to take full advantage of the knowledge available on these feedstuffs. Since mycotoxins in poultry feeds are a serious problem in hot climates where there are prolonged periods of high temperature accompanied at times with high humidity, Chapter 7 has been included to highlight this subject. This chapter focuses mainly on aflatoxins since these are the most widely spread in those regions. Mycotoxins such as citrinin, fumonisins, ochratoxins, oosporein, T-2 toxins, vomitoxin and zearalenone are also covered. Chapters 8 and 9 deal with selected aspects of feeding and management of broilers, replacement pullets and layers in hot climates. The reason that these two chapters present a combination of nutrition and management strategies is because the author believes that this is the most adequate approach to overcome problems of heat stress in poultry meat and egg stocks. Finally, Chapter 10 describes some of the important features of breeder management in hot climates as well as hatchery management and operation problems in such a climate.

This book is aimed at advanced undergraduate and graduate students studying animal production and specifically interested in production problems of hot and tropical countries. It will also serve as a reference for scientists working in the poultry industry in those regions. The book presents not only a review of published research on this subject during the past two decades, but also specific recommendations for poultry producers and farm supervisors.

It was not the intention of the authors to include in this book standard material that is usually present in a poultry production textbook. Only those areas that have actually been investigated in hot climates or studied at high temperatures have been covered.

The reference lists at the end of each chapter are not exhaustive, particularly in the case of certain chapters. However, they have been chosen to provide the reader with a suitable entry to the literature covering research reports on each facet of poultry science research in hot climates.

The authors of this book are indebted to research workers throughout the world who have conducted the research reported in this book. The editor is indebted to Dr Michel Picard for his review and criticism of Chapter 5, to Dr J.L. Sell for his valuable review of Chapter 6, to Dr M.L. Lukic for his review of Chapter 7 and to Dr J.H. Douglas for his review of Chapter 10. The editor would like to express his deepest thanks for the help, assistance and understanding of members of his family all through the preparation of this book.

N.J. Daghir

1

Present Status and Future of the Poultry Industry in Hot Regions

N.J. Daghir
Faculty of Agricultural Sciences, UAE University, Al-Ain, UAE.

Introduction	1
Present Status	2
Africa	3
Latin America	5
Middle East	5
South Asia (India, Pakistan, Bangladesh, Nepal, Sri Lanka and Bhutan)	6
Southeast Asia (Brunei, Cambodia, Hong Kong, Indonesia, Laos, Malaysia, Myanmar, Philippines, Singapore and Thailand)	6
Future Development	7
References	9

Introduction

Domestication of poultry is said to have started in Asia. The earliest record of poultry dates back to about 3200 BC in India. Chickens have been bred in captivity in Egypt since about 1400 BC. Domestication of poultry in China also dates back to about 1400 BC. The red jungle fowl, an Asian breed, is assumed to be the ancestor of our modern poultry breeds. Recently, there is some evidence indicating that the first domestication of the fowl took place much earlier and not in Southeast Asia but in China (Ketelaars and Saxena, 1992).

The warm regions of the world were the area from which all modern breeds of chicken have evolved. Poultry were kept by farmers in China, India and East Asia long before they were known to the Europeans and Americans (van Wulfeten Palthe, 1992). Poultry as a business, however, was not known before the twentieth century. It was not until 1882 that R.T. Maitland wrote his *Manual and Standards Book For the Poultry Amateur* in which he describes the husbandry, care and breeding of poultry with a short description of all poultry strains present at the time.

The World's Poultry Congresses have helped since their inception in 1921 to spread knowledge about poultry production. The International Association of Poultry Instructors and Investigators was founded in 1912. This organization was transformed later to become the World's Poultry Science Association and the first congress was held in The Hague in 1921. Out of 19 congresses held between 1921 and 1992, only three were held in the warm regions of the world (Barcelona, 1924, Madrid, 1970, and Rio de Janeiro, 1978). This may be one of the reasons why knowledge of modern poultry husbandry has not reached the warm regions of the world as rapidly as it has the temperate areas.

Present Status

The poultry industry has in recent years occupied a leading role among agricultural industries in many parts of the world. The potential for further growth is obvious in view of the value of eggs and poultry meat as basic protective foods in the human diet. Table 1.1 shows the world production of eggs and poultry meat and the increase in the production of these products in various continents. Although the increase in poultry meat production during the last decade is similar in all continents, there is a great deal of difference in egg production. Africa, Asia and South America show the greatest increase in egg production while North and Central America, Europe and Oceania show very little increase and in some cases a decrease in egg production. Table 1.2 presents production figures in the developing versus the developed regions of the world. During the past decade, the production of eggs continued to increase rapidly in the developing regions which include most of the hot regions of the world. On the other hand, egg production came practically to a standstill in the developed regions. As for poultry meat production, the increase during the past decade in the developing regions was on average 43%, while in the developed regions it was only 28.4%.

With the exception of a few countries like Saudi Arabia and Israel, the hot regions of the world have probably the greatest potential for further growth since the level of consumption is still very low. Table 1.3 shows the per capita consumption of eggs and broiler meat in selected countries from both the hot and the temperate regions of the world. Per capita egg consumption in countries located in the hot regions is still below 100 eggs and per capita poultry meat consumption is on average lower than in countries from the temperate regions.

Most governments in the hot regions are aware of the relative ease and

Table 1.1. World production of eggs and poultry meat in million metric tons. (From FAO, 1990.)

Continent	Eggs			Poultry meat		
	1980	1990	% Change	1980	1990	% Change
Africa	0.92	1.42	+35.2	1.17	1.79	+34.6
North and Central America	5.42	5.79	+6.4	7.99	12.83	+37.7
South America	1.56	2.31	+32.5	2.36	3.85	+38.7
Asia	7.58	14.27	+46.9	5.34	9.39	+43.1
Europe	7.22	7.17	−0.7	7.09	8.24	+14.0
Oceania	0.26	0.25	−3.2	0.34	0.48	+29.2
Former USSR	3.76	4.54	+17.2	2.14	3.28	+34.8
World	26.74	35.76	+25.2	26.44	39.86	+33.7

rapidity with which the industry can be developed in those areas. They are also aware of the contribution that the poultry industry can make towards improving the quality of human diets in their countries. There is therefore a very rapid expansion of the industry in many countries in these regions. This is very evident, for example, in Brazil in South America, Morocco and Nigeria in Africa and Saudi Arabia in the Middle East.

Poultry meat and eggs are among the highest-quality human foods. They can serve as important sources of animal protein in those areas of the world that have protein insufficiency. Most countries in the hot regions of the world have daily per capita animal protein consumption below that recommended by the Food and Agriculture Organization (FAO) and the World Health Organization (WHO). In the Middle East and North Africa, the average daily per capita intake of animal protein does not exceed 12.2 g as compared with 47 g in most of the developed countries (FAO, 1990).

In the following sections, a brief description of the known and reported status of the poultry industry is presented for Africa, Latin America, Middle East, South Asia and Southeast Asia.

Africa

The poultry industry is highly developed in South Africa and has seen a great deal of development in other African countries during the past two decades. Eggs and poultry meat are beginning to make a substantial

Table 1.2. Production of eggs and poultry meat in the developed vs. the developing world in million metric tons. (From FAO, 1990.)

Region	Eggs			Poultry meat		
	1980	1990	% Change	1980	1990	% Change
North America	4.45	4.34	−2.5	7.24	11.55	+37.3
Europe	7.22	7.17	−0.7	7.09	8.24	+14.0
Oceania	0.26	0.25	−3.2	0.34	0.48	+29.2
Former USSR	3.76	4.54	+17.2	2.14	3.28	+34.8
Others	2.25	2.69	+16.4	1.50	2.04	+26.5
Total developed	17.94	18.99	+5.6	18.31	25.59	+28.4
Africa	0.92	1.42	+35.2	1.17	1.79	+34.6
Latin America	2.53	3.76	+32.7	3.11	5.13	+39.4
Near East	0.76	1.28	+40.6	0.87	1.65	+47.3
Far East	4.85	10.70	+54.7	3.36	6.38	+47.3
Others	0.01	0.01	—	0.01	0.01	—
Total developing	9.07	17.17	+47.2	8.52	14.96	+43.0

contribution in relieving the protein insufficiency in many African countries. Adegbola (1988) reported that only 44 eggs were produced in the African continent per person per year.

Per capita consumption of broiler meat in South Africa increased from 12.0 kg to 15.5 kg between 1985 and 1990, while per capita egg consumption increased from 51.4 to 57.8 during the same period (Viljoen, 1991).

There is very little statistical information on the industry from other African countries, with few exceptions. In spite of the expansion that Nigeria has seen in its poultry meat industry, per capita consumption is still below 2.0 kg (Ikpi and Akinwumi, 1981). The Moroccan industry had undergone tremendous growth since the early 1970s and per capita consumption today is estimated at 50 eggs and 7.6 kg of poultry meat (Benabdeljalil, 1983). The top egg-producing countries of Africa (Nigeria, South Africa, Egypt and Algeria) have increased their production from 4.2 billion (million million) in 1960 to 16.5 billion in 1990 (Saxena, 1992).

Table 1.3. Per capita consumption of eggs (number) and broiler meat (kg) in selected countries during 1990. (From FAO, 1990.)

Country	Eggs	Broiler meat
Argentina	91.9	9.3
Brazil	93.4	12.7
Egypt	25.1	1.2
Hong Kong	67.9	24.8
South Africa	89.0	15.5
Average	73.5	12.7
Austria	212.5	8.2
Canada	162.1	22.5
France	256.3	10.5
Hungary	278.3	18.6
Former USSR	270.6	7.2
United States	208.7	32.0
United Kingdom	171.5	15.6
Average	222.9	16.4

Latin America

Commercial poultry development has been occurring in a number of Latin American countries, particularly Brazil, Chile, Colombia, Mexico and Venezuela. Per capita consumption of broiler meat in Brazil increased from 8.7 kg in 1985 to 12.7 kg in 1990, while egg consumption went up from 84.3 to 93.4 eggs during the same period (USDA, 1990). Broiler meat production showed very significant increases between 1985 and 1990 in both Brazil and Mexico, going up from 490,000 tons to 700,000 tons in Mexico and from 1,490,000 tons to 2,250,000 tons in Brazil.

Middle East

During the past decade, the fastest growth rate in poultry development has been achieved by the Middle East and North Africa. Massive investments were made in the development of environmentally controlled poultry houses equipped with evaporative cooling systems. The total chicken population of 21 Arab countries in the region rose from 234 million in 1980 to 467 million by 1988 (FAO, 1990). The highest increases were in Saudi Arabia (69 million), Iraq (76 million) and Jordan (60 million). These three countries constituted 43% of the total poultry population out of a total of 21 Arab countries. The value of eggs produced in Saudi Arabia rose from $58 million in 1980 to $315 million in 1989. Poultry meat value rose from $130 million to $652 million in 1989 (Anonymous, 1990).

It is noteworthy that this tremendous development of the poultry industry in the Arab Middle East all started in Lebanon in the late 1950s and early 1960s. In 1954, income from poultry amounted for less than 15% of the total income from animal production in that country whereas in 1967 it rose to 60% of the total (Taylor and Daghir, 1968). This growth had a significant effect on animal protein consumption rising from 14.6 to 28.3 g per person per day (McLaren and Pellett, 1970).

Potential for expansion in this area is still very great. The annual average per capita consumption of poultry meat and eggs in North Africa and Middle East is 4.4 kg and 49 eggs respectively, which is below world averages of 5.5 kg and 118 eggs. The potential for increased needs for poultry products in those countries is very obvious. This increased consumption is going to be hastened by population growth, increased income from oil and increased rural-to-urban migration.

South Asia (India, Pakistan, Bangladesh, Nepal, Sri Lanka, and Bhutan)

Although India ranks eighth in the world in egg production, per capita consumption is still very low in both eggs and poultry meat. Poultry meat consumption is estimated at only 150 g per person per year and egg consumption at about 20 eggs per person per year (Reddy, 1991). Saxena (1992) estimated egg production at about 20 billion in India and commercial broiler production at 250 million for the year 1990.

According to Panda (1988) the process of planned poultry development on scientific lines was initiated between the early 1960s and mid-1970s in South Asia. Despite sizeable growth of the poultry industry in those countries, per capita consumption varies from 5 to 35 eggs and from 0.25 kg to 1.25 kg of poultry meat.

Pakistan, with a population of 102 million people, has also gone through a sizeable growth in the production of poultry meat and eggs. Per capita availability went up from 23 eggs in 1977 to 47 eggs in 1988 and poultry meat availability increased from 0.64 kg to 1.48 kg during the same period.

Southeast Asia (Brunei, Cambodia, Hong Kong, Indonesia, Laos, Malaysia, Myanmar, Philippines, Singapore and Thailand)

Poultry-raising in Southeast Asia is characterized by traditional and small-scale systems of farming and operated predominantly by small and low-income farmers. The bulk of the national supplies of poultry meat and eggs in this region is derived from the small and subsistence-type farms rather than from large commercial operations. In Indonesia, for example, 90% of the poultry meat consumed comes from native scavenger flocks. In the Philippines, 80% of the estimated 46 million fowls are located in backyard operations and only 20% on commercial farms. Countries in this region have therefore a tremendous potential for increasing poultry production through the development of small and medium-scale commercial operations. Success of these operations depends on the adoption of modern practices of poultry production in an integrated approach. This development is going to be closely linked to availability of feed sources that do not compete with human food. Table 1.4, extracted from data presented by Leong and Jalaludin (1982), shows the total poultry numbers in those countries and the duck population, since duck farming constitutes a very significant part of the poultry sector in those countries. In Indonesia, for example, the duck population is about 13.5% of the total poultry population. Arboleda (1988) reported that Southeast Asia has the highest number of ducks per person of any other region in the world. Out of a total world duck population of 169 million, Southeast Asia has 77 million. Asia as a whole has 104 million ducks, Africa 16 million, Europe 27 million, North and Central America 13 million and South America 8 million.

Future Development

Based on the poultry industry's development during the past two decades and the need for increased animal protein sources in the hot regions of the world, there is general agreement that these areas are going to witness further expansion in the current decade. Richardson (1988) estimated that world production of eggs will reach 51 million tons by the year 2000, the greatest increase being in the developing countries. As for poultry meat,

Table 1.4. Total poultry and duck population in Southeast Asia. (Adapted from Leong and Jalaludin, 1982.)

Country	Total poultry (millions)	Eggs produced (millions)	Ducks (millions)	Duck eggs (millions)
Thailand	200	2500	13	2000
Indonesia	148	n/a	20	n/a
Malaysia	n/a	2000	n/a	n/a
Philippines	46	n/a	4	n/a
Singapore	15	n/a	2	n/a

he estimated that world production will reach 47 million tons, with demand in the developing countries reaching 16 million tons. Most researchers agree that, in view of the present rate of development, production might well exceed some of the projections for the year 2000. There is no doubt that this increase in the availability of eggs and poultry meat will contribute significantly to the improvement of the nutritional status of the people in the developing countries.

Although the need for more eggs and poultry meat is obvious and the availability of these products can go a long way to meet the protein needs of several populations in hot regions, there are several constraints to the future development of the poultry industry.

The first and foremost is the availability of capital. With the exception of the oil-rich countries, these regions are in general poor with low per capita incomes. If their governments are not able to provide loans and/or subsidies, they will not be able to go into commercial poultry production activities.

Another constraint in those areas is the availability of adequate supplies of grain and protein supplements necessary for the formulation of poultry feeds. Production of feed grains and oil-seeds to support a feed industry is a prerequisite for further growth in the poultry industry. It is true that many countries in those regions have plentiful supplies of agricultural and industrial by-products that can be used in poultry feed formulations, but, before these can be used at relatively high levels, they need to be evaluated both chemically and biologically to determine proper levels of inclusion in poultry feeds.

A third constraint on future poultry industry development in those areas is the need to develop the various supporting industries necessary for commercial poultry production. Production of poultry equipment, pharmaceuticals, packaging materials, housing materials, etc. is practically non-existent and needs to be developed alongside the development of commercial poultry production.

The lack of poultry-skilled people for middle management positions in those areas is a real hindrance to further growth in the industry. Training programmes in poultry management for secondary school graduates can go a long way in supplying these badly needed skills for the various integrated poultry operations.

Disease diagnosis and control are of primary importance in the development and continued growth of any animal production enterprise. It is of utmost importance for modern poultry enterprises because of the intensive nature of these enterprises. There is need in those areas for the establishment of poultry disease diagnostic laboratories as well as for the training of veterinarians in poultry pathology and poultry disease diagnosis. It is also important that these veterinarians receive some training in poultry management since disease control in commercial poultry production needs to be looked at as part of a total management package.

Finally, the most obvious constraint on poultry production in those regions is the climate. High temperature, especially when coupled with high humidity, imposes severe stress on birds and leads to reduced performance. Fortunately, during the past two decades, there has been a great deal of development in housing and housing practices for hot climates. Poultry equipment companies have come up with various devices that contribute to lowering house temperatures and reducing heat stress. These innovations in housing have probably been one of the most significant developments in poultry production practices for hot climates in recent years.

References

Adegbola, A.A. (1988) The structure and problems of the poultry industry in Africa. *Proceedings 18th World's Poultry Congress*, pp. 31–38.

Anonymous (1990) Poultry farming. *Arab World Agribusiness* 6, 22.

Arboleda, C.R. (1988) The structure and problems of the poultry industry in South East Asia. *Proceedings of the 18th World's Poultry Congress*, pp. 45–50.

Benabdeljalil, K. (1983) Poultry production in Morocco. *World's Poultry Science Journal* 39, 52–60.

FAO (1990) *Production Yearbook*, Vol. 44. Food and Agriculture Organization of the United Nations, Rome, Italy.

Ikpi, A. and Akinwumi, J. (1981) The future of the poultry industry in Nigeria. *World's Poultry Science Journal* 37, 39–43.

Ketelaars, E.H. and Saxena, H.C. (1992) *Management of Poultry Production in the Tropics*. Benekon, Holland, 11 pp.

Leong, E. and Jalaludin, S. (1982) The poultry industries of South East Asia – the need for an integrated farming system for small poultry producers. *World's Poultry Science Journal* 38, 213–219.

McLaren, D.S. and Pellett, P.L. (1970) Nutrition in the Middle East. *World Review of Nutrition and Dietetics* 12, 43–127.

Panda, B. (1988) The structure and problems of the poultry industry in South Asia. *Proceedings of the 18th World's Poultry Congress*, pp. 39–44.

Reddy, C.V. (1991) Poultry production in developing versus developed countries. *World Poultry* 7(1), 8–11.

Richardson, D.I.S. (1988) Trends and prospects for the world's poultry industry to year 2000. *Proceedings of the 18th World's Poultry Congress*, pp. 251–256.

Saxena, H.C. (1992) Evaluation of poultry development projects in India, Middle East and Africa. *Proceedings of the 19th World's Poultry Congress*, Vol. 2, pp. 647–651.

Taylor, D.C. and N.J. Daghir (1968) *Determining least-cost poultry rations in Lebanon*. Faculty of Agricultural Sciences, Publ. 32, American University of Beirut, Lebanon, 11 pp.

USDA (1990) World poultry situation. *Foreign Agricultural Service Circular Series*, FL and P 1–90, January, pp. 1–30.
van Wulfeten Palthe, A.W. (1992) *C.S. Th. van Gink's Poultry Paintings*. Dutch branch of the World's Poultry Science Association, Beekgergen, The Netherlands, pp. 10–13.
Viljoen, W.C.J. (1991) The poultry industry in the Republic of South Africa. *World's Poultry Science Journal* 47, 250–255.

Breeding for Resistance to Heat Stress

R.S. Gowe[1] and R.W. Fairfull[2]

[1]Centre for Genetic Improvement of Livestock, Department of Animal and Poultry Science, University of Guelph, Guelph, Ontario, Canada N1G 2W1; [2]Centre for Food and Animal Research, Research Branch, Agriculture Canada, Ottawa, Ontario, Canada K1A OC6.

Introduction	11
Population Differences in Resistance to Heat Stress	13
Major Genes that Affect Heat Tolerance	14
Naked neck (*Na*) – an incompletely dominant autosomal gene	14
Frizzle (*F*) – an incompletely dominant autosomal gene	16
Dwarf (*dw*) – a sex-linked recessive gene	16
Slow feathering (*K*) – a sex-linked dominant gene	17
Other genes	17
Interactions among major genes	18
Use of major genes in developing heat-resistant strains	18
Experiments on Selection for Heat Tolerance	19
Use of quantitative genes in developing heat-resistant strains	21
Feasibility of Developing Commercial Poultry Stocks with Heat Resistance	22
Genotype × location (tropical vs. temperate) interaction	22
Selecting under a controlled or a tropical environment	23
Summary	23
References	25

Introduction

Most of the major international poultry breeders are located in temperate countries (Canada, France, Germany, the Netherlands, the UK and the USA), although the southern part of the USA has a warmer climate than some of the other temperate countries. Nevertheless, much of the world's poultry production takes place under more extreme temperature conditions than even the southern USA. Often humidity conditions are very extreme in addition to the prolonged periods when the temperature is over 30°C.

The question then arises whether commercial stocks developed in more moderate climates are optimal for the high heat and humidity conditions of a very large segment of the poultry-producing areas in the world. Stocks developed in temperate climates are now being sold and used throughout the world, but are they the best genetic material for all conditions? Would it be possible to select strains that are resistant to heat stress and that have all the other economic characteristics and are, therefore, more profitable under these conditions?

This chapter will examine the evidence available on selecting for heat resistance. However, before looking at this subject in detail, it is important to point out that selection for resistance to heat stress alone will not lead to a profitable commercial bird. In laying stocks, all the other traits, such as egg production, sexual maturity, egg size, eggshell quality, interior egg quality, disease resistance, fertility, hatchability, body size and feed efficiency, must be improved or maintained. For meat birds, traits such as growth rate, meat yield, conformation and leanness must also be considered. Otherwise, the stocks will not be economically competitive with the widely sold commercial stocks because these traits are under continuous improvement by the major breeders. Exactly how resistance to heat stress, if genetically based and of sufficient economic importance, should be introduced to a comprehensive breeding programme will not be dealt with in this chapter. The subject of multi-trait selection is a very complex one, which has been partially dealt with in a recent book (Crawford, 1990).

For a more general discussion of the value of the new methods for estimating breeding values and some of the advantages and disadvantages, the reader is referred to the papers by Hill and Meyer (1988) and Hartmann (1992). The recent papers by Fairfull *et al.* (1991) and McAllister *et al.* (1990) outline one procedure to incorporate best linear unbiased prediction (BLUP) animal model estimates of breeding values for seven key egg stock traits. These authors use an economic function to weight the seven different traits. That still leaves several traits that these authors suggest are best incorporated into the selection programme by using separate independent culling levels. The use of independent culling levels for some traits and an index of some kind for others has been further elaborated by Gowe (1993) and Gowe *et al.* (1993). Even with such complex plans that use both indices and independent culling levels, selection in meat stocks occurs at more than one age. Individual selection at broiler age for growth, conformation, feed efficiency, viability and other traits, such as the level of abdominal fatness, is usually practised before selection based on adult characteristics is employed.

If the heat tolerance trait is inherited in a classical quantitative way, with many genes with small effects influencing the trait, then it may be necessary to incorporate the trait into several grandparent lines since most commercial stocks are two-, three- or four-way crosses to make use of heterosis. As a quantitative trait, it would be easier and perhaps more effective to test for heat tolerance at a young age, especially in broilers, while costs per bird are low and population size is higher and before adult characters are assessed.

Although this chapter is not designed to answer the many complex problems of how to incorporate each trait into a breeding programme, it is important to emphasize that demonstrating a genetic basis for a useful economic trait is only the beginning. It is very important to evaluate the cost of introducing each trait to an already complex programme. After evaluating the information available on the genetic basis for heat stress, the final section of this chapter will comment on the development of heat-resistant stocks for hot climates.

Population Differences in Resistance to Heat Stress

There is a large body of literature on breed and strain differences in resistance to heat stress dating back to the 1930s. Some of the results are based on field observations and some on critical laboratory experiments. Many of these experiments have been reviewed before (Hutt, 1949; Horst, 1985; Washburn, 1985) and will not be reviewed in detail here. In general, the early literature showed that there were significant breed and strain within breed differences. The White Leghorn (WL) has been shown to have a greater tolerance for high temperatures than heavier breeds such as Rhode Island Reds (RIR), Barred Plymouth Rocks (BPR), White Plymouth Rocks (WPR) and also Australorps. However, this generalization only held true where there was water available during the stress period (Fox, 1951). WL strains also differed significantly in their ability to withstand heat stress (Clark and Amin, 1965). The heritability of survival of WL under heat stress was high, varying from 0.30 to 0.45 depending on the method of calculation used (Wilson *et al.*, 1966).

To test whether a native breed (a desert Bedouin bird of the Sinai), developed under hot desert conditions, could tolerate very high temperatures better than WL, an experiment was run under controlled conditions by Arad *et al.* (1981). Although the numbers of birds used were not large, the Sinai bird withstood the higher temperatures used (up to 44°C) better than the WL. However, because of its much higher initial production level, the WL was still laying at a higher rate at the highest temperatures than the acclimatized Sinai bird. A series of later papers (Arad and Marder, 1982a, b, c) presented further evidence that the Sinai fowl is more tolerant of high temperatures than the Leghorn, and that crosses of the Sinai and Leghorn came close to the Leghorn in egg production under severe heat conditions, but they were much inferior in egg size. They suggested that the Bedouin (Sinai) bird could be used in breeding programmes to develop strains tolerant of extreme heat.

Others have tested indigenous strains of birds under general tropical conditions and have compared them with crosses utilizing indigenous and improved stocks (Horst, 1988, 1989; Zarate *et al.*, 1988; Abd-El-Gawad *et al.*, 1992; Mukherjee, 1992; Nwosu, 1992). Although the indigenous breeds performed better under higher levels of management than under village conditions, they still do not perform competitively under commercial conditions. Most workers expressed the opinion that the acclimatization traits of indigenous breeds can be best made use of in synthetic or composite strains and in cross-breeding programmes, since most indigenous stocks lack the productive capacity of the highly improved commercial stocks. However, as Horst (1989) has pointed out, a major contribution of native fowl may be in their contribution of major genes that have an effect on adaptability, which is to be discussed next.

Major Genes that Affect Heat Tolerance

There are several genes that affect heat tolerance. Some, such as the dominant gene for naked neck (*Na*), affect the trait directly by reducing feather cover, while others, such as the sex-linked recessive gene for dwarfism (*dw*), reduce body size and thereby reduce metabolic heat output. Each of the genes important in conferring heat tolerance will be discussed separately, and other potential genes will be mentioned briefly as a group.

*Naked neck (*Na*) - an incompletely dominant autosomal gene*

Although the naked neck gene (*Na*) is incompletely dominant, the heterozygote (*Na na*) can be identified by a tuft of feathers on the ventral side of the neck (Scott and Crawford, 1977). The homozygote (*Na Na*) reduces feather covering by about 40% and the heterozygote reduces feathers by about 30%. As described by Somes (1988), the *Na* gene reduces all feather tracts and some are absent. Feather follicles are absent from the head and neck except around the comb, the anterior spinal tract and two small patches on each side above the crop. The extensive literature describing the characteristics of both broiler and egg-type birds carrying the *Na* gene, when compared with normals (*na na*), has been summarized by Merat (1986, 1990), and the reader is referred to these excellent reviews for more details. In summary, meat-type birds carrying the *Na* gene and grown in a warm environment (usually over 30°C) have larger body-weight, better feed efficiency, a lower percentage of feathers, slightly more fleshing, higher viability (when temperatures are very high) and sometimes a lower rate of cannibalism. The homozygote is slightly superior in most tests to the heterozygote for body-weight and feed efficiency.

In egg-type birds tested at higher temperatures, the *Na* gene improves heat tolerance as indicated by higher egg production, better feed efficiency, earlier sexual maturity, larger eggs with possibly fewer cracks, and lower mortality when compared with *na na* birds with similar genetic backgrounds. The *Na* gene may have more positive effects in medium-weight layers than in light birds (Merat, 1990).

Recently, Cahaner *et al.* (1992, 1993) reported on experiments to determine the effect of the *Na* gene in a high-performing broiler sire line and crosses of this line. The *Na* gene was introduced through six back-cross generations to a commercial sire line so the stock carried the *Na* gene in a homozygous state, but most of the genome was made up of the genes of the high-performance commercial stock. It was then crossed to another commercial sire line (*na na*) to produce *Na na* stock. Several crosses were made to produce comparable *Na na* and *na na* progeny. These progeny were tested under Israel spring and summer conditions where the temperature ranged between 31 and 22°C. All five flocks tested showed that the 7-week broiler weight of the *Na na* birds was higher than that of the *na na* birds. Samples that were tested under a controlled, constant high temperature (32°C) showed the same results. These studies are important as they show that the wide range of benefits of the *Na* gene demonstrated previously for various relatively slow-growing stocks (see Merat, 1986, 1990) are applicable to stocks that grow in the range of the modern commercial broiler. The three genotypes, *Na Na*, *Na na* and *na na*, were also compared under a controlled, constant high-temperature experiment (32°C) and a controlled lower temperature of 23°C (after 3 weeks of age) to broiler market age. Since the birds grew to market size slower at the higher temperature, they were slaughtered at 8 weeks while those raised at 23°C were slaughtered at 6 weeks of age. The *Na na* and *Na Na* birds grew faster with better feed efficiency (although not significant at 23°C), lower feather production, higher breast meat yield and lower skin percentage at both temperatures, demonstrating the advantage of the *Na* gene even at the lower temperatures. Previous reports suggested there was either no advantage or a small disadvantage to *Na* at lower temperatures. The authors attribute this to the fact that their results were obtained with birds that grow close to the rate of modern commercial broiler stocks, while earlier studies used slower-growing birds.

The studies by Washburn and colleagues (Washburn *et al.*, 1992; 1993a, b) also showed that the effect of the *Na* gene in reducing heat stress as indicated by growth is greatest in more rapid-growing stocks than in slow-growing stocks. In addition, they demonstrated that small-body-weight birds had a higher basal body temperature at 32°C and a smaller change in body temperature when exposed to acute heat (40.5°C) for 45 minutes than larger-body-weight birds.

The many advantages of the *Na* gene are associated with an increase in embryonic mortality (lower hatchability), particularly in its homozygous state (see Merat, 1986, 1990; Deeb *et al.*, 1993). In any commercial use of this gene, this loss will have to be balanced against the positive effects of the gene under hot conditions. Selection for hatchability in *Na* stocks may reduce this disadvantage.

Frizzle (F) – incompletely dominant autosomal gene

The frizzle (*F*) gene causes the contour feathers to curve outward away from the body. In homozygotes, the curving is extreme and the barbs are extremely curled so that no feather has a flat vane. Heterozygotes have less extreme effects (Somes, 1988).

There is less information available on the effect of the *F* gene on heat tolerance than the *Na*, and most of it is recent. Nevertheless, there is some evidence now to indicate that this gene may be useful in stocks that have to perform under hot humid conditions. This gene will reduce the insulating properties of the feather cover (reduces feather weight) and make it easier for the bird to radiate heat from the body.

Haaren-Kiso *et al.* (1988) introduced the *F* gene into a dual-purpose brown egg layer sire line by repeated back-crossing for several generations. Heterozygous males (*Ff*) of this sire line were crossed with a high-yielding female line (*ff*). The *Ff* and *ff* progeny were compared under two temperatures, 18–20°C or 32°C, and egg production data obtained. In the hot (32°C) environment, the birds carrying the *F* gene laid 24 more eggs over a 364 d laying period, while in the cooler environment the *F* gene birds laid only three less eggs on average. They also reported that the *F* gene favourably affected egg weight, feed efficiency and viability under the 32°C environment, but gave no supporting data. Later, Haaren-Kiso *et al.* (1992) reported that the *F* gene (as a heterozygote) caused a 40% reduction of feather weight at slaughter and an increase in comb weight.

Deeb *et al.* (1993) reported that the *F* gene (as *Ff*) reduced the feather weight of broilers in addition to the reduction caused by the *Na* gene.

Dwarf (dw) – a sex-linked recessive gene

The main effect of the dwarf gene (*dw*) is to reduce the body-weight of the homozygous males by about 43% and that of homozygous females by 26 to 32%. There are many other associated physiological and biochemical effects of the gene. The reader is referred to the reviews by Merat (1990) and Somes (1990a) for details.

Some reports show an advantage to the small-body-weight dwarf (*dw w*) hen at high temperatures over comparable normal (*Dw w*) hens. There is less depression of egg size and egg production. Other reports show no advantage (Merat, 1990). Industrially, the *dw* gene has been used with some success (particularly in Europe) in the female parent of the commercial broiler since there are substantial savings in feed and housing for this smaller broiler mother. The slight loss in growth of the heterozygote male broiler (*Dw dw*) that results from matings of *Dw Dw* male × *dw w* female may be offset by the lower cost of producing broiler hatching eggs. This seems to hold true in situations where feed costs for parents are high or the broilers are slaughtered at a small size and the cost of the hatching egg is relatively more important to the overall costs.

Horst (1988, 1989), however, in a test carried out in Malaysia, reported that Dahlem Red breeding birds (an egg-type synthetic cross) with the *dw* gene were substantially poorer in egg production and egg size than the normals (*Dw*) of the same stock. There is a body of evidence to show that under heat stress conditions the optimum body size of egg-type birds is found in the smaller breeds or strains and that within strain the optimum size is intermediate (Horst, 1982, 1985).

Slow feathering (K) - a sex-linked dominant gene

The slow feathering (*K*) gene has been widely used to 'auto-sex' strain and breed crosses. At hatching, the primary and secondary feathers of the recessive birds (*kw* or *kk*) project well beyond the wing coverts while those of slow-feathered chicks (*Kk* or *Kw*) do not. There are two other alleles in the series, both dominant to the wild-type or *k* gene, that are not used commercially (Somes, 1990b). Horst (1988) also credited the *K* gene with the indirect effects: (i) reduced protein requirement; (ii) reduced fat deposit during juvenile life; and (iii) increased heat loss during early growth, all of which may assist the bird in resisting heat stress. Merat (1990) reviewed other effects of this gene.

Other genes

It has been suggested by Horst (1988, 1989) that several other genes may be useful in making fowl tolerant of tropical conditions.

The recessive gene for silky (*h*), which affects the barbules on the feathers, may improve the ability to dissipate heat. The dominant gene for peacomb (*P*) reduced feather tract width, reduced comb size and changed skin structure. These may improve the ability to dissipate heat. The recessive sex-linked multiple allelic locus for dermal melanin (*id*) may improve radiation from the skin. However, to date no serious investigations were reported on the use of these genes to develop heat tolerance in commercial-type birds.

Interactions among major genes

There was evidence that the *Na* and *F* genes can interact to improve the performance of egg stocks under heat stress (Horst, 1988, 1989). In a brief report, Mathur and Horst (1992) claimed that the three genes *Na*, *F* and *dw* interact so that the combined effects of one or two genes are lower than the sum of their individual and additive effects, but still higher than the individual gene effects. The cross of the Dahlem Red naked neck strain with the Dahlem White frizzle strain (both developed at Berlin University) has competed successfully in the Singapore random sample test (Mukherjee, 1992), suggesting the interaction of the genes *Na* and *F* had a positive effect on performance.

In broiler stocks, there is one report that 6-week-old broilers with the *Ff Na na* genotype had fewer feathers than the *ff Na na* comparable stocks, which had fewer feathers than the *ff na na* stock. The effects which are not fully additive indicate that it might be advantageous to introduce both genes (*F* and *Na*) into a broiler sire line (Deeb *et al.*, 1993). Birds from strains selected for slow feathering (*S*) within the *K* genotype and carrying the *Na* gene had lower feather cover than *S* birds not carrying *Na*, indicating that both *Na* and selecting for slow feathering within the *K* genotype were improving performance in a warm environment (Lou *et al.*, 1992).

Selecting for quantitative genes for slower feather growth in a broiler line breeding true for the *K* gene (Edriss *et al.*, 1988; Ajang *et al.*, 1993) reduced feathers at 48 d, reduced carcass fat, increased carcass protein and carcass meat and increased growth and feed efficiency. Although these results were obtained at a moderate test temperature of 20°C, selection for slow feather growth in birds with the *K* gene should increase their heat tolerance by enhancing their ability to dissipate heat.

Use of major genes in developing heat-resistant strains

There is little doubt now that the advantages of using the naked neck gene for birds to be grown under high temperature (30°C+) conditions outweigh the disadvantage of the slightly lowered hatchability associated with this gene (Merat, 1986, 1990). Besides the advantage of fewer feathers on the neck and other parts of the body, there are other positive effects of this gene for broiler stocks such as increased carcass weight and meat yield, higher body-weights, lower fat content and better feed efficiency (Merat, 1986; Cahaner *et al.*, 1993). Also, the positive effects of this gene have recently been shown to be present when the birds were grown under more temperate conditions (i.e. 23 ± 2°C from 3 weeks to slaughter), even though the benefits of the gene are greater at higher ambient temperatures (Cahaner *et al.*, 1993). The positive effects of the *Na* gene were clearly demonstrated for high growth rate in modern-type broilers when raised as commercial broilers under the spring and summer conditions of Israel, where temperatures ranged between 31°C in the day and 22°C at night (Cahaner *et al.*, 1992). The *Na* gene in this study was introduced by back-crossing for six generations to a commercial sire line.

Although there is less evidence that the *Na* gene, incorporated into high-producing egg stocks, will improve performance under high temperatures, the report of Horst (1988) on the Dahlem Red stock with this gene shows it to be superior in egg production, egg weight and body size to the comparable stock not carrying the gene. Production levels of this test were comparable to commercial performance in this region.

The *Na* gene can be successfully incorporated by back-crossing into high-performing meat or egg stocks. One procedure was outlined in detail by Horst (1989). If the gene can be obtained in a stock already improved, fewer back-cross generations will be required. If it is from a relatively unimproved stock, from five to eight generations would be needed.

Although there is less conclusive evidence for the use of the frizzle gene, the limited evidence available suggests that breeders could seriously consider using this gene along with the *Na* gene to develop stocks specifically for the hot humid tropics (Horst, 1988; Haaren-Kiso *et al.*, 1992; Mathur and Horst, 1992; Deeb *et al.*, 1993). Also, there appear to be advantages to the use of the double heterozygote (*Na na Ff*) for stocks to be reared in the hot humid tropics. Both *Na* and *F* could be back-crossed into sire lines at the same time. It would take larger populations and a few more generations if the two genes came from different sources.

Whether the sex-linked recessive dwarf gene (*dw*) will be useful in female parent broiler stocks for the tropics, beyond its well-known characteristics of reduced body size and bird space requirements and improved feed efficiency, depends on the economics of broiler production of the region. Perhaps it may be even more useful in combination with the *Na* gene, the *F* gene or both since the *Na* gene tends to increase egg size of stocks carrying that gene (Merat, 1986, 1990). Critical experiments are required before recommending the use of this gene in combination with *Na* and *F*.

Experiments on Selection for Heat Tolerance

The early research on the genetics of heat tolerance was mainly concerned with preventing losses when the chickens were exposed to high temperatures, often accompanied by high humidity during a heat wave in temperate climates (Hutt, 1949).

In an early experiment, Wilson *et al.* (1966, 1975) placed 4-week-old Leghorn chicks in a chamber at 41°C with a relative humidity of 75%. Survival time of families during a 2-hour exposure was recorded and used to select parents for breeders to form two lines, one with a long survival period and another with the shortest survival period. Although the heritability of survival time of the 4-week-old chick was quite high (about 0.4) and the two lines diverged significantly, there was no evidence presented that the line that survived longer in this short-term test was able to perform better under high ambient temperatures normally found where chickens are grown. Even if this procedure did result in chickens able to perform better under high ambient temperatures, the high levels of mortality that resulted from this kind of test would be unacceptable now in most countries of the world.

Despite the legitimate welfare objections that might be raised when chickens are exposed to temperature levels that result in over 50% mortality, a seven-generation selection experiment, to develop heat tolerance in a Leghorn strain, was recently reported by Yamada and Tanaka (1992). They exposed adult Leghorn hens to 37°C and 60% relative humidity for 10 days. Survival rate varied from 16% for generation zero to 69% for generation seven, a very positive genetic increase in tolerance to the 10-day heat treatment. Each generation, the survivors were reproduced after a recovery phase of 60 days. The survivors of generation seven recovered faster as indicated by the reduced days to the start of lay after the 10-day period at 37°C. Although the authors claimed superior egg production for the thermotolerant strain, all the survivors apparently regained their normal egg production and egg quality when returned to a 21°C environment, as was shown to be the case when laying birds were exposed to 32°C and returned to 21°C (de Andrade, 1976). When the thermotolerant strain birds of Yamada and Tanaka (1992) were exposed to 38°C for 38 days, the authors claimed these birds were also superior physiologically, but the exact mechanisms were not made clear.

Selection under 38°C for heat tolerance (HR) and susceptibility to heat (HS) in the low growth rate Athens-Canadian random-bred (AC) and in a faster-growing broiler (BR) strain has been briefly reported by El-Gendy and Washburn (1989, 1992). In the AC population, after five generations of selection in both directions for 6-week body-weight in the 38°C environment, body-weight of the HR line was significantly heavier (39%), while the HS line was significantly smaller (46%) than the genetic controls. The realized heritability of 6-week body-weight in the HR line was 0.4 compared with 0.9 in the HS line. It is important to note the authors reported that mortality over the 6-week heat test was not increased in either line. The HR line had lower body temperatures than the HS line under 38°C. After only two generations of selection for 4-week body-weight and heat tolerance (HR) or heat susceptibility (HS) in the fast-growing broiler lines (BR), the HR line was 7% heavier and the HS line 6% smaller than controls. The realized heritability of body-weight for the HR line was 0.4 and 0.3 in the HS line. Both the selected lines had lower mortality than the control population. Although the detailed data were not yet available, this study showed that it is possible to successfully select for growth rate in broiler-type stocks under heat stress (38°C).

Bohren et al. (1981) selected Leghorn lines for fast and slow growth in a hot (32°C) environment and a normal environment (21°C) and then compared the lines under both environments. They also tested the lines under a higher heat stress (40°C) and found there was no significant difference in survival between lines selected in the two different environments. They postulated that the same growth genes were being selected for under both environments. This may be true for short-term selection for growth genes in Leghorns, but it may not be true for selection over many generations in large broiler-type birds (Washburn et al., 1992).

Selection for genetically lean broilers reduces fatness and also gives these leaner birds the ability to perform better under higher temperatures than birds selected for fatness (MacLeod and Hocking, 1993). Another two populations of meat birds successfully selected in two directions for leanness or fatness were tested under hot conditions (32°C) and a moderate environment (22°C). The lean birds grew to a greater weight than the fat birds at 32°C, also demonstrating a greater resistance to hot conditions (Geraert et al., 1993). Since selection for feed efficiency reduces body fat (Cahaner and Leenstra, 1992), the physiological mechanisms are probably the same. Selecting for feed efficiency and/or leanness would give broiler stock the additional advantage of greater heat tolerance.

Use of quantitative genes in developing heat-resistant strains

The research of Yamada and Tanaka (1992) demonstrates that it is feasible to successfully select for heat tolerance in a White Leghorn strain. Similarly, Washburn and colleagues (El-Gendy and Washburn, 1989, 1992) have shown that it is possible to select for heat tolerance in broiler stocks by selecting for body-weight under heat stress. These studies have shown there is a high heritability for selection for body-weight or survival under heat stress. Evidence is not yet available whether selection for better performance under heat stress can be effectively combined with selection for all the other necessary economic traits of egg and meat stocks.

Recent research discussed previously shows that selection for feed efficiency and/or leanness conveys an increased ability to withstand heat stress. Since these traits are independently valuable, it would seem important to emphasize these traits for stocks to be used in hot climates, more than might be indicated by their economic value in temperate climates.

Feasibility of Developing Commercial Poultry Stocks with Heat Resistance

Genotype × location (tropical vs. temperate) interaction

There were no scientific reports that these authors are aware of on the comparison of commercial stocks that were developed by the major poultry breeders, and tested in both temperate and tropical environments to evaluate the magnitude of the interaction. Similarly, there were no published studies on the comparison of different commercial genotypes tested under different controlled temperature conditions. However, some comparisons of strains developed by major breeding companies with indigenous stocks made in tropical environments have shown that the indigenous stocks generally cannot compete, even in the climate to which they are acclimatized. There is a need for much more definitive research in this area (see Arad et al., 1981; Arad and Marder, 1982b; Horst, 1988; Mukherjee, 1992).

A comparison of sire progeny groups of a brown egg cross that were tested in both a temperate climate (Berlin) and a tropical climate (Kuala Lumpur) showed that there were highly significant sire × location interactions for sexual maturity, egg production, egg weight and feed consumption (Mukherjee et al., 1980). There was a negative correlation (-0.39) between the breeding value for body-weight and egg production in Kuala Lumpur and a small positive correlation (0.10) for these traits at Berlin. This led Mukherjee et al. (1980) to suggest that smaller-bodied birds might improve egg output in the tropics, presumably because of their ability to withstand heat better than larger birds, and supports the thesis of Horst (1985) that optimal body-weight is critical for heat tolerance of egg stocks.

In a later experiment, Mathur and Horst (1988) tested pedigreed laying birds in three environments, a controlled warm temperature at Berlin (32°C), a partially controlled temperate climate (20° ± 2°C) at Berlin and a natural tropical environment in Malaysia (28° ± 6°C). The genetic correlations of breeding values of sires whose progeny were tested in all three locations was low for egg production, egg mass, laying intensity and persistency, but relatively high for egg weight and body-weight. These authors concluded that if the aim is to improve productivity in the tropics that selection should take place under tropical conditions; however, there was no direct evidence to support this conclusion.

After testing half-sib groups of broilers to 8 weeks of age in two locations that differed in temperature and other conditions (Federal German Republic and Spain), Hartman (1990) found the genetic correlation of progeny performance for body-weight was 0.68 and 0.56 for conformation and grade. He concluded that the correlation was high enough to justify selection at one location for use in both locations since simple errors in recording would account to some degree for the correlation not being 1.0; therefore, most of the selection would be directed at the same families in both locations.

Selecting under a controlled environment or a tropical environment

Most of the studies on the genetics of heat tolerance have made use of controlled temperature chambers, with temperatures maintained constantly at about 32°C (for example, see Cahaner *et al.*, 1993; Eberhart and Washburn, 1993a, b). Although high temperatures of 40°C or higher have been used, particularly in early experiments (Wilson *et al.*, 1966), they are not generally used now because of the high levels of mortality. Nevertheless, recently Yamada and Tanaka (1992) used a controlled temperature of 37°C for 10 days to successfully select for heat tolerance in adult hens, but their losses were very high.

There is no doubt that selection for heat tolerance under controlled heat conditions is feasible. The heritability of heat tolerance is quite high and rapid progress can be expected (Wilson *et al.*, 1966; El-Gendy and Washburn, 1992; Yamada and Tanaka, 1992). However, there is no evidence that strains selected for multiple performance traits as well as for heat tolerance in a constant temperature chamber do better under the variable conditions found in most tropical environments than the strains that were selected under variable temperate conditions. Experiments are needed to clarify whether the heat tolerance selected for under short-term controlled temperature conditons is the kind of heat stress tolerance that is needed in tropical and subtropical countries.

If the physiological mechanisms for resisting heat stress were better understood, it might be possible to more directly select for the trait required. It is possible as the physiological role of the heat-shock proteins are better understood that breeders will be able to select directly for the specific protein(s) that enables the bird to withstand the high temperatures (see Chapter 3 for a discussion of heat-shock proteins).

Summary

There is no doubt that it is feasible to select for resistance to heat stress, although all the practical problems have not yet been resolved. Nevertheless, the evidence is rapidly accumulating that those breeders, whether in tropical or temperate countries, that add heat tolerance or heat resistance to the many traits already needed for successful commercial egg or meat stock breeding will have an advantage in the tropical and semitropical regions.

It is particularly important for meat stock breeders to add heat tolerance to the breeding programme, since there is an antagonism between rapid growth, large-bodied birds and heat resistance. The rapid-growing bird needs a high feed consumption rate and, in turn, this generates metabolic heat which these large rapidly growing birds find difficult to dissipate rapidly enough following meals. This leads to reduced feed intake, to reduced growth and, if the temperatures are high enough, to poorer feed conversion ratios (Washburn and Eberhart, 1988; Cahaner and Leenstra, 1992). In egg stocks, it is also essential for high-production hens to maintain feed consumption under heat stress or egg production will fall drastically. The large-bodied female parent of the broiler must also be able to dissipate heat.

The breeder has many approaches possible. Selection can be directed at the quantitative genes responsible for heat resistance by challenging under controlled conditions or natural conditions. The controlled temperature approach was demonstrated as feasible for egg stocks (Yamada and Tanaka, 1992) and for meat stocks (El-Gendy and Washburn, 1989, 1992). It is not yet clear from the research to date whether most progress would be made by selecting under controlled high-temperature conditions, or by selecting under the variable ambient temperatures of a tropical or semitropical climate. Mathur and Horst (1988) thought it would be best to select under the prevailing climatic conditions where the birds are to be used. Also, if a controlled temperature environment is used for testing birds, studies are needed to determine the optimal temperature, humidity and length of exposure.

Reports indicate that the heritability of the trait is quite high (El-Gendy and Washburn, 1992; Yamada and Tanaka, 1992), and rapid progress has been made in uni-trait selection studies using constant temperature chambers. There is no published evidence of selecting for heat tolerance in a multiple trait selection programme as yet. Although there is no reason why the trait couldn't be added to a selection programme, it would probably reduce selection pressure on other economic traits.

It is now clear that selection for weight gain under variable heat stress (as in Israel) results in stocks that have better adaptability than stocks developed in the Netherlands to conditions that are 'normal' for the Netherlands (from 15 to 47 days of age, the temperature varied from 28 to 20°C) than to lower temperature conditions (i.e. from 15 to 47 days of age, the temperature varied from 22 to 15°C). Stocks developed in the Netherlands did better than the Israel stocks at the above low temperatures (Leenstra and Cahaner, 1991). It is apparent that some general adaptability to the specific climates of the two countries was occurring in the broiler stocks. This research supports the concept of selection under the variable conditions in hot climates for performance in hot climates.

As discussed earlier, there would be a positive effect on heat tolerance for strains selected either for feed efficiency or leanness or both. These traits are valuable in any climate.

Although the quantitative genetics approach discussed above may prove useful in the long run, the introduction of major genes such as *Na* and *F* into high-producing lines (both egg and meat), or perhaps only into the sire or the female parent of the commercial product, would appear to provide a quicker way to introduce heat tolerance. By back-crossing six to eight generations into high-performing sire lines, the benefit of the heat tolerance can be achieved with little or no loss of other essential performance traits (Horst, 1988, 1989; Cahaner *et al.*, 1993).

Native or indigenous chickens produce more than 50% of the poultry meat and eggs of the tropical countries (Horst, 1989; Mukherjee, 1992). Many of these breeds or strains must possess some tolerance to heat stress and adaptability to tropical conditions. These indigenous village chickens could be upgraded by mating them to males from improved strains of indigenous stock, or by crossing them to exotic stock that have higher productivity and possibly one or two major genes for heat tolerance (Horst, 1989). A small increase in the performance of these scavenging birds will bring large returns to many people. Some native stocks might be used in composites or synthetics that include high-performance strains. The adaptability genes of the indigenous stock would be selected for along with the performance genes of the commercial stocks in the composite, if selection is carried out under tropical conditions. The addition of major genes such as *Na* or *F* into these composites (if not already there) might help ensure there is greater heat tolerance as well as general adaptability. Such new strains developed from synthetic strains might then be useful in crosses to be used under tropical conditions. However, this would be a long-range project with no guarantee that a useful strain would emerge.

The challenge for the poultry breeder is to introduce heat stress tolerance while retaining and improving the wide array of other economic traits needed in commercial chickens.

References

Abd-El-Gawad, E.M., Khalifah, M. and Merat, P. (1992) Egg production of a dwarf (*dw*) F1 cross between an experimental line and local lines in Egypt, especially in small scale production. *Proceedings 19th World's Poultry Congress*, Amsterdam, the Netherlands, Vol. 2, pp. 48-52.

Ajang, O.A., Prijono, S. and Smith, W.K. (1993) Effect of dietary protein content on growth and body composition of fast and slow feathering broiler chickens. *British Poultry Science* 34, 73-91.

Arad, Z. and Marder, J. (1982a) Differences in egg shell quality among the Sinai

Bedouin fowl, the commercial White Leghorn and their crossbreds. *British Poultry Science* 23, 107–112.

Arad, Z. and Marder, J. (1982b) Comparison of the productive performance of the Sinai Bedouin fowl, the White Leghorn and their crossbreds: study under laboratory conditions. *British Poultry Science* 23, 329–332.

Arad, Z. and Marder, J. (1982c) Comparison of the productive performances of the Sinai Bedouin fowl, the White Leghorn and their crossbreds: study under natural desert condition. *British Poultry Science* 23, 333–338.

Arad, Z., Marder, J. and Soller, M. (1981) Effect of gradual acclimatization to temperatures up to 44°C on productive performance of the desert Bedouin fowl, the commercial White Leghorn and the two reciprocal crossbreds. *British Poultry Science* 22, 511–520.

Bohren, B., Carsen, J.R. and Rogler, J.C. (1981) Response to selection in two temperatures for fast and slow growth to nine weeks of age. *Genetics* 97, 443–456.

Cahaner, A. and Leenstra, F. (1992) Effects of high temperature on growth and efficiency of male and female broilers from lines selected from high weight grain, favourable feed conversion and high or low fat content. *Poultry Science* 71, 1237–1250.

Cahaner, A., Deeb, N. and Gutman, M. (1992) Improving broiler growth at high temperatures by the naked neck gene. *Proceedings 19th World's Poultry Congress*, Amsterdam, Vol. 2, pp. 57–60.

Cahaner, A., Deeb, N. and Gutman, M. (1993) Effects of the plumage reducing Naked Neck (*Na*) gene on the performance of fast growing broilers at normal and high ambient temperatures. *Poultry Science* 72, 767–775.

Clark, C.E. and Amin, M. (1965) The adaptability of chickens to various temperatures. *Poultry Science* 44, 1003–1009.

Crawford, R.D. (ed.) (1990) *Poultry Breeding and Genetics*. Elsevier, Amsterdam.

de Andrade, A.N. (1976) Influence of constant elevated temperature and diet on egg production and shell quality. *Poultry Science* 55, 685–693.

Deeb, N., Yunis, R. and Cahaner, A. (1993) Genetic manipulation of feather coverage and its contribution to heat tolerance of commercial broilers. In: Gavora, J.S. and Boumgartner, J. (eds) *Proceedings of the 10th International Symposium on Current Problems in Avian Genetics*, Nitra, Slovakia, p. 36.

Eberhart, D.E. and Washburn, K.W. (1993a) Variation in body temperature response of naked neck and normally feathered chickens to heat stress. *Poultry Science* 72, 1385–1390.

Eberhart, D.E. and Washburn, K.W. (1993b) Assessing the effects of the naked neck gene on chronic heat stress resistance in two genetic populations. *Poultry Science* 72, 1391–1399.

Edriss, M., Smith, K. and Dun, P. (1988) Divergent selection for feather growth in broiler chickens. *Proceedings 18th World's Poultry Congress*, Nagoya, Japan, pp. 561–563.

El-Gendy, E.A. and Washburn, K.W. (1989) Selection for extreme heat stress in young chickens. 3. Response of the S_1 generation. *Poultry Science* 68 (Suppl. 1), 49 (Abstract).

El-Gendy, E.A. and Washburn, K.W. (1992) Selection for heat tolerance in young chickens. *Proceedings 19th World's Poultry Congress*, Amsterdam, the Netherlands, Vol. 2, p. 65.

Fairfull, R.W., McAllister, A.J. and Gowe, R.S. (1991) A profit function for White Leghorn layer selection. In: Zelenka, D. (ed.) *Proceedings Fortieth Annual National Breeders Roundtable*. St Louis Poultry Breeders of America, Tucker, Georgia, pp. 36–49.

Fox, T.W. (1951) Studies on heat tolerance in the domestic fowl. *Poultry Science* 30, 477–483.

Geraert, P.A., Guillaumin, S. and LeClerq, B. (1993) Are genetically lean broilers more resistant to hot climate? *British Poultry Science* 34, 643–653.

Gowe, R.S. (1993) Egg genetics: conventional approaches: should all economic traits be included in an index? In: Gavora, J.S. and Boumgartner, J. (eds) *Proceedings 10th International Symposium on Current Problems in Avian Genetics*. Nitra, Slovakia, pp. 33–40.

Gowe, R.S., Fairfull, R.W., McMillan, I. and Schmidt, G.S. (1993) A strategy for maintaining high fertility and hatchability in a multiple trait egg stock selection program. *Poultry Science* 72, 1433–1448.

Haaren-Kiso, A.V., Horst, P. and Zarate, A.V. (1988) The effect of the Frizzle gene (F) for the productive adaptability of laying hens under warm and temperate environmental conditions. *Prooceedings 18th World's Poultry Congress*, Nagoya, Japan, pp. 386–388.

Haaren-Kiso, A.V., Horst, P. and Zarate, A.V. (1992) Genetic and economic relevance of the autosomal incompletely dominant Frizzle gene (F). *Proceedings 19th World's Poultry Congress*, Amsterdam, the Netherlands, Vol. 2, p. 66.

Hartman, W. (1990) Implications of genotype–environment interactions in animal breeding: genotype–location interactions in poultry. *World's Poultry Science Journal* 46, 197–210.

Hartmann, W. (1992) Evaluations of the potentials of new scientific developments for commercial poultry breeding. *World's Poultry Science Journal* 48, 16–27.

Hill, W.G. and Meyer, K. (1988) Developments in methods for breeding value and parameter estimation on livestock. In: Land, R.B., Bulfield, G. and Hill, W.G. (eds) *Animal Breeding Opportunities*. British Society of Animal Production Occasional Publication No. 12, Edinburgh.

Horst, P. (1982) General perspectives for poultry breeding on improved productive ability to tropical conditions. *2nd World Congress on Genetics Applied to Livestock Production*, Madrid, Spain, Vol. 8, pp. 887–892.

Horst, P. (1985) Effects of genotype × environment interactions on efficiency of improvement of egg production. In: Hill, W.G., Manson, J.M. and Hewitt, D. (eds) *Poultry Genetics and Breeding*. British Poultry Science Ltd., Longman Group, Harlow, UK, pp. 147–156.

Horst, P. (1988) Native fowl for a reservoir of genomes and major genes with direct and indirect effects on productive adaptability. *Proceedings 18th World's Poultry Congress*, Nagoya, Japan, pp. 99–105.

Horst, P. (1989) Native fowl as a reservoir for genomes and major qenes with direct and indirect effects on the inadaptability and their potential for tropically oriented breeding plans. *Archiv fur Geflugelkunde* 53, 93–101.

Hutt, F. B. (1949) *Genetics of the Fowl*. McGraw-Hill, New York, Toronto and London.

Leenstra, F. and Cahaner, A. (1991) Genotype by environment interactions using fast growing lean or fat broiler chickens, originating from the Netherlands or Israel, raised at normal or low temperatures. *Poultry Science* 70, 2028-2039.

Lou, M.L., Quio, O.K. and Smith, W.K. (1992) Effects of naked neck gene and feather growth rate on broilers in two temperatures. *Proceedings 19th World's Poultry Congress*, Amsterdam, the Netherlands, Vol. 2, p. 62.

McAllister, A.J., Fairfull, R.W. and Gowe, R.S. (1990) A preliminary comparison of selection by multiple trait culling levels and best linear unbiased prediction. In: Hill, W.G., Thompson, R. and Woolliams, J.A. (eds) *Proceedings of the 4th World Congress on Genetics Applied to Livestock Production*, Vol. 16, University of Edinburgh, Edinburgh.

MacLeod, M.G. and Hocking, P.M. (1993) Thermoregulation at high ambient temperature in genetically fat and lean broiler hens fed *ad libitum* or on a controlled feeding regime. *British Poultry Science* 34, 589-596.

Mathur, P.K. and Horst, P. (1988) Efficiency of warm stall tests for selection on tropical productivity in layers. *Proceedings 18th World's Poultry Congress*, Nagoya, Japan, pp. 383-385.

Mathur, P.K. and Horst, P. (1992) Improving the productivity of layers in the tropics through additive and non-additive effects of major genes. *Proceedings 19th World's Poultry Congress*, Amsterdam, the Netherlands, Vol. 2, p. 67.

Merat, P. (1986) Potential usefulness of the *Na* (Naked Neck) gene in poultry production. *World's Poultry Science Journal* 42, 124-142.

Merat, P. (1990) Pleiotropic and associated effects of major genes. In: Crawford, R.D. (ed.) *Poultry Breeding and Genetics*. Elsevier, Amsterdam, pp. 429-467.

Mukherjee, T.K. (1992) Usefulness of indigenous breeds and imported stocks for poultry production in hot climates. *Proceedings 19th World's Poultry Congress*, Amsterdam, the Netherlands, Vol. 2, pp. 31-37.

Mukherjee, T.K., Horst, P., Flock, D.K. and Peterson, J. (1980) Sire × location interactions from progeny tests in different regions. *British Poultry Science* 21, 123-129.

Nwosu, C.C. (1992) Genetics of local chickens and its implications for poultry breeding. *Proceedings 19th World Poultry Congress*, Amsterdam, the Netherlands, Vol. 2, pp. 38-42.

Scott, T. and Crawford, R.D. (1977) Feather number and distribution in the throat tuft of naked neck chicks. *Poultry Science* 56, 686-688.

Somes, R.G. (1988) International registry of poultry genetic stocks. *Bulletin* 476, Storrs Agricultural Experiment Station, University of Connecticut, Storrs.

Somes, R.G., Jr. (1990a) Mutations and major variants of muscles and skeleton in chickens. In: Crawford, R.D. (ed.) *Poultry Breeding and Genetics*. Elsevier, Amsterdam, the Netherlands, pp. 209-237.

Somes, R.G., Jr. (1990b) Mutations and major variants of plumage and skin in chickens. Crawford, R.D. (ed.) *Poultry Breeding and Genetics*. Elsevier, Amsterdam, pp. 169-208.

Washburn, K.W. (1985) Breeding of poultry in hot and cold environments. In: Yousef, M.K. (ed.) *Livestock Physiology*, Vol. 3, *Poultry*. C.R.C. Publications, Boca Raton, FL, pp. 111-122.

Washburn, K.W. and Eberhart, D. (1988) The effect of environmental temperature on fatness and efficiency of feed utilization. *Proceedings 8th World's Poultry Congress*, Nagoya, Japan, pp. 1166–1167.

Washburn, K.W., El-Gendy, E. and Eberhart, D.E. (1992) Influence of body weight and response to a heat stress environment. *Proceedings 19th World's Poultry Congress*, Amsterdam, Vol. 2, pp. 53–56.

Wilson, H.R. Armas, A.E., Ross, J.J., Dominey, R.W. and Wilcox, C.J. (1966) Familial differences of Single Comb White Leghorn Chickens in tolerance to high ambient temperatures. *Poultry Science* 45, 784–788.

Wilson, H.R., Wilson, C.S., Voitle, R.A., Baird, C.S. and Dominey, R.W. (1975) Characteristics of White Leghorn chickens selected for heat tolerance. *Poultry Science* 54, 126–130.

Yamada, M. and Tanaka, M. (1992) Selection and physiological properties of thermotolerant White Leghorn hen. *Proceedings 19th World's Poultry Congress*, Amsterdam, the Netherlands, Vol. 2, pp. 43–47.

Zarate, A.V., Horst, P., Harren-Kiso, A.V. and Rahman, A. (1988) Comparing performance of Egyptian local breeds and high yielding German medium heavy layers under controlled temperature and warm environmental conditions. *Proceedings 18th World's Poultry Congress*, Nagoya, Japan, pp. 389–391.

3

Behavioural, Physiological, Neuroendocrine and Molecular Responses to Heat Stress

R.J. Etches[1], T.M. John[2] and A.M. Verrinder Gibbins[1]
Departments[1] of Animal and Poultry Science and[2] of Zoology,
University of Guelph, Guelph, Ontario, Canada N1G 2W1.

Introduction	31
Heat Stress and the Maintenance of Body Temperature	32
Behavioural Responses to Heat Stress	34
Physiological Responses to Heat Stress	36
Acclimatization to high ambient temperature	36
Consumption of feed and water	37
Sensible heat loss through specialized heat exchange mechanisms	37
Sensible heat loss and feather cover in poultry	38
Changes in respiration rate and blood pH	38
Changes in plasma concentrations of ions	40
Heart rate, cardiac output, blood-pressure and total peripheral resistance	41
Hormonal Involvement in Thermoregulation	42
Neurohypophyseal (posterior pituitary) hormones	42
Arginine vasotocin (AVT)	42
Mesotocin (MT)	43
Growth hormone (GH)	44
The hypothalamic–pituitary adrenal axis	44
Corticosterone	44
Catecholamines	46
Melatonin	47
Reproductive hormones	47
Thyroid hormones	48
Heat-shock Proteins and Heat Stress	49
References	53

Introduction

The net energy stored in the tissues of a bird equals the difference between energy intake and energy loss. Metabolism of food and high environmental temperature are potential sources of energy while low environmental temperatures and the maintenance of normal body temperature are potential expenditures of energy. Excessive flow of energy into the body and excessive depletion of energy from the body both lead to death although many birds can survive conditions in which the potential for energy flux is extreme, by invoking various adaptive mechanisms that increase or decrease the flow of energy to or from the environment. The extreme examples of very cold environmental temperature and very hot environmental temperature both lead to death because the animal cannot cope with the excessive flow of energy out of or into, respectively, their body mass. In many parts of the world, particularly in warm tropical and subtropical regions, poultry are maintained in environmental temperatures which require the involvement of intricate molecular, physiological and behavioural changes that enable domestic birds to cope with the flux of energy into their tissues at high ambient temperatures. This chapter describes the range and complexity of molecular, physiological, neuro-endocrine and behavioural responses that are invoked to maintain body temperature within the normal range at high ambient temperatures.

Heat Stress and the Maintenance of Body Temperature

Body temperature of domestic chickens is maintained within a relatively narrow range that is usually reflected by the upper and lower limits of a circadian rhythm in deep body temperature. In well-fed chickens that are neither dissipating heat to the environment nor gaining heat from the environment, the upper limit of the circadian rhythm is usually about 41.5°C and the lower limit is about 40.5°C. When exposed to a hot environment and/or performing vigorous physical activity, body temperature might rise by 1 or 2°C as heat is stored. Heat storage cannot continue for extended periods before body temperature increases past the limit that is compatible with life. Conversely, when birds are exposed to very cold environments, heat escapes from the bird and, unless it is replenished by energy from the metabolism of food, body temperature will decline until the bird is incapacitated and dies. These general considerations of the effect of environmental temperature have been synthesized into terminology that is commonly used to discuss the response of homoeothermic animals to changes in environmental temperature and they are illustrated in Fig. 3.1. In this diagram, body temperature is approximated as a constant that is maintained over a wide range of environmental temperatures, indicated as the *zone of normothermia*. The lower critical temperature ([a] in Fig. 3.1) is the minimum environmental temperature which, even if maintained over a period of days, is compatible with life. When environmental temperature is less than the lower critical temperature, body temperature

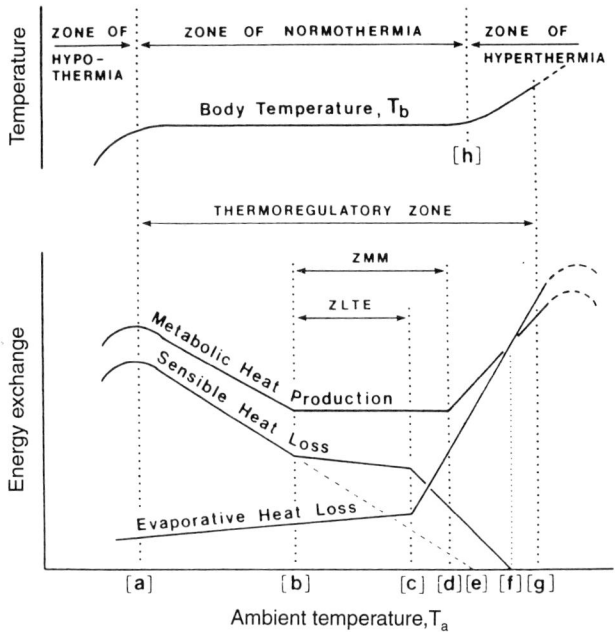

Fig. 3.1. Generalized schematic diagram illustrating T_b (body temperature) and partitioning of energy exchange through a wide range of T_a (ambient temperature). The zones and points are: ZMM, zone of minimum metabolism; ZLTE, zone of least thermoregulatory effort; point [a] lower critical temperature; point [b], critical temperature; point [c], temperature at which intense evaporative heat loss begins; point [d], upper critical temperature; point [e], T_a which equals normothermic T_b; point [f] where sensible heat loss is zero because metabolic heat production equals evaporative heat loss; point [g], critical thermal maximum; point [h], point of incipient hyperthermia. (From Hillman et al., 1985.)

begins to decline and death occurs within the *zone of hypothermia*. At the upper end of the zone of normothermia, body temperature increases (indicated by [h] in Fig. 3.1) in the *zone of hyperthermia* until the critical thermal maximum ([g] in Fig. 3.1) is reached, above which the bird expires. Between the lower critical temperature and the critical thermal maximum, thermoregulatory processes are initiated to cope with the ambient temperature. Within the *zone of least thermoregulatory effort* (ZLTE in Fig. 3.1), metabolic heat production is at a minimum, sensible heat loss is relatively constant because physiological and behavioural responses limit the escape of heat throughout this range of environmental temperatures, and evaporative heat losses are limited to those occurring as a by-product of normal respiration and exposure of non-insulated areas of the body. The *zone of minimum metabolism* (ZMM in Fig. 3.1) extends throughout the range of the ZLTE and includes higher ambient temperatures (from [c] to [d] in Fig. 3.1) that can be accommodated by increasing both evaporative and sensible heat loss. Metabolic heat production increases as environmental temperature declines below the ZMM and ZLTE ([b] in Fig. 3.1) to provide energy to maintain body temperature, and increases above the ZMM ([d] in Fig. 3.1) to provide energy for panting. Evaporative heat loss is minimal at low ambient temperatures and increases rapidly as soon as thermoregulation is required to alleviate an increase in ambient temperature, i.e. at the transition from the ZLTE to the ZNM (indicated by [c] in Fig. 3.1). Sensible heat transfer (designated as sensible heat loss in Fig. 3.1), which is cumulative heat transfer from the bird by radiation, conduction and convection, is negative when the environment is colder than the bird and positive when the bird is colder than the environment (indicated by [e] in Fig. 3.1). As metabolic heat production increases in the zone of hyperthermia, body temperature rises and therefore sensible heat losses can increase if the ambient temperature is lower than body temperature. The point at which hyperthermic body temperature is equal to the ambient temperature, and above which sensible heat transfer increases body temperature further into the zone of hyperthermia, is indicated by [f] in Fig. 3.1.

As ambient temperature rises and falls, a wide variety of physiological, behavioural, neuroendocrine and molecular responses are initiated to maintain body temperature within the normal limits. In some instances, the responses are short-term measures that are invoked to withstand a brief period of extreme temperature. The responses can also be invoked to develop a longer-term response aimed at acclimatizing the bird to ambient temperatures that fall within the upper regions of the thermoregulatory zone. Finally, the responses can be terminal reactions that can be sustained for only brief periods; these responses are initiated to cope with extreme and life-threatening environmental conditions.

Behavioural Responses to Heat Stress

During thermal stress, birds alter their behaviour to help maintain body temperature within the normal limits. Behavioural adjustments can occur rapidly and at less cost to the bird than most physiological adjustments (Lustick, 1983), although they are preceded by the molecular response to heat stress that is mediated by the heat-shock proteins (see below).

As ambient temperature increases above the comfort zone, chickens devote less time to walking and standing (see Mench, 1985; McFarlane *et al.*, 1989). During exposure to high temperature, chickens consume less feed and more water (May and Lott, 1992) to compensate for water lost through evaporative cooling (see Mench, 1985) although a reduction in drinking time was observed when heat stress was applied concurrently with other stressors (McFarlane et al., 1989). When exposed to high temperatures, domestic fowl may splash water on their combs and wattles in order to increase evaporative cooling from these surfaces (see Whittow, 1986). Heat-stressed birds also spend relatively less time engaging in social behaviour and in changing their posture. When maintained in cages, heat-stressed chickens tend to distance themselves from each other, pant, and often stand with their wings drooped and lifted slightly from the body to maximize sensible heat loss (see Mench, 1985).

In recent years, it has been shown that chickens will select their thermal environment to increase heat gain if the effective environmental temperature induces heat loss or to increase heat loss if the effective environmental temperature induces heat gain. In a natural environment, the hen would move to a shady area or seek a microenvironment that avoided the environmental extreme. In confinement, chickens will choose a preferred environment using operant control mechanisms (Richards, 1976; Morrison and Curtis, 1983; Morrison and McMillan, 1985, 1986; Morrison et al., 1987a,b; Laycock, 1989; Hooper and Richards, 1991). Operant control over the environment is achieved when the bird learns to perform a simple behaviour (usually pecking a switch) to choose a preferred environmental condition. Operant conditioning can be used to determine the optimum effective environmental temperature, which may be a combination of several factors such as temperature, air speed and humidity. In a study on young chicks, Morrison and McMillan (1986) observed that the birds responded rapidly to changing ambient temperatures when given the opportunity to press a microswitch to provide themselves with supplementary heat from a 250 W infrared bulb. As environmental temperature was reduced in increments of 1°C below an ambient temperature of 20°C, the chicks increased the request for supplementary heat by $1.6 \min h^{-1}$. Although it is often suggested that evaporative cooling is unimportant to birds because they have no sweat glands, when the ambient temperature is 40°C mature domestic hens will request a 30 s stream of air at a dry bulb temperature of 22°C using an operant control (Richards, 1976).

Operant control of the environment has been used to evaluate the thermal environment of mature laying hens (Laycock, 1989). Cooler cage temperatures are requested by hens during the night and during periods of low feeding activity, indicating that their thermal requirements are diminished during these periods. A similarity between the diurnal rhythm of operant-controlled supplemental heat and metabolic heat production was also observed in these birds, which might indicate that circadian rhythms of activity, feeding, basal metabolic rate and thermal requirements are interrelated.

Physical factors in the environment can also influence the effective environmental temperature and operant control of supplemental heating sources can be used to evaluate these factors. For example, the influence of flooring was evaluated by Morrison and Curtis (1983) and Morrison and McMillan (1985), who observed that chicks raised on wire flooring requested 4.22 min h^{-1} more supplemental heat than those on floors with shavings. These data indicate that heat loss is greater on wire and consequently chicks experience cooler temperature on wire, although the ambient temperature was identical on both types of flooring. The influence of feather loss has also been investigated using operant control of environmental temperature, and, in the absence of their natural thermal insulation, domestic hens increase their request for supplemental heat (Horowitz *et al.*, 1978). Studies with hens (Horowitz *et al.*, 1978) as well as with pigeons (Necker, 1977) have demonstrated that the greatest demand for supplemental heat was produced when the bird's back was exposed, indicating that this region is the most thermally sensitive part of the bird.

From their studies on domestic fowl, Hooper and Richards (1991) concluded that the relative contribution of operant behaviour to overall temperature regulation was different both qualitatively and quantitatively under heat load and cold load temperatures. Under heat load, the preferential method of thermoregulation included an operant response, whereas under cold exposure the autonomic response (as indicated by increases in oxygen consumption and heart rate) appeared to be the major contributor to thermoregulation.

Physiological Responses to Heat Stress

The physiological responses to heat stress in birds involve the functional integration of several organs to meet the metabolic needs of birds that are trying to dissipate heat and maintain homoeostasis. Responses that can be described in anatomical terms and involve the whole animal are discussed below and integration of the physiological responses by endocrine systems are described in the following section.

Acclimatization to high ambient temperature

Exposure of chickens to high environmental temperatures produces an initial increase in the temperature of peripheral tissues and subsequently in core body temperature (Fronda, 1925; Heywang, 1938; Thornton, 1962; Kamar and Khalifa, 1964; Boone and Hughes, 1971a; Wang *et al.*, 1989). Boone (1968) observed that the body temperature of chickens began increasing when the ambient temperature rose above 30°C if the rate of increase in ambient temperature was rapid. On the other hand, if the ambient temperature rose more slowly (Boone and Hughes, 1971b), the birds maintained their normal body temperature until the ambient temperature reached 33°C. Using the length of time required for body temperature to become constant as a measure of acclimatization, Hillerman and Wilson (1955) reported that adult chickens required 3 to 5 days to acclimatize to both hot and cold environments. Ambient temperatures above 32°C produce transient hyperthermia in turkeys (Wilson and Woodard, 1955) that can last for up to 21 days if the birds are exposed to 38°C (Parker et al., 1972).

Consumption of feed and water

As birds accumulate heat in their tissues, several responses to increase the dissipation of heat are invoked to reduce the heat load. Water consumption increases when chickens are exposed to high ambient temperature (North and Bell, 1990; Deyhim and Teeter, 1991; May and Lott, 1992) and survival in a hot environment is dependent upon the consumption of large volumes of water (Fox, 1951). Voluntary feed consumption is diminished in response to high environmental temperatures (Otten et al., 1989) and fasting for 1–3 days has been shown to progressively increase survival time of chicks exposed to heat stress (McCormick et al., 1979). However, since heavier broilers are more susceptible to heat stress (Reece et al., 1972), it is uncertain whether the effect of depressed feed consumption is due to the reduction in feed consumption or to a reduction in body-weight resulting from low food intake. The increase in water consumption occurs immediately, whereas the reduction in food consumption is delayed until several hours after the birds have experienced high temperatures (May and Lott, 1992). The immediate increase in water consumption meets the immediate demands of evaporative cooling from respiratory surfaces and the associated decline in food consumption reduces the contribution of metabolic heat to the total heat load that requires dispersion.

Sensible heat loss through specialized heat exchange mechanisms

Heat is dispersed through anatomical specializations in birds that provide increased blood flow to surfaces that can effectively transfer heat by radiation and conduction. The vascular system in the legs and feet of many birds, including domestic fowl, contains arteriovenous heat exchange mechanisms that facilitate the dispersal of heat through these uninsulated surfaces. The volume of blood that flows through the arteriovenous network, which serves as a heat exchanger, is regulated by shunts in the vascular system (Midtgard, 1989). At high ambient temperatures, these shunts bring cool venous blood in close proximity to arterial blood to dissipate the maximum amount of heat to the environment. The loss of heat by sensible transfer through feet is adopted quickly by hens given the opportunity to roost on pipes in which cold water is circulating (Otten et al., 1989). While cooled roosts have not been extensively tested in poultry production systems, it would appear that they have considerable potential in the management of hens under heat stress.

In many birds, heat exchange can be increased through the rete opthalmicum, an arteriovenous heat exchanger situated between the optic cavity and the brain (Midtgard, 1989). This anatomical specialization can be used to dissipate heat through the cornea, the eye, the buccal cavity, the beak and the nasal passages. The extent to which domestic birds can disperse heat through this mechanism is not yet established.

Sensible heat loss and feather cover in poultry

Partial feather loss is not uncommon in hens, particularly in those housed in battery cages during the laying cycle. Feather loss occurs mainly from the neck, back and breast regions, and is believed to be related to cage shape, cage size, crowding of hens (Hill and Hunt, 1978) and feather pecking (Hughes and Wood-Gush, 1977; Hughes, 1978). In some breeds, the absence of feathers in the neck is controlled by a single autosomal dominant gene designated as naked neck (*Na*). In addition to eliminating plumage from the neck, the naked neck gene suppresses 30–40% of the plumage in all of the other feather tracts. Regardless of the cause of poor feather covering, sensible heat loss is substantially increased (Richards, 1977). In normal environments, food consumption is increased to offset the increase in heat dissipation (Emmans and Charles, 1977; Gonyou and Morrison, 1983) and, consequently, the feed efficiency of poorly feathered birds is decreased (Leeson and Morrison, 1978). At ambient temperatures above 25°C, however, where the ability to dissipate heat is an asset, naked neck (*Na Na*) chickens possessed superior growth rate, viability, egg weight and female reproductive performance (Merat, 1990).

Changes in respiration rate and blood pH

Panting is one of the visible responses of poultry during exposure to heat. This specialized form of respiration dissipates heat by evaporative cooling at the surfaces of the mouth and respiratory passageways. Hens may begin panting at an ambient temperature of 29°C (North, 1978) after 60 min of exposure to 37°C and 45% relative humidity (Wang et al., 1989) or when their body temperature reaches 42°C (Hillman et al., 1985). Panting enables hens to increase the rate of water evaporation from 5 to $18 \, \text{g} \, \text{h}^{-1}$ in response to a change in ambient temperature from 29 to 35°C with relative humidity of 50–60% (Lee et al., 1945). However, it is believed that, at an ambient temperature of 32°C and relative humidity of 50–60%, hens reach the maximal ability to lose heat through evaporation (Barrot and Pringle, 1941; Wilson, 1948).

Panting increases the loss of carbon dioxide from the lungs, which leads to a reduction in the partial pressure of carbon dioxide (Wang *et al.*, 1989), and thus bicarbonate, in blood plasma. In turn, the lowered concentration of hydrogen ions causes a rise in plasma pH (Mongin, 1968; Richards, 1970), a condition generally referred to as alkalosis. In laying hens, the reduction in the plasma concentrations of bicarbonate compromises egg-shell formation by limiting the availability of the anion required during formation of $CaCO_3$ crystals in the shell (Mongin, 1968).

The occurrence of respiratory alkalosis in response to thermal stress has not been consistently observed in all studies in poultry. For example, acute hyperthermia in female turkeys produced alkalosis (Kohne and Jones, 1975a), chronic hyperthermia in female turkeys had no effect on plasma pH (Kohne and Jones, 1975b) and thermal stress (37.8°C, 7 days) in male turkeys was associated with a reduction in blood pH (Parker and Boone, 1971). In Leghorn hens exposed to increasing heat, Darre *et al.* (1980) noted a curvilinear increase in blood pH, whereas, in broilers reared under continuous heat (35°C), Siegel *et al.* (1974) and Vo and Boone (1975) were unable to detect any significant change. On the other hand, Bottje *et al.* (1983) and Raup and Bottje (1990) reported that blood pH in cockerels and broilers was increased at higher ambient temperature. Such discrepancies are presumed to be due to variations in the degree of thermal stress, the length of thermal stress period and the degree to which the birds had been acclimatized to the conditions (see Teeter *et al.*, 1985). Variation in plasma pH noted during thermal stress may also have been a consequence of the time at which blood was sampled relative to bouts of panting and normal breathing that accompany exposure to chronically high ambient temperatures, since blood pH of panting chicks was elevated while that of non-panting birds was not significantly altered (Teeter *et al.*, 1985).

The physiological mechanisms that are invoked by birds exposed to high temperatures must meet the opposing demands of thermoregulation and respiratory alkalosis. Dissipation of heat by evaporative cooling demands an increase in respiration, while respiratory alkalosis demands a decrease in respiration. Gular flutter and utilization, during panting, of respiratory passages that are not involved in gas exchange (e.g. nasal cavities, nasopharynx, larynx and trachea) are known to reduce respiratory alkalosis in heat-stressed birds (see Hillman *et al.*, 1985). Gular flutter involves the rapid and sometimes resonant vibration of the upper throat passages driven by the hyoid apparatus (Welty and Baptista, 1988). In pigeons, panting is superimposed on slower deeper breathing, a system which may minimize respiratory alkalosis (Ramirez and Bernstein, 1976). The air sacs of birds are also utilized during panting to move air over surfaces that limit the exchange of gas between blood and air while facilitating evaporative heat loss (Whittow, 1986).

Respiratory alkalosis can also be combated nutritionally by providing a source of anion via feed or water. For example, Teeter and Smith (1986) have shown that supplemental ammonium chloride in drinking water of chronically heat-stressed birds can return blood pH to normal and enhance weight gain. During acute heat stress, the provision of ammonium chloride (Branton et al., 1986) or carbonated water (Bottje and Harrison, 1985) has been found to decrease blood pH.

Changes in plasma concentrations of ions

The normal functions of tissues are dependent upon the stability of the total osmolarity of intracellular and extracellular fluids. The major ions of the plasma are sodium, chloride, potassium, calcium, phosphate, sulphate and magnesium. The plasma concentration of each ion normally varies only within a remarkably small range; a substantial shift in their concentration can cause serious disturbance to cells since these ions and the plasma proteins play a major role in establishing the osmotic balance between plasma and fluids bathing the cells. The concentrations of the major ions are also important in determining the pH of the body fluids.

Elevation of body temperature to 44.5–45.0°C by exposing chickens to 41°C ambient temperature has been associated with increased plasma sodium and chloride and decreased plasma potassium and phosphate (Ait-Boulahsen et al., 1989). In normally hydrated fowls, however, heat stress (35–45°C for 10–12 h) produced no significant changes in the serum concentrations of sodium, potassium, chloride and calcium, or in serum osmolality, although serum phosphate declined (Arad et al., 1983). Subjecting domestic fowl to 37°C and approximately 45% relative humidity for a period of 150 min did not significantly alter the plasma osmolality (Wang et al., 1989). The different responses observed in these studies may be attributed to the differences in the extent and duration of heat exposure, and to the fact that the birds used by Arad et al. (1983) were acclimatized to high ambient temperature for a long period prior to the experiment.

The availability of water to heat-stressed hens is essential to support evaporative cooling from the respiratory surfaces. Whereas laying hens maintained within the thermoneutral zone will drink approximately 200 ml of water per day, hens at 40°C will consume approximately 500 ml per day (North and Bell, 1990). Arad et al. (1983) observed that water deprivation for 48 h, which included 24 h without food, of fowls maintained at 25°C was associated with an increase in serum concentrations of sodium and in serum osmolality. The increase in serum concentration of sodium and in serum osmolality were exacerbated and associated with a decline in serum concentrations of phosphate when dehydrated birds were subjected to 35–45 °C for 10–12 h. In pigeons subjected to heat stress and water deprivation, serum concentrations of sodium increased while serum potassium levels declined (John and George, 1977). This response has been attributed to the release of neurohypophyseal hormones in heat-stressed pigeons (John and George, 1977; George, 1980), since it is known that injection of neurohypophyseal hormones into hens results in an elevation of blood sodium level and a decrease of blood potassium level (Rzasa and Neizgoda, 1969).

Heart rate, cardiac output, blood-pressure and total peripheral resistance

Exposure to high ambient temperature is associated with a decline in blood-pressure, an increase in cardiac output and a decrease in peripheral resistance (Weiss et al., 1963; Whittow et al., 1964; Sturkie, 1967; Darre and Harrison, 1987). As birds become acclimatized to elevated ambient temperature, however, cardiac output decreases, blood-pressure increases and peripheral resistance returns to normal (Vogel and Sturkie, 1963; Sturkie, 1967).

Although, in some experiments, the heart rate of chickens exposed to acute heat stress declined (Darre and Harrison, 1981, 1987), an increase in heart rate was apparent in other experiments (Darre and Harrison, 1987; Wang et al., 1989). In pigeons exposed to ambient temperatures between 6 °C and 34 °C, John and George (1992) observed that heart rate was significantly lower at ambient temperatures above 28 °C when compared to temperatures below 28 °C. In part, the variation in heart rate following exposure to heat stress may be the consequence of the trauma of repeated blood sampling, which would increase heart rate in some instances, overriding the inhibitory effect of thermal stress (Darre and Harrison, 1987). In both of the above experiments with chickens (Darre and Harrison, 1987; Wang et al., 1989) in which an increase in heart rate was observed following heat exposure, the birds had been subjected to repeated blood samplings.

The flow of blood from the body core to the periphery plays a significant role in the transfer of heat from deep body tissues to the peripheral tissues that are capable of dissipating heat to the environment (see Darre and Harrison, 1987). In chickens exposed to high ambient temperature, blood flow through the comb, wattles and shanks is increased due to peripheral vasodilation, and excess heat is dissipated to the surrounding air (Whittow et al., 1964; Darre and Harrison, 1981). During acute heat stress, the cardiovascular system distributes blood to functions related to thermoregulation, giving only secondary importance to other functions such as those related to the exchange of respiratory gases and digestion (see Darre and Harrison, 1987). In heat-exposed chickens, for example, Bottje and Harrison (1984) demonstrated that blood flow to the viscera was reduced by 44%.

Based upon their observation of cardiovascular response of chickens to acute mild hyperthermia, Darre and Harrison (1987) proposed that the thermoregulatory response starts with a decreased heart rate and peripheral vasodilation, which leads to decreased blood-pressure, decreased peripheral resistance and compensatory increases in stroke volume and cardiac output. It is suggested that the large increase in cardiac output during heat stress demonstrates the intense demand placed upon the cardiovascular system to dissipate heat from the bird. Darre and Harrison (1987) concluded that the fine-tuning of the body temperature is accomplished primarily by cardiovascular adjustments to prevent major overshoots or undercontrol.

Hormonal Involvement in Thermoregulation

Hormones are produced by endocrine tissues and transported through the circulatory system to their target tissues. They provide an important link in the flow of information among cells and tissues in an animal to initiate and maintain the physiological and behavioural responses to heat stress. As a bird attempts to cope with heat stress, an intricate series of changes that is mediated by many, if not all, hormonal systems is initiated. The relative importance of each of these systems and the extent to which they are called upon depend on the severity of the heat stress. The following paragraphs describe the major physiological adaptations to heat stress that require the participation of endocrine systems.

Neurohypophyseal (posterior pituitary) hormones

Arginine vasotocin (AVT)

The principal neurohypophyseal hormone in birds, AVT, is an antidiuretic hormone in non-mammalian vertebrates (Munsick *et al.*, 1960; Ames *et al.*, 1971). AVT is released in response to dehydration and stimulates the resorption of water by the kidney. However, AVT is believed to play a role in heat dissipation that is independent of its role in osmoregulation in chickens (Robinzon *et al.*, 1988; Wang *et al.*, 1989) and pigeons (John and George, 1992). In non-heat-acclimatized fowl, plasma concentrations of AVT increased after 90 min of exposure to 32°C (Wang, 1988) and after 60 min of exposure to 37°C (Wang *et al.*, 1989), without a significant change in plasma osmolality. Similar observations have been reported by Azahan and Sykes (1980) although, in some cases, plasma AVT levels increased only after 48 h of dehydration (Arad *et al.*, 1985). These observations have led to the suggestion that, in non-heat-acclimatized fowl, heat stress alone could increase AVT levels, while in heat-acclimatized birds, an increase in plasma osmolality could also be necessary (Wang et al., 1989). The thermoregulatory role of AVT is shown further by the observation that injections of AVT decrease shank and comb temperatures in fowl (Robinzon et al., 1988) and induce a drop in cloacal and foot temperature in heat-stressed pigeons (John and George, 1992). In pigeons, but not in chickens, AVT has also been implicated in the control of basal metabolic rate and respiratory rate (Robinzon et al., 1988; John and George, 1992).

The release of AVT in response to heat stress has also been implicated in the mobilization of free fatty acids (FFA). In pigeons subjected to heat stress and dehydration, serum triglyceride levels dropped to less than half those in controls (John and George, 1977), whereas the plasma FFA levels showed a more than twofold increase (John et al., 1975), suggesting that the increase in FFA is due to the breakdown (lipolysis) of blood triglycerides. In normothermic pigeons, intravenous injection of AVT (400 mU per pigeon) brought about a highly significant increase in FFA at 30 min post-injection (John and George, 1973) although injection of AVT into immature female chickens was without effect (Rzasa et al., 1971). In vitro studies of pigeon adipose tissue have revealed that AVT induces FFA release (John and George, 1986). It is well established that fat is the major substrate for sustained muscular activity in birds (George and Berger, 1966) and a ready supply of FFA could meet the increased energy requirement of the respiratory muscles as panting is initiated in heat-stressed birds.

In a recent study with pigeons (John, T.M. and George, J.C., unpublished), increases in AVT were associated with increases in plasma thyroxine (T_4) level and concomitant decreases in triiodothyronine (T_3) level. Basal metabolic rate is determined by plasma T_3 in birds (see 'Thyroid hormones' below), and therefore a drop in T_3 level would reduce metabolic heat production to alleviate heat stress.

Mesotocin (MT)

Mesotocin, the avian analogue of oxytocin, has recently been implicated in thermoregulation in domestic fowl (Robinzon et al., 1988; Wang et al., 1989). Heat stress suppressed the circulating level of MT (Wang et al., 1989), but it is not clear what role MT plays in thermoregulation. Since AVT could suppress MT release (Robinzon et al., 1988), the decrease in MT may be the result of elevated concentrations of AVT. As in amphibians (Stiffler et al., 1984), MT could be a diuretic hormone in birds (Wang et al., 1989). This suggestion is supported by the observation that MT release is stimulated by hypotonic saline infusion (Koike et al., 1986) and is positively correlated with avian renal blood flow (Bottje et al., 1989). It has been suggested that, if MT has a diuretic function in fowl, then the suppression of MT with a concomitant increase in AVT would be a useful mechanism to aid in the conservation of body fluids during heat stress (Wang et al., 1989). Infusion of MT has been shown to produce a dramatic increase in respiratory rate (Robinzon et al., 1988), which could enhance evaporative heat loss during heat stress. It is suggested that the thermoregulatory function of MT is carried out via either the central nervous system or peripheral mechanisms (Robinzon et al., 1988).

Growth hormone (GH)

In pigeons deprived of drinking water and subjected to high ambient temperature for 3 days (28°C, 31°C and 36.5°C, respectively), the plasma levels of growth hormone (GH) increased significantly (John et al., 1975). This increase in GH levels is believed to play a role in fatty acid mobilization since GH is a major lipolytic hormone in birds (John et al., 1973). Although the sequence of events that lead to the release of GH is not clearly understood, the importance of GH in diverting metabolism to provide a high-energy substrate for muscle metabolism is believed to contribute to the support of panting-related muscular activity during heat exposure.

The hypothalamic–pituitary adrenal axis

Corticosterone

Corticosterone is the principal steroid hormone of the avian adrenal cortex (Holmes and Phillips, 1976). Heat stress stimulates the release of corticosterone from the adrenal glands (Edens, 1978) and increases plasma concentrations of corticosterone in chickens (Edens and Siegel, 1975; Ben Nathan et al., 1976), turkeys (El Halawani et al., 1973) and pigeons (Pilo et al., 1985). Heat stress has also produced adrenal enlargement in ducks (Hester et al., 1981) and quail (Bhattacharyya and Ghosh, 1972). Since increased levels of circulating corticosterone have been observed under various stress situations including cold exposure (Etches, 1976; Pilo et al., 1985), the response to heat exposure is considered primarily as a reaction to stress.

The release of corticosterone from the adrenal cortex is mediated by the hypothalamus and the pituitary gland. Both neural and endocrine inputs to the central nervous system stimulate the production of corticotrophin-releasing factor (CRF) from neurons within the median eminence of the hypothalamus. CRF is released into the hypothalamic portal vascular system and transported to the pituitary gland where it stimulates the production of adrenocorticotrophic hormone (ACTH). ACTH is released into the general circulatory system and is transported to its major target tissue, the adrenal cortex. Under ACTH stimulation, the adrenal cortex increases the production and release of all of the adrenocortical hormones, although the major hormones are corticosterone and aldosterone. As the primary mediators of stress responses, ACTH, aldosterone and corticosterone have widespread effects on many target tissues throughout the body. None of these effects are specific to heat stress, but all are initiated as a bird mounts a coordinated physiological response to cope with increasing environmental temperature.

Plasma concentrations of corticosterone increase in response to heat stress but high levels of corticosterone can only be maintained for short periods to cope with acute exposure to the high temperatures. When birds are chronically exposed to high temperatures, plasma concentrations of corticosterone will decline after the initial surge and, unless other physiological and/or behavioural responses can be implemented to alleviate the heat stress, the bird will become hyperthermic and die. For example, in young chickens exposed to high ambient temperature (43°C), plasma corticosterone increased within 30 min, but dropped below pre-exposure levels within 120 min (Edens, 1978). This drop in circulating levels of corticosterone was accompanied by low plasma concentrations of glucose, phosphate and Na^+ and elevated plasma pH. In response to this severe acute thermal stress, therefore, the chickens exhibited acute adrenal cortical insufficiency within 120 min, which, in conjunction with a massive secretion of catecholamines, resulted in cardiovascular failure and death. Prevention of a decline in corticosterone by pretreatment of birds with reserpine, propanolol or dihydroergotamine sustained the adrenal response and reduced mortality during heat stress (Edens and Siegel, 1976). Hydrocortisone and cortisone therapy have also been found to reduce mortality in birds exposed to high temperatures (Burger and Lorenz, 1960; Sammelwitz, 1967), providing further evidence that corticosteroids can protect against the lethal effects of high ambient temperature.

Although the acute effects of aldosterone, corticosterone and ACTH are believed to initiate or support a number of physiological changes that delay incipient hyperthermia, the details of these effects are remarkably sparse. Increased plasma levels of aldosterone act in concert with AVT to promote renal absorbtion of water to prevent dehydration as evaporative cooling is utilized for thermoregulation.

Changes in the plasma concentrations of corticosteroids and ACTH affect the lymphoid tissues and consequently the ability of chickens to mount an immune response. For example, a diminution in the mass of the thymus, spleen and bursa of Fabricius (Garren and Shaffher, 1954, 1956; Glick, 1957, 1967; Siegel and Beane, 1961), and a decrease in the number of circulating lymphocytes and an increase in neutrophilic or heterophilic granulocytes (Dougherty and White, 1944; Gross and Siegel, 1983) have been reported following administration of corticosteroids.

Corticosteroids bind to specific cytoplasmic receptors in lymphatic cells to redirect differentiation and metabolism of this cell lineage (Thompson and Lippman, 1974; Sullivan and Wira, 1979), and both ACTH administration and exposure to high temperature increase corticosteroid binding to cells in the lymphoid system (Gould and Siegel, 1981). The precise effect of corticosteroids, however, on this diverse cell lineage depends on both the severity of persistence of the stress and the role of the affected cells in both humoral and cell-mediated immunity.

Catecholamines

The catecholamines, epinephrine (E) and norepinephrine (NE), are synthesized and released from adrenal chromaffin cells. Their secretion in response to stress is similar to the corticosterone response since both adrenal cortical hormones and ACTH stimulate the release of both E and NE (see Harvey *et al.*, 1986). For example, a substantial increase in the circulating levels of both E and NE were observed over a 140 min period in 8-week-old broilers exposed to 45°C (Edens and Siegel, 1975). Chronic exposure of adult roosters to an ambient temperature of 31°C produced little change in blood concentrations of E and NE (Lin and Sturkie, 1968). The increase in plasma concentrations of E and NE in response to acute heat stress are probably as transitory as that of corticosterone since, in 9-week-old male turkeys, the adrenal concentrations of E and NE were not significantly altered when examined 6 h after exposure to 32°C (El Halawani *et al.*, 1973). Catecholamines may also exert an influence on body temperature more directly than through their role as mediators of a general response to heat stress since infusion of catecholamines into the hypothalamus of both young (Marley and Stephenson, 1970) and adult (Marley and Nistico, 1972) chickens lowers body temperature.

The turnover rate in brain tissue of NE increased and of E decreased upon acute exposure of turkeys to 32°C (El Halawani and Waibel, 1976). However, chronic exposure to the same temperature reduced the increase in NE turnover while not affecting the decrease in E turnover. Since similar observations have been reported for Japanese quail subjected to heat stress, Branganza and Wilson (1978a) suggested that acute heat stress increases central noradrenergic neuronal activity which returns to normal following acclimatization. Branganza and Wilson (1978b) also reported that, while high ambient temperature (34°C) for 6 h increased brain NE level and NE turnover rate, it decreased heart NE levels and enhanced the turnover rate of NE in the heart. Although chronic exposure for 5 weeks to high temperatures did not increase NE turnover in the brain, NE turnover in the heart was increased. Thus, NE appears to play a role in heat stress in at least two organs and the effects in each of them appears to be dissimilar.

Melatonin

Melatonin (*N*-acetyl-5-methoxytryptamine) is an indoleamine that has been recognized as the major pineal hormone (Quay, 1974). In birds, as in many other vertebrates, melatonin is also produced in several extrapineal tissues (Pang *et al.*, 1977; Cremer-Bartels *et al.*, 1980; Ralph, 1981) such as the retina and the Harderian gland. Melatonin has been implicated in thermoregulation in birds (Binkley *et al.*, 1971; John *et al.*, 1978; John and George, 1984, 1991; George and John, 1986) and may regulate the circadian rhythm in body temperature since, in pigeons, body temperature is relatively low in the night when both plasma and pineal levels of the melatonin are high (John *et al.*, 1978). Conversely, body temperature is higher in the day when the melatonin levels are low. In the presence of high concentrations of melatonin, heat dissipation by peripheral tissues is enhanced by vasodilatation and blood flow, particularly to the foot, which is an important site for heat dissipation in birds (Jones and Johansen, 1972). Furthermore, melatonin may act centrally by lowering the set-point of the main 'thermostat', which is believed to be present in the hypothalamus (see John and George, 1991). A more general role for the pineal gland in the regulation of body temperature in chickens was suggested by Cogburn *et al.* (1976), who observed that the return to normal body temperature was delayed in pinealectomized birds.

Reproductive hormones

The effect of heat stress on reproductive performance in chickens has been indicated by reduced egg production. The diminished egg production is suspected to be at least partly influenced by the ovulatory hormones. In the hen, heat stress reduces serum luteinizing hormone (LH) levels, hypothalamic content of luteinizing hormone-releasing hormone (LHRH) (Donoghue *et al.*, 1989) and the preovulatory surges of plasma LH and progesterone (Novero *et al.*, 1991). Since the preovulatory surges of LH and progesterone are controlled in a positive feedback loop (Etches and Cunningham, 1976; Wilson and Sharp, 1976; Williams and Sharp, 1978; Johnson *et al.*, 1985) and since both hormone levels are depressed concomitantly (Novero *et al.*, 1991), it is difficult to identify the site or sites of action of heat stress. However, the hypothalamus could be a primary target for heat stress because it receives both neural and endocrine inputs that could be translated into general inhibition of the reproductive system. For example, corticosterone is known to have a suppressive effect on circulating LH levels in birds (Wilson and Follett, 1975; Deviche *et al.*, 1979; Etches *et al.*, 1984; Petitte and Etches, 1988) and the high concentration of circulating corticosterone during thermal stress (Donoghue *et al.*, 1989) could promote the decline of LHRH. Brain monoamines have also been suspected of being involved in inhibiting hypothalamic function (Donoghue *et al.*, 1989). Since heat stress has been reported to alter brain monoamine levels and turnover rate (El Halawani and Waibel, 1976; Branganza and Wilson, 1978a, b) and since monoamines are considered to be putative hypothalamic regulators of gonadotrophin secretion (El Halawani *et al.*, 1982), monoamine turnover could be involved in bringing about the reduction in hypothalamic LHRH content that is associated with heat stress.

Thyroid hormones

The importance of the thyroid gland in adaptation to heat stress is related to the central role that thyroid hormones play in the regulation of metabolic rate of birds (Bellabarba and Lehoux, 1981, 1985; McNicholas and McNabb, 1987; McNabb, 1987). This effect has been demonstrated by surgical or chemical thyroidectomy of chickens, which produces a decrease in metabolic rate (Winchester, 1939; Mellen and Wentworth, 1962) and body temperature (Nobukumi and Nishiyama, 1975; Davison *et al.*, 1980; Lam and Harvey, 1990), and by thyroid hormone administration, which stimulates heat production (Mellen and Wentworth, 1958; Singh *et al.*, 1968; Arieli and Berman, 1979). In chickens, thyroid hormone secretion is depressed as ambient temperature increases (Reineke and Turner, 1945; Hahn *et al.*, 1966) and heat tolerance improves as thyroid function is reduced. For example, radiothyroidectomy (Bowen *et al.*, 1984) or chemical thyroidectomy by thiouracil (Fox, 1980; Bowen *et al.*, 1984) increased survival time in heat-stressed chickens, while thyroid hormone administration decreased the survival time (Fox, 1980; May, 1982; Bowen *et al.*, 1984).

The two active forms of thyroid hormones are T_4 and T_3, and the inactive form is reverse triiodothyronine ($r\text{-}T_3$). The selective peripheral conversion of T_4 to T_3 or $r\text{-}T_3$ is believed to play an important role in thermoregulation in domestic fowl (Kuhn and Nouwen, 1978; Decuypere *et al.*, 1980; Rudas and Pethes, 1984); when chickens are exposed to warm temperatures, T_4 is inactivated by conversion into $r\text{-}T_3$, whereas during cold exposure T_4 is converted into T_3, which stimulates metabolic rate.

While it is generally accepted that T_3 stimulates metabolic rate and that both T_3 and T_4 are depressed following heat stress, this pattern is not universally observed. In Japanese quail and pigeons, for example, plasma T_3 and T_4 concentrations have been reported to increase, decrease or remain unchanged following heat stress (John and George, 1977; Bobek *et al.*, 1980; Bowen and Washburn, 1985; Pilo *et al.*, 1985). Heat stress experiments conducted on heat-acclimatized or non-acclimatized broilers of different lines also failed to elicit a consistent pattern in thyroid hormone response to heat stress (May *et al.*, 1986), leading to the conclusion that the complex physiological response to heat stress does not consistently affect plasma concentrations of thyroid hormones.

Heat-shock Proteins and Heat Stress

A response of all organisms – animal, plant or microbe – to elevated temperature is the increased synthesis of a group of proteins known as the heat-shock proteins (reviewed by Lindquist and Craig, 1988; Pardue et al., 1989; Morimoto et al., 1990; Craig and Gross, 1991; Nover, 1991). The universal response to heat stress is the most highly conserved genetic system known, and some of the heat-shock proteins are not only the most abundant proteins found in nature but are the most highly conserved proteins that have been analysed. This degree of conservation indicates the fundamental role that these proteins must play in restoring normal function to cells or whole organisms that are exposed to potentially damaging stimuli. As discussed below, heat-shock proteins play an essential role by associating with a variety of proteins and affecting their conformation and location. In a heat-shocked cell, the heat-shock proteins may bind to heat-sensitive proteins and protect them from degradation, or may prevent damaged proteins from immediately precipitating and permanently affecting cell viability. Heat-shock proteins are members of a larger family of stress proteins, some of which may be synthesized because of nutrient deprivation, oxygen starvation or the presence of heavy metals, oxygen radicals or alcohol. These stimuli can also cause the synthesis of some of the heat-shock proteins. An important aspect of these stressful circumstances is that cells or organisms that have recovered from a mild stressful episode and are expressing elevated levels of stress proteins can exhibit tolerance to doses of the stress-causing agent, including heat, that would normally cause developmental abnormalities or death (for example, see Mizzen and Welch, 1988; Welch and Mizzen, 1988).

The most commonly found forms of heat-shock proteins have relative molecular masses of approximately 10,000–30,000, 70,000, 90,000 and 100,000–110,000, and so are referred to as HSP70, HSP90 and so on. Some of these proteins, or their close relatives, apparently fulfil vital functions in normal cells, whereas some are required for growth at the upper end of the normal temperature range, and some are required to withstand the toxic effects of extreme temperatures. Strictly speaking, the term 'heat-shock' protein should apply only to those proteins that are synthesized in a cell in response to a heat shock, but currently it is unclear whether these proteins carry out any specialized function that cannot be performed by their constitutively expressed relatives present under normal conditions. These latter proteins are sometimes named 'heat-shock cognates' (HSCs). For the purposes of this chapter, and as is usually the case, the 'heat-shock' protein designations will be used broadly to refer to any members of these families, whether inducible or constitutive.

In general, heat-shock proteins are involved in the assembly or disassembly of proteins or protein-containing complexes during the life and death of a normal cell (the role of heat-shock proteins in protein folding is reviewed by Gething and Sambrook, 1992). Each type of heat-shock protein may interact with a specific group of molecules – for example, members of the HSP70 family bind to immunoglobulin heavy chains, clathrin baskets and deoxyribonucleic acid (DNA) replication complexes, and may be involved either in maintaining these structures in partially assembled form until they are required or in degrading them after use. All members of the HSP70 family bind adenosine triphosphate (ATP) with high affinity, have adenosine triphosphatase (ATPase) domains (Flaherty *et al.*, 1990) and probably undergo conformational changes using energy released on ATP hydrolysis; the conformation of structures with which the heat-shock proteins are interacting may then also change (Pelham, 1986). HSP90, whose family members are abundant at normal temperatures (up to 1% of the total soluble protein in the cytoplasm), interacts with steroid hormone receptors and apparently masks the DNA-binding region of the receptor until the receptor has bound to the appropriate steroid hormone; the receptor can then bind to DNA and activate expression of specific genes. Another profound influence of HSP90 is its modulation of phosphorylation activity within the cell as a consequence of it binding to cellular kinases. These molecular chaperon characteristics of HSP90 apparently depend on the ability of HSP90 to recognize and bind certain subsets of non-native proteins and influence their folding to the native state, which they can do in the absence of nucleoside triphosphates (Wiech *et al.*, 1992). Members of the HSP70 family also act as molecular chaperons and can conduct certain proteins through intracellular membranes (Chirico *et al.*, 1988: Deshaies *et al.*, 1988); perhaps HSP70 binds to newly synthesized proteins that are just being released from ribosomes and prevents the proteins from aggregating and precipitating before they are properly folded, transported and incorporated into complexes or organelles. Members of the HSP70 family can certainly distinguish folded from unfolded proteins (Flynn *et al.*, 1991). This activity might provide a clue to the role of the large amount of HSP70 that is synthesized following heat shock – HSP70 may bind to cellular proteins that have been denatured by the heat and may prevent their catastrophic precipitation (Finley *et al.*, 1984). On return to normal environmental conditions, sufficient cellular protein might be released in an undamaged state to allow the cell to resume its activity, and protein that has been irreparably damaged can be removed gradually by the ubiquitin salvage pathway. Ubiquitin expression is also induced by heat, and ubiquitin is conjugated through its terminal glycine residue to cellular proteins prior to their selective degradation. A further feature of heat-shock proteins that is yet to be fully explored is the role that they might play in immunity and immunopathology (Young, 1990). Although excessive heat can have many deleterious effects on the structure and physiology of a cell, including impairment of transcription, ribonucleic acid (RNA) processing, translation, post-translational processing, oxidative metabolism, membrane structure and function, cytoskeletal structure and function, etc., it is not clear which of these is the most harmful (reviewed by Rotii-Roti and Laszlo, 1987) or which is alleviated by any particular heat-shock protein.

The response of cells or a whole organism to heat shock is extremely rapid, but transient, and involves the redistribution of preformed heat-shock proteins within the cell, as well as immediate translation of preformed messenger RNA (mRNA) into heat-shock proteins, immediate transcription of genes encoding heat-shock proteins and cessation of transcription or translation of other genes or mRNA (Yost *et al.*, 1990). An example of the redistribution of preformed proteins on heat shock is provided by some members of the HSP70 family that immediately migrate from their normal location in the cytoplasm to the nucleus, where they associate with preribosomes in the nucleoli (Welch and Suhan, 1985). Interestingly, the damage to nucleolar morphology that can be seen in cells that have undergone a heat shock can be repaired more rapidly if the cells are artificially induced to overexpress HSP70 by transformation with an exogenous HSP70-encoding gene (Pelham, 1984). During recovery from heat shock, HSP70 migrates back to the cytoplasm.

Transcription of heat-shock protein-encoding genes is regulated by a heat-shock factor (HSF), which interacts with a conserved DNA sequence, the heat-shock element (HSE), located in the 5' flanking regions of the genes. In vertebrates, binding of significant amounts of HSF to HSE sequences only occurs after a heat shock, and may result from the conversion of small, inactive oligomeric forms of the protein to an active multimer (Westwood *et al.*, 1991). The HSE consists of three repeats of a five-base sequence, arranged in inverted orientation, and multiple copies of the HSE may result in a cooperative increase in levels of transcription (Tanguay, 1988); high-affinity binding of the HSF to this complex may only occur if the HSF is in the multimeric form. The HSF appears to be prelocalized in the nucleus, perhaps to minimize the time for response to a heat shock, and the chromatin surrounding some heat-shock protein-encoding genes is in an 'open' conformation, allowing immediate association with transcription factors following a heat shock. Also, RNA polymerase may in some cases already be interacting with the gene but is blocked from transcribing until the heat shock occurs. In addition to associating with HSE in heat-shock protein-encoding genes, there is evidence that the activated multimeric HSF may bind to other regions of the genome, and may play a role in suppressing transcription of other genes during the period of stress (Westwood *et al.*, 1991).

Although selective transcription of heat-shock protein-encoding genes takes place at elevated temperatures, various aspects of RNA metabolism under these circumstances also result in the preferential synthesis of heat-shock proteins (Yost et al., 1990). Most eukaryotic genes are transcribed to give complex RNA structures that require processing by removal of introns before functional mRNA results. This splicing reaction is inhibited under conditions of heat stress, perhaps to prevent the production of a range of incorrectly spliced, mutant mRNAs. Most heat-shock protein genes do not contain introns, however, and so produce mRNAs that are functional under heat stress. Indeed, there is evidence that heat-shock protein may contribute to the thermotolerance of cells or organisms that have been previously heat-shocked by protecting the splicing machinery and thus enabling the production of functional mRNAs from a range of genes – a primed protection of normal function of the cell. A further aspect of heat-shock protein mRNA metabolism that is of interest is that some classes of these molecules appear to be stable in heat-stressed cells but are highly unstable in unstressed cells, indicating that a significant increase in heat-shock protein synthesis may result from stabilization of pre-existing or newly formed mRNA. Other evidence (Theodorakis et al., 1988), however, indicates that the levels of some classes of heat-shock protein mRNA may not be significantly different between stressed and unstressed cells, but that the elongation rate is much lower during protein synthesis from these templates in unstressed cells than in stressed cells.

Much of the work on the molecular biology of heat-shock protein expression in vertebrates has been done using the chicken as a model system. Several of the chicken heat-shock protein-encoding genes have been cloned and characterized, including *hsp70* (Morimoto et al., 1986), *hsp90* (Catelli et al., 1985), *hsp108* (Kulomaa et al., 1986; Sargan et al., 1986) and *UbI* and *UbII*, which are ubiquitin-encoding genes (Bond and Schlesinger, 1986). The *hsp108* gene was first cloned because of its expression in the chick oviduct and was subsequently shown to have sequence homology with heat-shock proteins. HSP108 is found constitutively in all chicken tissues examined, but its levels are enhanced following heat shock. Surprisingly, the *hsp108* gene is steroid-regulated in chick oviduct (Baez et al., 1987), and HSP108, like HSP90, is found associated with all types of steroid hormone receptors. The induction of heat-shock protein synthesis has been studied in a range of avian cell types, including reticulocytes (Atkinson et al., 1986; Theodorakis et al., 1988), lymphoid cells (Banerji et al., 1986; Miller and Qureshi, 1992) and macrophages (Miller and Qurechi, 1992), generally by exposing the cells to 43–45°C for 30–60 min, and usually resulting in the synthesis of proteins of approximately 23,000, 70,000 and 90,000 Da. An additional protein of 47,000 Da is synthesized in chicken embryo fibroblasts that are exposed to heat shock, and appears to have collagen-binding activity (Nagata et al., 1986). Most protein synthesis in the chick embryo recovered from the unincubated egg is still dependent on recruitment of maternal mRNA rather than on embryonic gene expression, but expression of several heat-shock protein-encoding genes can apparently be induced by heat shock at this stage (Zagris and Matthopoulos, 1988). By stage XIII of embryonic development (Eyal-Giladi and Kochav, 1976), heat-shock protein expression is regionally specific within the blastoderm (Zagris and Matthopoulos, 1986) and heat shock (isolated blastoderms subjected to a temperature of 43–44°C for 2.5 h) disrupts normal embryonic development – the blastoderm becomes approximately twice as large and the primitive streak fails to form.

Certain breeds or strains of poultry appear to survive heat stress more successfully than others, and birds that are acclimatized gradually to elevated temperatures are more resilient than those experiencing a sudden heat shock. Jungle fowl survive heat stress more successfully than do normal commercial strains of chicken, as do Bedouin fowl of the Israeli desert (Marder, 1973). A cross of Leghorn and Bedouin fowl produced offspring with improved heat tolerance relative to the Leghorn parents (Arad *et al.*, 1975), indicating a genetic component in this tolerance. It is tempting to speculate that part of the difference in heat tolerance of various breeds could be attributable to different alleles of heat-shock protein-encoding genes, resulting in differing ability to respond rapidly to a heat stress, differing final concentrations of the relevant heat-shock proteins in stressed birds or differing ability of various heat-shock protein isoforms to interact with their normal ligands in the cell. A thorough molecular characterization of the heat-shock response in a range of poultry would provide the basis for future genetic manipulation of the heat-shock response in a way that has not been possible before. In the long term, perhaps, levels of certain heat-shock proteins could be raised, as indicated by Pelham (1984), bearing in mind that ancillary proteins such as the HSF might also have to be present in increased amounts (see Johnston and Kucey, 1988). These manipulations might be achieved either by introgression and marker-assisted selection, or by more radical means through the development of transgenic birds. An intriguing hint of the genetic plasticity of the heat-shock response is provided by experiments that resulted in significant improvement of the temperature-specific fitness of lines of the bacterium, *Escherichia coli* (which has a heat-shock response that is very similar to that of birds), maintained at 42°C for 200 generations (Bennett *et al.*, 1990).

References

Ait-Boulahsen, A., Garlich, J.D. and Edens, F.W. (1989) Effect of fasting and acute heat-stress on body temperature, blood acid–base and electrolyte status in chickens. *Comparative Biochemistry and Physiology* 94A, 683–687.

Ames, E., Steven, K. and Skadhauge E. (1971) Effects of arginine vasotocin on renal excretion of Na^+, K^+, Cl^- and urea in the hydrated chicken. *American Journal of Physiology* 221, 1223–1228.

Arab, Z., Moskovits, E. and Marder, J. (1975) A preliminary study of egg production and heat tolerance in a new breed of fowl (Leghorn × Bedouin). *Poultry Science* 54, 780–783.

Arad, Z., Marder, J. and Eylath, U. (1983) Serum electrolyte and enzyme responses to heat stress and dehydration in the fowl (*Gallus domesticus*). *Comparative Biochemistry and Physiology* 74A, 449–453.

Arad, Z., Arnason, S.S., Chadwick, A. and Skadhauge, E. (1985) Osmotic and hormonal responses to heat and dehydration in the fowl. *Journal of Comparative Physiology* 155B, 227–234.

Arieli, A. and Berman, A. (1979) The effect of thyroxine on thermoregulation in the mature fowl (*Gallus domesticus*). *Journal of Thermal Biology* 4, 247–249.

Atkinson, B.G., Dean, R.L. and Blaker, T.W. (1986) Heat shock induced changes in the gene expression of terminally differentiating avian red blood cells. *Canadian Journal of Genetics and Cytology* 28, 1053–1063.

Azahan, E. and Sykes, A.H. (1980) The effects of ambient temperature on urinary flow and composition in the fowl. *Journal of Physiology, London* 302, 389–396.

Baez, M., Sargan, D.R., Elbrecht, A., Kulomaa, M.S., Zarucki-Schulz, T., Tsai, M.-J. and O'Malley, B.W. (1987) Steroid hormone regulation of the gene encoding the chicken heat shock protein Hsp 108. *Journal of Biological Chemistry* 262, 6582–6588.

Banerji, S.S., Berg, L. and Morimoto, R.I. (1986) Transcription and post-transcriptional regulation of avian HSP70 gene expression. *Journal of Biological Chemistry* 261, 15740–15745.

Barrot, H.G. and Pringle, E.M. (1941) Energy and gaseous metabolism of the hen as affected by temperature. *Journal of Nutrition* 22, 273–286.

Bellabarba, D. and Lehoux, J.-G. (1981) Triiodothyronine nuclear receptor in chick embryo: nature and properties of hepatic receptor. *Endocrinology* 109, 1017–1025.

Bellabarba, D. and Lehoux, J.-G. (1985) Binding of thyroid hormones by nuclei of target tissues during the chick embryo development. *Mechanisms of Ageing and Development* 30, 325–331.

Ben Nathan, D., Heller, E.D. and Perek, M. (1976) The effect of short heat stress upon leucocyte count, plasma corticosterone level, plasma and leucocyte ascorbic acid content. *British Poultry Science* 17, 481–485.

Bennett, A.F., Dao, K.M. and Lenski, R.E. (1990) Rapid evolution in response to high-temperature selection. *Nature* 346, 79–81.

Bhattacharyya, T.K. and Ghosh, A. (1972) Cellular modification of interrenal tissue induced by corticoid therapy and stress in three avian species. *American Journal of Anatomy* 133, 483–494.

Binkley, S., Kluth, E. and Menaker, M. (1971) Pineal function in sparrows: circadian rhythms and body temperature. *Science* 174, 311–314.

Bobek, S., Niezgoda, J., Pietras, M., Kacinska, M. and Ewy, Z. (1980) The effect of acute cold and warm ambient temperatures on the thyroid hormone concentration in blood plasma, blood supply, and oxygen consumption in

Japanese quail. *General Comparative Endocrinology* 40, 201–210.
Bond, U. and Schlesinger, M.J. (1986) The chicken ubiquitin gene contains a heat shock promoter and expresses an unstable mRNA in heat-shocked cells. *Molecular and Cellular Biology* 6, 4602–4610.
Boone, M.A. (1968) Temperature at six different locations in the fowl's body as affected by ambient temperatures. *Poultry Science* 47, 1961–1962.
Boone, M.A. and Hughes, B.L. (1971a) Effect of heat stress on laying and non-laying hens. *Poultry Science* 50, 473–477.
Boone, M.A. and Hughes, B.L. (1971b) Wind velocity as it affects body temperature, water consumption and feed consumption during heat stress of roosters. *Poultry Science* 50, 1535–1537.
Bottje, W.G. and Harrison, P.C. (1984) Mean celiac blood flow (MBF) and cardiovascular response to α-adrenergic blockade or elevated ambient CO_2 (%CO_2) during acute heat stress. *Poultry Science* 63 (Suppl. 1), 68–69. (Abstract).
Bottje, W.G. and Harrison, P.C. (1985) The effect of tap water, carbonated water, sodium bicarbonate, and calcium chloride on blood acid–base balance in cockerels subjected to heat stress. *Poultry Science* 64, 107–113.
Bottje, W.G., Harrison, P.C. and Grishaw, D. (1983) Effect of an acute heat stress on blood flow in the coeliac artery of Hubbard cockerels. *Poultry Science* 62, 1386–1387 (Abstract).
Bottje, W.G., Holmes, K.R., Neldon, H.L. and Koike, T.I. (1989) Relationships between renal hemodynamics and plasma levels of arginine vasotocin and mesotocin during hemorrhage in the domestic fowl (*Gallus domesticus*). *Comparative Biochemistry and Physiology* 92A, 423–427.
Bowen, S.J. and Washburn, K.W. (1985) Thyroid and adrenal response to heat stress in chickens and quail differing in heat tolerance. *Poultry Science* 64, 149–154.
Bowen, S.J., Washburn, K.W. and Huston, T.M. (1984) Involvement of the thyroid gland in the response of the young chicken to heat stress. *Poultry Science* 63, 66–69.
Branganza, A. and Wilson, W.O. (1978a) Elevated temperature effects on catecholamines and serotonin in brains of male Japanese quail. *Journal of Applied Physiology* 45, 705–708.
Branganza, A. and Wilson, W.O. (1978b) Effect of acute and chronic elevated air temperatures, constant (34°) and cyclic (10–34°), on brain and heart norepinephrine of male Japanese quail. *General and Comparative Endocrinology* 36, 233–237.
Branton, S.L., Reece, F.N. and Deaton, J.W. (1986) Use of ammonium chloride and sodium bicarbonate in acute heat exposure. *Poultry Science* 65, 1659–1663.
Burger, R.E. and Lorenz, F.W. (1960) Pharmacologically induced resistance to heat shock. II. Modifications of activity of the central-nervous and endocrine systems. *Poultry Science* 39, 477–482.
Catelli, M.G., Binart, N., Feramisco, J.R. and Helfman, D.M. (1985) Cloning of the chick *hsp90* cDNA in expression vector. *Nucleic Acids Research* 13, 6035–6047.
Chirico, W.J., Waters, M.G. and Blobel, G. (1988) 70K heat shock related

proteins stimulate protein translocation into microsomes. *Nature* 332, 805–810.

Cogburn, L.A., Harrison, P.C. and Brown, D.E. (1976) Scotophase-dependent thermoregulatory dysfunction in pinealectomized chickens. *Proceedings of the Society of Experimental Biology and Medicine* 153, 197–201.

Craig, E.A. and Gross, C.A. (1991) Is hsp70 the cellular thermometer? *Trends in Biochemical Science* 16, 135–140.

Cremer-Bartels, G., Kuchle, H.J., Ludtmann, W. and Malten, K. (1980) Melatonin biosynthesis in avian pineal gland and retina. *Advances in Bioscience* 29, 47–56.

Darre, M.J. and Harrison, P.C. (1981) The effects of heating and cooling localized areas of the spinal cord and brain stem on thermoregulatory responses of domestic fowl. *Poultry Science* 60, 1644. (Abstract).

Darre, M.J. and Harrison, P.C. (1987) Heart rate, blood pressure, cardiac output, and total peripheral resistance of single comb white leghorn hens during an acute exposure to 35°C ambient temperature. *Poultry Science* 66, 541–547.

Darre, M.J., Odom, T.W., Harrison, P.C. and Staten, F.E. (1980) Time course of change in respiratory rate, blood pH, and blood PCO_2 of SCWL hens during heat stress. *Poultry Science* 59, 1598 (Abstract).

Davison, T.F., Misson, B.H. and Freeman, B.M. (1980) Some effects of thyroidectomy on growth, heat production and the thermoregulatory ability of the immature fowl (*Gallus domesticus*). *Journal of Thermal Biology* 5, 197–202.

Decuypere, E., Hermans, C., Michels, H., Kuhn, E.R. and Verheyen, J. (1980) Thermoregulatory response and thyroid hormone concentration after cold exposure of young chicks treated with iopanoic acid or saline. In: Pethes, G., Peczely, P. and Rudas, P. (eds) *Recent Advances of Avian Endocrinology*. Pergamon Press, Oxford/Akademiai Kiado, Budapest, pp. 291–298.

Deshaies, R.J., Koch, B.D., Werner-Washburne, M., Craig, E.A. and Schekman, R. (1988) A subfamily of stress proteins facilitates translocation of secretory and mitochondrial precursor polypeptides. *Nature* 332, 800–805.

Deviche P., Heyns, W., Balthazart, J. and Hendick, J. (1979) Inhibition of LH plasma levels by corticosterone administration in the male duckling. *IRCS Medical Science* 7, 622–630.

Deyhim, F. and Teeter, R.G. (1991) Research note: sodium and potassium chloride drinking water supplementation effects on acid–base balance and plasma corticosterone in broilers reared in thermoneutral and heat-distressed environments. *Poultry Science* 70, 2551–2553.

Donoghue, D., Krueger, B.F., Hargis, B.M., Miller, A.M. and El Halawani, M.E. (1989) Thermal stress reduces serum luteinizing hormone and bioassayable hypothalamic content of luteinizing hormone releasing hormone in the hen. *Biology and Reproduction* 41, 419–424.

Dougherty, T.F. and White, A. (1944) Influence of hormones on lymphoid tissue structure and function: the role of pituitary adrenotrophic hormone in the regulation of lymphocytes and other cellular elements of blood. *Endocrinology* 35, 1–12.

Edens, F.W. (1978) Adrenal cortical insufficiency in young chickens exposed to a high ambient temperature. *Poultry Science* 57, 1746–1750.

Edens, F.W. and Siegel, H.S. (1975) Adrenal responses in high and low ACTH response lines of chicken during acute heat stress. *General and Comporative Endocrinology* 25, 64–73.

Edens, F.W. and Siegel, H.S. (1976) Modification of corticosterone and glucose responses by sympatholytic agent in young chickens during acute heat exposure. *Poultry Science* 55, 1704–1712.

El Halawani, M.E. and Waibel, P.E. (1976) Brain indole and catecholamines of turkeys during exposure to temperature stress. *American Journal of Physiology* 230, 110–114.

El Halawani, M.E., Waibel, P.E., Appel, J.R. and Good, A.L. (1973) Effects of temperature stress on catecholamines and corticosterone of male turkeys. *American Journal of Physiology* 224, 384–388.

El Halawani, M.E., Burke, W.H. Dennison, P.T. and Silsby, J.L. (1982) Neuropharmacological aspects of neural regulation of avian endocrine function. In: Scanes, C.G., Ottinger, M.A., Kenny, A.D., Balthazart, J., Cronshaw, J. and Chester Jones, I. (eds) *Aspects of Avian Endocrinology. Practical and Theoretical Implications*. Texas Tech Press, Lubbock, pp. 33–40.

Emmans, G.C. and Charles, D.R. (1977) Climatic environment and poultry feeding in practice. In: Haresign, W., Swan, H. and Lewis, D. (eds) *Nutrition and the Climatic Environment*. Butterworths, London, pp. 31–48.

Etches, R.J. (1976) A radioimmunoassay for corticosterone and its application to the measurement of stress in poultry. *Steroids* 28, 763–773.

Etches, R.J. and Cunningham, F.J. (1976) The interrelationship between progesterone and luteinizing hormone during the ovulation cycle of the hen (*Gallus domesticus*). *Journal of Endocrinology* 71, 51–58.

Etches, R.J., Williams, J.B. and Rzasa, J. (1984) Effects of corticosterone and dietary changes in the domestic hen on ovarian function, plasma LH and steroids and the response to exogenous LHRH. *Journal of Reproduction and Fertility* 70, 121–130.

Eyal-Giladi, H. and Kochav, S. (1976) From cleavage to primitive streak formation: a complementary normal table and a new look at the first stages of the development of the chick. I. General morphology. *Developmental Biology* 49, 321–337.

Finley, D., Ciechanover, A. and Varshavsky, A. (1984) Thermolability of ubiquitin-activating enzyme from the mammalian cell cycle mutant ts85. *Cell* 37, 43–55.

Flaherty, K.M., Deluca-Flaherty, C. and McKay, D.B. (1990) Three-dimensional structure of the ATPase fragment of a 70K heat-shock cognate protein. *Nature* 346, 623–628.

Flynn, G.C., Pohl, J., Flocco, M.T. and Rothman, J.E. (1991) Peptide-binding specificity of the molecular chaperone BiP. *Nature* 353, 726–730.

Fox, T.W. (1951) Studies on heat tolerance in domestic fowl. *Poultry Science* 30, 477–483.

Fox, T.W. (1980) The effects of thiouracil and thyroxine on resistance to heat shock. *Poultry Science* 59, 2391–2396.

Fronda, F.M. (1925) Some observations on the body temperature of Poultry. *Cornell Vet* 15, 8–20.

Garren, H.W. and Shaffner, C.S. (1954) Factors concerned in the response of young New Hampshires to muscular fatigue. *Poultry Science* 33, 1095–1104.

Garren, H.W. and Shaffner, C.S. (1956) How the period of exposure to different stress stimuli affects the endocrine and lymphatic gland weights of young chicks. *Poultry Science* 35, 266–272.

George, J.C. (1980) Structure and physiology of posterior lobe hormones. In: Epple, A. and Stetson, M.H. (eds) *Avian Endocrinology*. Academic Press, New York, pp. 85–115.

George, J.C. and Berger, A.J. (1966) *Avian Myology*. Academic Press, New York.

George, J.C. and John, T.M. (1986) Physiological responses to cold exposure in pigeons. In: Heller, H.C., Musacchia, X.J. and Wang, C.H. (eds) *Living in the Cold: Physiological and Biochemical Adaptations*. Elsevier Science, New York, pp. 435–443.

Gething, M.J. and Sambrook, J. (1992) Protein folding in the cell. *Nature* 355, 33–45.

Glick, B. (1957) Experimental modification of the growth of the bursa of Fabricius. *Poultry Science* 36, 18–24.

Glick, B. (1967) Antibody and gland studies in cortisone and ACTH-injected birds. *Journal of Immunology* 98, 1076–1084.

Gonyou, H.W. and Morrison, W.D. (1983) Effects of defeathering and insulative jackets on production by laying hens at low temperatures. *British Poultry Science* 24, 311–317.

Gould, N.R. and Siegel, H.S. (1981) Viability of and corticosteroid binding in lymphoid cells of various tissues after corticotropin injection. *Poultry Science* 60, 891–893.

Gross, W.B. and Siegel, H.S. (1983) Evaluation of heterophile : lymphocyte ratio as a measure of stress in chickens. *Avian Diseases* 27, 972–979.

Hahn, D.W., Ishibashi, T. and Turner, C.W. (1966) Alteration of thyroid secretion rate in fowls changed from a cold to a warm environment. *Poultry Science* 45, 31–33.

Harvey, S., Scanes, C.G. and Brown, K.I. (1986) Adrenals. In: Sturkie, P.D. (ed.) *Avian Physiology*. Springer-Verlag, New York, pp. 479–493.

Hester, P.Y., Smith, S.G., Wilson, E.K. and Pierson, F.W. (1981) The effect of prolonged heat stress on adrenal weight, cholesterol and corticosterone in white pekin ducks. *Poultry Science* 60, 1583–1586.

Heywang, B.W. (1938) Effect of some factors on the body temperature of hens. *Poultry Science* 17, 317–323.

Hill, A.T. and Hunt, J.R. (1978) Layer cage depth effects on nervousness, feathering, shell breakage, performance and net egg returns. *Poultry Science* 57, 1204–1216.

Hillerman, J.P. and Wilson, W.O. (1955) Acclimation of adult chickens to environmental temperature changes. *American Journal of Physiology* 180, 591–595.

Hillman, P.E., Scott, N.R. and van Tienhoven, A. (1985) Physiological responses and adaptations to hot and cold environments. In: Yousef, M.K. (ed). *Stress Physiology in Livestock*, Vol. 3, *Poultry*. CRC Press, Boca Raton, Florida, pp. 1–71.

Holmes, W.N. and Phillips, J.G. (1976) The adrenal cortex of birds. In: Chester-

Jones, I. and Henderson, I.W. (eds) *General and Clinical Endocrinology of the Adrenal Cortex.* Academic Press, London, pp. 293–420.

Hooper, P. and Richards, S.A. (1991) Interaction of operant behaviour and autonomic thermoregulation in the domestic fowl. *British Poultry Science* 32, 929–938.

Horowitz, K.A., Scott, N.R., Hillman, P.E. and Van Tienhoven, A. (1978) Effects of feathers on instrumental thermoregulatory behaviour in chickens. *Physiology and Behaviour* 21, 233–238.

Hughes, B.O. (1978) The frequency of neck movements in laying hens and the improbability of cage abrasion causing feather wear. *British Poultry Science* 19, 389–393.

Hughes, B.O. and Wood-Gush, D.G.M. (1977) Agonistic behaviour in domestic hens: the influence of housing method and group size. *Animal Behaviour* 25, 1056–1062.

John, T.M and George, J.C. (1973) Influence of glucagon and neurohypophysial hormones on plasma free fatty acid levels in the pigeon. *Comparative Biochemistry and Physiology* 45A, 541–547.

John, T.M. and George, J.C. (1977) Blood levels of cyclic AMP, thyroxine, uric acid, certain metabolites and electrolytes under heat-stress and dehydration in the pigeon. *Archives Internationales de Physiologie et Biochimie* 85, 571–582.

John, T.M. and George, J.C. (1984) Diurnal thermal response to pinealectomy and photoperiod in the pigeon. *Journal of Interdisciplinary Cycle Research* 15, 57–67.

John, T.M. and George, J.C. (1986) Arginine vasotocin induces free fatty acid release from avian adipose tissue *in vitro. Archives Internationales de Physiologie et Biochimie* 94, 85–89.

John, T.M. and George, J.C. (1991) Physiological responses of melatonin-implanted pigeons to changes in ambient temperature. In: Riklis, E. (ed.) *Photobiology.* Plenum Press, New York, pp. 597–605.

John, T.M. and George, J.C. (1992) Effects of arginine vasotocin on cardiorespiratory and thermoregulatory responses in the pigeon. *Comparative Biochemistry and Physiology* 102C, 353–359.

John, T.M., McKeown, B.A. and George, J.C. (1973) Influence of exogenous growth hormone and its antiserum on plasma free fatty acid level in the pigeon. *Comparative Biochemistry and Physiology* 46A, 497–504.

John, T.M., McKeown, B.A. and George, J.C. (1975) Effect of thermal stress and dehydration on plasma levels of glucose, free fatty acids and growth hormone in the pigeon. *Archives Internationales de Physiologie et Biochimie* 83, 303–308.

John, T.M., Itoh, S. and George, J.C. (1978) On the role of pineal hormones in the thermoregulation in the pigeon. *Hormone Research* 9, 41–56.

Johnson, P.L., Johnson, A.L. and van Tienhoven, A. (1985) Evidence for a positive feedback interaction between progesterone and luteinizing hormone in the induction of ovulation in the hen, *Gallus domesticus. General and Comparative Endocrinology* 58, 478–485.

Johnston, R.N. and Kucey, B.L. (1988) Competitive inhibition of *hsp70* gene expression causes thermosensitivity. *Science* 242, 1551–1554.

Jones, D.R. and Johansen, K. (1972) The blood vascular system of birds. In:

Farner, D.S. and King, J.R. (eds) *Avian Biology*, Vol. 2. Academic Press, New York, pp. 257–285.

Kamar, G.A.R. and Khalifa, M.A.S. (1964) The effect of environmental conditions on body temperature of fowls. *British Poultry Science* 5, 235–244.

Kohne, H.J. and Jones, J.E. (1975a) Changes in plasma electrolytes, acid–base balance and other physiological parameters of adult female turkeys under conditions of acute hyperthermia. *Poultry Science* 54, 2034–2038.

Kohne, H.J. and Jones, J.E. (1975b) Acid–base balance, plasma electrolytes and production performance of adult turkey hens under conditions of increasing ambient temperature. *Poultry Science* 54, 2038–2045.

Koike T.I., Neldon, H.L., McKay, D.W. and Rayford, P.L. (1986) An antiserum that recognizes mesotocin and isotocin: development of a homologous radioimmunoassay for plasma mesotocin in chickens (*Gallus domesticus*). *General Comparative Endocrinology* 63, 93–103.

Kuhn, E.R. and Nouwen, E.J. (1978) Serum levels of triiodothyronine and thyroxine in the domestic fowl following mild cold exposure and injection of synthetic thyrotropin releasing hormone. *General and Comparative Endocrinology* 34, 336–342.

Kulomaa, M.S., Weigel, N.L., Kleinsek, D.A., Beattie, W.G., Conneely, O.M., March, C., Zarucki-Schulz, T., Schrader, W.T. and O'Malley, B.W. (1986) Amino acid sequence of a chicken heat shock protein derived from the complementary DNA nucleotide sequence. *Biochemistry* 25, 6244–6251.

Lam, S.K. and Harvey, S. (1990) Thyroid regulation of body temperature in anaesthetized chickens. *Comparative Biochemistry and Physiology* 95A, 435–439.

Laycock, S.R. (1989) Operant conditioning techniques and their use in evaluating the thermal environment of the mature laying hen. MSc Thesis, University of Guelph, Ontario, Canada.

Lee, D.H., Robinson, K.W., Yeates, N.T.M. and Scott, M.I.R. (1945) Poultry husbandry in hot climates – experimental enquiries. *Poultry Science* 24, 195–207.

Leeson, S.A. and Morrison, W.D. (1978) Effect of feather cover on feed efficiency in laying birds. *Poultry Science* 57, 1094–1096.

Lin, Y.C. and Sturkie, P.D. (1968) Effect of environmental temperatures on the catecholamines of chickens. *American Journal of Physiology* 214, 237–240.

Lindquist, S. and Craig, E.A. (1988) The heat-shock proteins. *Annual Review of Genetics* 22, 631–677.

Lustick, S.I. (1983) Cost–benefit of thermoregulation in birds: influences of posture, microhabitat selection, and color. In: Aspey, W. and Lustick, S.I. (eds) *Behavioral Energetics*. Ohio State University Press, Columbus, Ohio, pp. 265–294.

McCormick, C.C., Garlich, J.D. and Edens, F.W. (1979) Fasting and diet affect the tolerance of young chickens exposed to acute heat stress. *Journal of Nutrition* 109, 1797–1809.

McFarlane, J.M., Curtis, S.E., Shanks, R.D. and Carmer, S.G. (1989) Multiple concurrent stressors in chicks. 1. Effect on weight gain, feed intake, and behaviour. *Poultry Science* 68, 501–509.

McNabb, F.M.A. (1988) Peripheral thyroid hormone dynamics in precocial and

altricial avian development. *American Zoology* 28, 427–440.
McNicholas, M.J. and McNabb, F.M.A. (1987) Influence of dietary iodine availability. *Journal of Experimental Zoology* 244, 263–268.
Marder, J. (1973) Temperature regulation in the bedouin fowl (*Gallus domesticus*). *Physiology and Zoology* 46, 208–217.
Marley, E. and Nistico, G. (1972) Effects of catecholamines and adenosine derivative given into the brain of fowls. *British Journal of Pharmacology* 46, 629–636.
Marley, E. and Stephenson, J.D. (1970) Effects of catecholamines infused into the brain of young chickens. *British Journal of Pharmacology* 40, 639–658.
May, J.D. (1982) Effect of dietary thyroid hormone on survival time during heat stress. *Poultry Science* 61, 706–709.
May, J.D. and Lott, B.D. (1992) Feed consumption patterns of broilers at high environmental temperatures. *Poultry Science* 71, 331–336.
May, J.D., Deaton, J.W., Reece, F.N. and Branton, S.L. (1986) Effect of acclimation and heat stress on thyroid hormone concentration. *Poultry Science* 65, 1211–1213.
Mellen, W.J. and Wentworth, B.C. (1958) Studies with thyroxine and triiodothyronine in chickens. *Poultry Science* 37, 1226.
Mellen, W.J. and Wentworth, B.C. (1962) Observations on radiothyroidectomized chickens. *Poultry Science* 41, 134–141.
Mench, J. (1985) Behaviour and stress. *Maryland Poultryman*, July, 1–2.
Merat, P. (1990) Gènes que majeurs chez la poule (*Gallus gallus*): autres gènes que ceux affectant la taille. *INRA Production Animal* 3, 355–368.
Midtgard, U. (1989) Circulatory adaptations to cold in birds. In: Bech, C. and Reinertsen, R.E. (eds) *Physiology of Cold Adaptation in Birds*. Plenum Press, New York, pp. 211–222.
Miller, L. and Qureshi, M.A. (1992) Molecular changes associated with heat-shock treatment in avian mononuclear and lymphoid lineage cells. *Poultry Science* 71, 473–481.
Mizzen, L.A. and Welch, W.J. (1988) Characterization of the thermotolerant cell. I. Effects on protein synthesis activity and the regulation of heat-shock protein 70 expression. *Journal of Cell Biology* 106, 1105–1116.
Mongin, P.E. (1968) Role of acid–base balance in the physiology of egg-shell formation. *World's Poultry Science Journal* 24, 200–230.
Morimoto, R.I., Hunt, C., Huang, S.-Y., Berg, K.L. and Banerji, S.S. (1986) Organization, nucleotide sequence, and transcription of the chicken HSP70 gene. *Journal of Biological Chemistry* 261, 12692–12699.
Morimoto, R.I., Tissieres, A. and Georgopoulos, C. (eds) (1990) *Stress Proteins in Biology and Medicine*. Cold Spring Harbor Laboratory, Cold Spring Harbor, New York.
Morrison, W.D. and Curtis, S. (1983) Observations of environmental thermoregulation by chicks. *Poultry Science* 62, 1912–1914.
Morrison, W.D. and McMillan, I. (1985) Operant control of the thermal environment by chicks. *Poultry Science* 64, 1656–1660.
Morrison, W.D. and McMillan, I. (1986) Response of chicks to various environmental temperatures. *Poultry Science* 65, 881–883.
Morrison, W.D., McMillan, I. and Amyot, E. (1987a) Operant control of the

Thermal environment and learning time of young chicks and piglets. *Canadian Journal of Animal Science* 67, 343–347.

Morrison, W.D., McMillan, I. and Bate, L.A. (1987b) Effect of air movement on operant heat demand of chicks. *Poultry Science* 66, 854–857.

Munsick, R.S., Sawyer, W.H. and Van Dyke, H.B. (1960) Avian neurohypophysial hormones: pharmacological properties and tentative identification. *Endocrinology* 66, 860–871.

Nagata, K., Saga, S. and Yamada, K.M. (1986) A major collagen-binding protein of chick embryo fibroblasts is a novel heat shock protein. *Journal of Cell Biology* 103, 223–229.

Necker, R. (1977) Thermal sensitivity of different skin areas in pigeons. *Journal of Comparative Physiology* 116, 239–246.

Nobukumi, K. and Nishiyama, H. (1975) The influence of thyroid hormone on the maintenance of body temperature in male chicks exposed to low ambient temperature. *Japan Journal of Zootechnical Science* 46, 403–407.

North, M.O. (1978) *Commercial Chicken Production Manual*. AVI Publishing Company, Westport, Connecticut.

North, M.O. and Bell, D.D. (1990) *Commercial Chicken Production Manual*, 4th edn. Van Nostrand Reinhold, New York, 262 pp.

Nover, L. (1991) *Heat Shock Response*. CRC Press, Boca Raton, Florida.

Novero, R.P., Beck, M.M., Gleaves, E.W., Johnson, A.L. and Deshazer, J.A. (1991) Plasma progesterone, luteinizing hormone concentrations and granulosa cell responsiveness in heat-stressed hens. *Poultry Science* 70, 2335–2339.

Otten, L., Morrison, W.D., Braithwaite, L.A. and Smith, J.H. (1989) Development of cooled roosts for heat-stressed poultry. Presentation at the Meeting of the Canadian Society of Agricultural Engineering and American Society of Agricultural Engineers held in Quebec, PQ, Canada, 25–28 June 1989. Paper No. 89-4081, 13 pp.

Pang, S.F., Brown, G.M., Grota, L.J., Chambers, J.W. and Rodman, R.L. (1977) Determination of N-acetylserotonin and melatonin activities in the pineal gland, retina, Harderian gland, brain and serum of rats and chickens. *Neuroendocrinology* 23, 1–13.

Pardue, M.L., Feramisco, J.R. and Lindquist, S. (eds) (1989) *Stress-induced Proteins*. Alan R. Liss, New York.

Parker, J.T. and Boone, M.A. (1971) Thermal stress effects on certain blood characteristics of adult male turkeys. *Poultry Science* 50, 1287–1295.

Parker, J.T., Boone, M.A. and Knechtges, J.F. (1972) The effect of ambient temperature upon body temperature, feed consumption, and water consumption, using two varieties of turkeys. *Poultry Science* 51, 659–664.

Pelham, H.R.B. (1984) Hsp70 accelerates the recovery of nucleolar morphology after heat shock. *EMBO Journal* 3, 3095–3100.

Pelham, H.R.B. (1986) Speculations on the functions of the major heat shock and glucose-regulated proteins. *Cell* 46, 959–961.

Petitte, J.N. and Etches, R.J. (1988) The effect of corticosterone on the photoperiodic response of immature hens. *General and Comparative Endocrinology* 39, 424–430.

Pilo, B., John, T.M., George, J.C. and Etches, R.J. (1985) Liver Na^+K^+-ATPase activity and circulating levels of corticosterone and thyroid hormones following

cold and heat exposure in the pigeon. *Comparative Biochemistry and Physiology* 80A, 103–106.
Quay, W.B. (1974) *Pineal Chemistry*. Charles C. Thomas, Illinois.
Ralph, C.L. (1981) Melatonin production by extrapineal tissues. In: Birau, N. and Schloot, W. (ed.) *Melatonin – Current Status and Perspectives*. Pergamon Press, New York, pp. 35–46.
Ramirez, J.M. and Bernstein, M.H. (1976) Compound ventilation during thermal panting in pigeons: a possible mechanism for minimizing hypocapnic alkalosis. *Federation Proceedings, Federation of the American Society of Experimental Biology* 35, 2562–2565.
Raup, T.J. and Bottje, W.G. (1990) Effect of carbonated water on arterial pH, P_{CO_2}, and plasma lactate in heat-stressed broilers. *British Poultry Science* 31, 377–384.
Reece, F.N., Deaton, J.W. and Kubena, L.F. (1972) Effects of high temperature and humidity on heat prostration of broiler chickens. *Poultry Science* 51, 2021–2025.
Reineke, E.P. and Turner, C.W. (1945) Seasonal rhythm in thyroid hormone secretion of the chick. *Poultry Science* 24, 499–504.
Richards, S.A. (1970) Physiology of thermal panting in birds. *Annals of Biology, Animal Biophysics* 10, 151–168.
Richards, S.A. (1976) Behavioural temperature regulation in the fowl. *Journal of Phvyiology* 258, 122P–123P.
Richards, S.A. (1977) The influence of loss of plumage on temperature regulation in laying hens. *Journal of Agriculture Science, Cambridge* 89, 393–398.
Robinzon, B., Koike, T.I., Neldon, H.L., Kinzler, S.L., Hendry, I.R. and El Halawani, M.E. (1988) Physiological effects of arginine vasotocin and mesotocin in cockerels. *British Poultry Science* 29, 639–652.
Rotii-Roti, J.L. and Laszlo, A. (1987) The effects of hyperthermia on cellular macromolecules. In: Urano, M. and Douple, A. (eds) *Hyperthermia and Oncology*, Vol. 1. VNU Scientific Publishers, the Netherlands, pp. 13–56.
Rudas, P. and Pethes, G. (1984) The importance of peripheral thyroid hormone deiodination in adaptation to ambient temperature in the chicken (*Gallus domesticus*). *Comparative Biochemistry and Physiology* 77A, 567–571.
Rzasa, J. and Neizgoda, J. (1969) Effects of the neurohypophysial hormones on sodium and potassium level in the hen's blood. *Bulletin of the Academy of Polish Science Cl. II, Ser. Science Biology* 17, 585–588.
Rzasa, J. Skotinicki, J. and Niezgoda, J. (1971) Metabolic effects of neurohypophysial hormones in the chicken. *Bulletin of the Academy of Polish Science Cl. II, Ser. Science Biology* 19, 431–434.
Sammelwitz, P.H. (1967) Adrenocortical hormone therapy of induced heat stress mortality in broilers. *Poultry Science* 46, 1314.
Sargan, D.R., Tsai, M.J. and O'Malley, B.W. (1986) hsp108, a novel heat shock inducible protein of chicken. *Biochemistry* 25, 6252–6258.
Siegel, H.S. and Beane, W.L. (1961) Time responses to single intramuscular doses of ACTH. *Poultry Science* 40, 216–219.
Siegel, H.S., Drury, L.N. and Patterson, W.C. (1974) Blood parameters of broilers grown in plastic coops and on litter at two temperatures. *Poultry Science* 53, 1016–1024.

Singh, A., Reineke, E.P. and Ringer, R.K. (1968) Influence of thyroid status of the chick on growth and metabolism, with observations on several parameters of thyroid function. *Poultry Science* 47, 212–219.

Stiffler, D.F., Roach, S.C. and Pruett, S.J. (1984) A comparison of the responses of the amphibian kidney to mesotocin, isotocin, and oxytocin. *Physiological Zoology* 57, 63–69.

Sturkie, P.D. (1967) Cardiovascular effects of acclimation to heat and cold in chickens. *Journal of Applied Physiology* 22, 13–15.

Sullivan, D.A. and Wira, C.R. (1979) Sex hormone and glucocorticoid receptor in the bursa of Fabricius of immature chickens. *Journal of Immunology* 122, 2617–2623.

Tanguay, R.M. (1988) Transcriptional activation of heat-shock genes in eukaryotes. *Biochemistry and Cell Biology* 66, 584–593.

Teeter, R.G. and Smith, M.O. (1986) High chronic ambient temperature stress effects on broiler acid–base balance and their response to supplemental ammonium chloride, potassium chloride, and potassium carbonate. *Poultry Science* 65, 1777–1781.

Teeter, R.G., Smith, M.O., Owens, F.N., Arp, S.C., Sangiah, S. and Breazile, J.E. (1985) Chronic heat stress and respiratory alkalosis: occurrence and treatment in broiler chicks. *Poultry Science* 64, 1060–1064.

Theodorakis, N.G., Banerji, S.S. and Morimoto, R.I. (1988) HSP70 mRNA translation in chicken reticulocytes is regulated at the level of elongation. *Journal of Biological Chemistry* 263, 14579–14585.

Thompson, E.B. and Lippman, M.E. (1974) Mechanism of action of glucocortoids. *Metabolism* 23, 159–202.22.

Thornton, P.A. (1962) The effect of environmental temperature on body temperature and oxygen uptake by the chicken. *Poultry Science* 41, 1053–1062.

Vo, K.V. and Boone, M.A. (1975) The effect of high temperatures on broiler growth. *Poultry Science* 54, 1347–1348 (Abstract).

Vogel, J.A. and Sturkie, P.D. (1963) Cardiovascular responses of the chicken to seasonal and increased temperature changes. *Science* 140, 1404–1406.

Wang, S. (1988) Arginine vasotocin and mesotocin response to acute heat stress in domestic fowl. MS thesis, University of Arkansas.

Wang, S., Bottje, W.G., Kinzler, S., Neldon, H.L. and Koike, T.I. (1989) Effect of heat stress on plasma levels of arginine vasotocin and mesotocin in domestic fowl (*Gallus domesticus*). *Comparative Biochemistry and Physiology* 93A(4), 721–724.

Weiss, H.S., Frankel., H. and Hollands, K.G. (1963) The effect of extended exposure to a hot environment on the response of the chicken to hyperthermia. *Canadian Journal of Biochemistry* 41, 805–815.

Welch, W.J. and Mizzen, L.A. (1988) Characterization of the thermotolerant cell. II. Effects on the intracellular distribution of heat-shock protein 70, intermediate filaments, and small nuclear ribonucleoprotein complexes. *Journal of Cell Biology* 106, 1117–1130.

Welch, W.J. and Suhan, J.P. (1985) Morphological study of the mammalian stress response: characterization of changes in cytoplasmic organelles, cytoskeleton, and nucleoli, and appearance of intranuclear actin filaments in rat fibroblasts after heat-shock treatment. *Journal of Cell Biology* 101, 1198–1211.

Welty, J.C. and Baptists, L. (1988) *The Life of Birds*, 4th edn. Saunders College Pubishing, Toronto, p. 138.

Westwood, J.T., Clos, J. and Wu, C. (1991) Stress-induced oligomerization and chromosomal relocalization of heat-shock factor. *Nature* 353, 822–827.

Whittow, G.C. (1986) Regulation of body temperature. In: Sturkie, P.D. (ed.) *Avian Physiology*. Springer-Verlag, New York, pp. 221–252.

Whittow, G.C., Sturkie, P.D. and Stein, G., Jr (1964) Cardiovascular changes associated with thermal polypnea in the chicken. *American Journal of Physiology* 207, 1349–1353.

Wiech, H., Buchner, J., Zimmerman, R. and Jakob, U. (1992) Hsp90 chaperones protein folding *in vitro*. *Nature* 358, 169–170.

Williams, J.B. and Sharp, P.J. (1978) Control of the preovulatory surge of luteinizing hormone in the hen (*Gallus domesticus*): the role of progesterone and androgens. *Journal of Endocrinology* 77, 57–65.

Wilson, F.E. and Follett, B.K. (1975) Corticosterone-induced gonadosuppression in photostimulated tree sparrows. *Life Science* 17, 1451–1456.

Wilson, S.C. and Sharp, P.J. (1976) Induction of luteinizing hormone release by gonadal steroids in the ovariectomized domestic hen. *Journal of Endocrinology* 71, 87–98.

Wilson, W.O. (1948) Some effects of increasing environmental temperatures on pullets. *Poultry Science* 27, 813–817.

Wilson, W.O. and Woodard, A. (1955) Some factors affecting body temperature of turkeys. *Poultry Science* 34, 369–371.

Winchester, C.F. (1939) Influence of thyroid on egg production in poultry. *Endocrinology* 27, 697–703.

Yost, H.J., Petersen, R.B. and Lindquist, S. (1990) RNA metabolism: strategies for regulation in the heat shock response. *Trends in Genetics*, 6, 223–227.

Young, R.A. (1990) Stress proteins and immunology. *Annual Review of Immunology* 8, 401–420.

Zagris, N. and Matthopoulos, D. (1986) Differential heat-shock gene expression in chick blastula. *Roux's Archives of Developmental Biology* 195, 403–407.

Zagris, N. and Matthopoulos, D. (1988) Gene expression in chick morula. *Roux's Archives of Developmental Biology* 197, 298–301.

4

Housing for Improved Performance in Hot Climates

R.A. Ernst

Extension Poultry Specialist, Department of Avian Sciences,
University of California, Davis, California 95616-8532, USA.

Introduction	68
Definitions	68
Conversion of Temperature or Units of Length	70
Principles Related to Housing Design	70
Methods of heat loss or gain	70
Operative temperature	71
Thermoneutral zone	72
Upper and lower lethal temperatures	72
Acclimatization	72
Effects of high temperature and humidity on performance	72
Effects of diet on metabolic heat production	73
Poultry House Design	74
Factors affecting design choice	74
Basics of building design	75
Insulation	76
Amount of insulation	79
Vapour barriers	79
Design of naturally ventilated housing	81
Design of closed, fan-ventilated houses	86
Flex housing	96
Poultry House Maintenance	96
Direct-drive fans	96
Belt-drive fans	96
Pad maintenance	96
Evaluating house and fan performance	97
Monitoring House Performance	98
Emergency Systems	98
References	98

Introduction

The purpose of a poultry house is to confine the birds; to protect them from predators and environmental extremes which would cause mortality or reduce growth, feed efficiency, immunocompetence, fertility or egg production; to facilitate light control; and to facilitate bird management. When chicks are hatched they are incomplete homoeotherms and must have supplemental heat until they are fully feathered. At 1 day old, chicks require an ambient temperature of 29.4–30.5°C (85–87°F) or a cooler ambient temperature with supplemental radiant heat for optimum health and growth. Chicks can also be raised successfully in a cool room adjacent to an area (or hover) with a warmer temperature (33–35°C or 91–95°F at 1 day old). This is sometimes referred to as cool-room brooding. As chicks grow and feather out, their ambient temperature requirement declines until about 6 weeks of age. At this age they become complete homoeotherms within the temperature ranges normally experienced in tropical areas.

This chapter is written for use by farmers who want to design practical poultry houses. It will not address details of design for a specific structure. Detailed plans need to be developed by engineers or other qualified professionals who can calculate structural requirements for the building.

Definitions

Anemometer: a device for measuring air velocity (see Fig. 4.1).
British thermal unit (BTU): a measure of heat; it takes 1 BTU to raise the temperature of 1 lb of water 1°F (0.56°C); 1 BTU = 0.252 kcal; 1 kcal = 1 BTU × 0.252.
Dry-bulb temperature: the temperature measured by a regular thermometer.
Wet-bulb temperature: temperatures measured by a thermometer that has a wetted wick or sock around it. Air must be moving past the sock for accurate wet-bulb temperature measurement. Wet-bulb temperature is affected by dry-bulb temperature and air moisture content.
Psychrometer: a device for measuring dry- and wet-bulb temperatures (see Fig. 4.2).
Evaporative cooling: cooling in which the dry-bulb temperature is reduced while the wet-bulb temperature remains the same. Water is evaporated, and the heat absorbed by the water as it evaporates cools the air. This cooled air will contain more water than it did originally; therefore, its relative humidity is higher than before it was cooled. Theoretically, evaporative cooling can reduce the dry-bulb temperature of the air to the wet-bulb temperature. In actual practice, air is usually cooled to within about 1°C (2°F) of the wet-bulb temperature.

Fig. 4.1. Pinwheel anemometer used to measure air speed and estimate fan output (photo courtesy of Parsons and Bell, 1968).

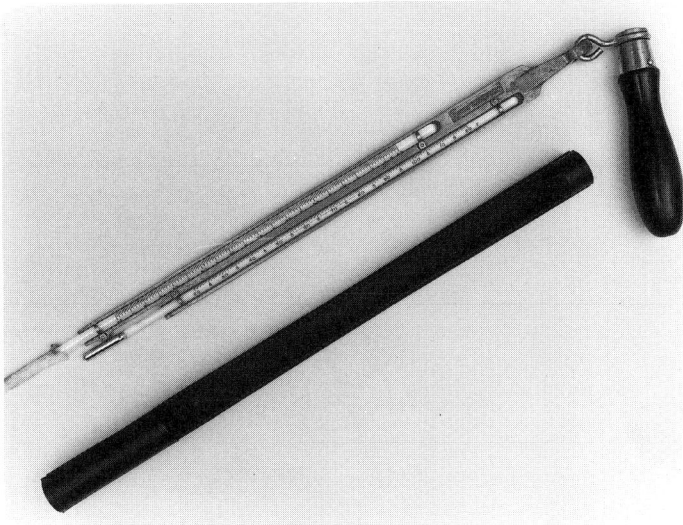

Fig. 4.2. Sling psychrometer used to measure wet- and dry-bulb temperature (photo courtesy of Parsons and Bell, 1968).

Latent heat: heat exchanged when water changes its state. An example is the cooling of a bird by evaporation of water from the respiratory system while the air temperature is not changed.

Sensible heat: heat that causes a temperature change in a material. Air is the material in which temperature is changed when poultry house ventilation is being considered.

Conversion of Temperature or Units of Length

Both Fahrenheit and Celsius temperature scales are used in this chapter. Units of length are expressed in feet or metres. Conversion can be made using the following equations:

$°F = (1.8 \times °C) + 32$
$°C = 0.555 \times (°F - 32)$
Feet = metres $\times$ 3.281
Metres = $\dfrac{\text{feet}}{3.281}$ (or feet $\times$ 0.3048)

Principles Related to Housing Design

Methods of heat loss or gain

Radiation

Solar radiation in open pens or radiation from a hot roof can be a significant source of heat for birds. Whenever two bodies differ in temperature, there will be a net heat gain by the cooler body (heat flows from the warmer object to the cooler object). Radiant heat passes through air without changing air temperature.

Conduction

Conduction is generally defined as the transfer of heat through a solid medium. Examples are the transfer of heat through wood or concrete. Materials that are poor conductors are useful as insulation.

Convection

Convection is heat transfer associated with a moving fluid. Convective heat loss or gain occurs when the temperature of the air surrounding a bird differs from its body temperature. Heat exchange by convection is greatly affected by air movement in the environment. Figure 4.3 shows the effect of air movement on heat stress of chickens.

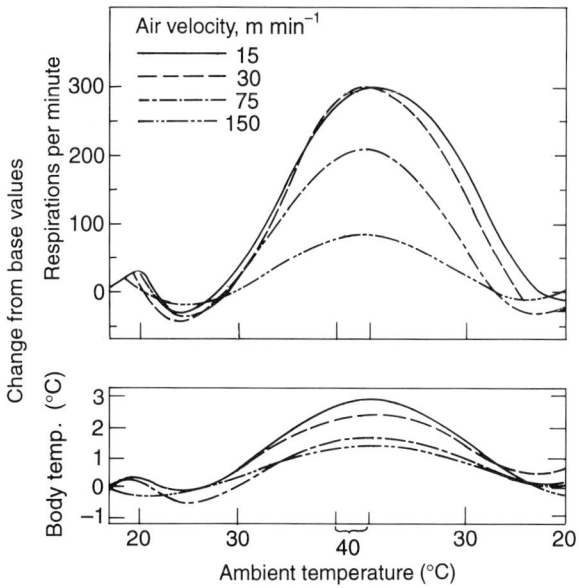

Fig. 4.3. Effects of air velocity and temperature on respiratory rate and body temperature of chickens. The curves are derived from fourth-degree polynomial equations computed from hourly oscillographic data showing changes from base values (based on Siegel and Drury, 1968).

Evaporation

Heat is required to convert liquid water to water vapour. When ambient temperatures exceed the body temperature of poultry, evaporative heat loss is the only way birds can dissipate metabolic heat. Evaporative heat loss is an extremely important method of removing heat from a poultry building. This occurs by evaporation of water from the skin and respiratory membranes of the bird and by evaporation of water from the manure or litter within the poultry house.

Operative temperature

Operative temperature is a term sometimes used to indicate the temperature in any environment as it is perceived by a bird. A bird is affected simultaneously by all forms of heat transfer (radiation, convection, conduction and evaporation) in a particular environment. The summation of these determines whether the bird is gaining or losing heat, not just the ambient temperature. For example, a well-feathered chicken may be quite comfortable at an ambient temperature of 10°C if there is little air movement and it is receiving significant amounts of solar radiation. On the other hand, it may be uncomfortable at 20°C if it is wet and there is significant air velocity and no net gain in radiant energy.

Thermoneutral zone

The thermoneutral zone is a range in temperature within which poultry can maintain their body temperature by adjusting heat loss or gain without energy expenditure (Sturkie, 1976). The upper and lower limits of this zone are referred to as critical temperatures. When temperatures are within this zone, feed requirements for growth and egg production are minimized. Critical temperatures vary with age, previous temperature experience of the bird and to a lesser extent, by other factors such as diet, feathering, etc.

Upper and lower lethal temperatures

These refer to the deep body temperature at which birds will begin to die (Sturkie, 1976).

Acclimatization

Poultry adapt to hot environments following previous exposure to high temperatures. Thus exposure to high temperatures for 1 or more days will result in an increase in both the upper and lower critical temperatures which define their thermoneutral zone (Sturkie, 1976, pp. 150–154). Poultry that have been recently exposed to high temperatures will begin panting at a higher temperature and will also have a higher lethal temperature. This means that a gradual shift from cool to hot temperatures is much less stressful than a sudden shift. The length of time that heat tolerance will persist following high temperature exposure is not well defined, but certainly intervening cool temperatures will acclimatize poultry in the opposite direction.

Effects of high temperature and humidity on performance

High humidity reduces evaporative heat loss from the skin and respiratory membranes of poultry and thereby increases the effects of high temperatures (Esmay, 1978, p. 100).

Feed consumption

Feed consumption is reduced when ambient temperatures rise above the upper critical temperature, apparently because poultry are attempting to reduce their metabolic heat load (Leeson and Summers, 1991).

Growth

Weight gain is reduced during high temperature stress as a result of reduced feed intake and because feed energy must be used for muscle contraction associated with panting (Lesson and Summers, 1991).

Egg production, egg size

Like growth, these decline as feed consumption drops.

Fertility and mating

Poultry respond to severe heat stress by minimizing their muscular activity, presumably because this results in less heat which must be dissipated. Reduced muscular activity results in reduced mating activity. If cooler temperatures can be maintained during part of the daylight period, both feeding and mating activity can be improved.

Acid–base balance

The acid–base balance of a bird is regulated by an internal control system. When high temperatures (in excess of the upper critical temperature) are encountered, the body temperature of the bird increases. This activates a response by the bird in the form of panting. Panting increases water evaporation from the respiratory surfaces, but it also results in an increased air movement over the respiratory surfaces resulting in a loss of carbon dioxide from the blood via the lungs. This loss of carbon dioxide results in alkalosis. This alkalotic condition is slowly compensated by the kidneys, but when heat stress is prolonged a chronic alkalosis results. This condition interferes with eggshell calcification and with bone deposition. It can therefore contribute to an increased incidence of leg problems (e.g. tibial dyschondroplasia) and eggshell thinning (Leeson and Summers, 1991, p. 108).

Effects of diet on metabolic heat production

Protein or carbohydrate metabolism for energy results in more metabolic heat production than fat metabolism. Therefore, fat is sometimes added to poultry diets during heat-stress periods (Lesson and Summers, 1991, p. 106).

Poultry House Design

Factors affecting design choice

Climate

Expected temperature and humidity extremes are important factors that influence poultry house design. Prolonged periods with daily maximum temperatures over 42°C (108°F) make it very difficult (if not impossible) to achieve good flock performance with naturally ventilated housing. High-humidity environments reduce the effectiveness of evaporative cooling. However, relative humidity drops as environmental temperatures increase because the quantity of water which air can hold increases with an increase in air temperature.

When outside or ambient temperature approaches the body temperature of poultry it rapidly becomes more difficult to successfully remove heat from the building. When a flock of commercial poultry is housed at typical space allowances (e.g. 25 kg m^{-2}), the birds produce approximately 15 times more heat than enters through the roof and walls of an insulated poultry house. All of this heat must be removed to prevent the inside temperature from rising. This heat can only be removed by increasing the temperature of the air moving through the building or by evaporation of water into this air. When the outside temperature reaches 41°C (106°F), increased air movement through the building is not helpful in removal of heat. Instead the building should be closed to allow the minimum amount of ventilation needed to remove moisture. Evaporation inside the building should be increased to the maximum possible by increased fogging of water or direct wetting of the poultry (e.g. with a hose) and operating internal fans to increase water evaporation. Poultry under these conditions are subjected to severe stress and they cannot be sustained for long periods of time under these conditions without evaporatively cooled housing.

Power availability

Controlled-environment housing requires a reliable source of power. If frequent, prolonged power interruptions are expected, it will be necessary to provide supplemental power generation equipment large enough to operate most of the electrical equipment. This will increase the initial cost of housing and operating costs.

Available funds

A well-designed controlled-environment house will cost 20–50% more than a naturally ventilated facility.

Power costs

Electrical costs increase significantly when controlled-environment housing is used. Costs will vary considerably with climate and relate directly to the amount of time that cooling is needed to maintain a productive environment. In a controlled-environment facility, fans are used continuously for ventilation (exceptions with flex housing will be discussed later) and therefore electrical costs for air movement are higher than with natural ventilation.

Management preference

Successful flock management is generally easier when controlled-environment housing is used in a tropical climate. When properly maintained, the mechanical system will adjust air movement and cooling automatically to achieve the desired conditions. In naturally ventilated houses, the manager is usually responsible for ensuring proper ventilation, although the control in these houses can be automated, if desired.

Basics of building design

Several types of construction can be used successfully for poultry houses. In the USA many poultry houses are of a wood-frame or steel rigid-frame truss construction. Pole structures and buildings with masonry walls are also used successfully. Clear-span truss roof construction is often preferred for cage houses because there are no posts to interfere with cage location.

Roof type

A gable roof or a shed roof is commonly used on poultry buildings. Ridge vents are often added to a gable roof to provide an escape for hot air at the roof peak. Research in North Carolina (Timmons *et al.*, 1986) has shown that ridge vents are of limited value in open-type poultry houses. The use of a ridge vent would be most beneficial in locations where hot wheather is accompanied by extended periods with little wind. Shed-type roof designs are most effctive on buildings less than 10 m (33 ft) wide.

Foundation

Other than pole buildings, all buildings should have an adequate foundation to provide the support needed for the prevailing soil type. Expansive soils require a more substantial foundation. Walls should always be anchored to the foundation.

Floors

Concrete floors are highly desirable in both floor and cage poultry houses for ease of cleaning and effective disinfection between flocks. However, cost often prohibits their use. Houses with packed earthen floors are widely and successfully used.

Walls

The solid portion of walls can be constructed of any building material which is of adequate strength. Materials which are impervious to moisture, or which can be sealed to provide a waterproof surface, are highly desirable. These will withstand poultry house conditions and can be disinfected more effectively than permeable surfaces. Wall materials should also be resistant to insect damage and, in floor houses, must be of material that poultry cannot damage by picking.

Insulation

Insulation of poultry houses is highly desirable in hot climates to reduce the influx of heat into the building through the roof and walls. Several types of insulation are available for specific applications. The choice should be made based on the cost, durability, effectiveness and the nature of the area to be insulated. Table 4.1 shows the R or resistance value for some common insulating materials. Reflective materials are not shown because R values are not an appropriate measure of their effectiveness.

Calculating resistance values

The R value of wall, ceiling or roof units can be calculated by summing the various parts of the unit (North and Bell, 1990, pp. 185–186). An example is given in Table 4.2. In this example, if the polystyrene board is placed directly below the metal roofing and above the rafters, as is commonly done in poultry houses, the R value of 2.33 for the horizontal air space will not be included, and the R value for the roof will be reduced to 5.87.

Table 4.1. R values of various building materials.

Item	Thickness (cm (in.))	Resistance rating
Insulation		
Mineral wool blanket	2.54 (1)	3.12
Wood fibre blanket	2.54 (1)	4.00
Cellulose fibre batt	2.54 (1)	4.16
Expanded polystyrene (bead board)	2.54 (1)	3.50
Expanded polystyrene (extruded)	2.54 (1)	5.00
Urethane foam	2.54 (1)	6.60
Fibreglass (glass wool)	2.54 (1)	3.70
Rock wool (loose fill)	2.54 (1)	3.33
Rock wool blanket	2.54 (1)	3.33
Glass fibre blanket	2.54 (1)	3.33
Mineral wool	2.54 (1)	3.33
Vermiculite (expanded)	2.54 (1)	2.05
Sawdust or shavings	2.54 (1)	2.22
Straw	2.54 (1)	1.75
Other materials		
Air space, horizontal	1.9 (0.75+)	2.33
Air space, vertical	1.9 (0.75+)	0.91
Building paper		0.15
Concrete	20 (8)	0.61
Concrete block	20 (8)	1.11
Hardboard	0.64 (0.25)	0.18
Plywood	0.64 (0.25)	0.18
Surface, inside		0.61
Surface, outside		0.17
Siding (drop)	1.9 (0.75)	0.94
Metal siding		0.09
Glass (single pane)		0.61
Shingles (composition)		0.18
Shingles wood		0.78
Roll roofing (25 kg or 55 lb)		0.15
Vapour barrier		0.15

Reflective materials

Several reflective roof coatings are available which can be applied to the exterior of the roof to reduce roof temperatures and thereby reduce the radiant heat reaching the birds. In California tests, good reflective coatings have produced temperature reductions approximately equivalent to an inch of expanded polystyrene board insulation located below the roof

Table 4.2. Calculating R value

Component	R value
Outside surface	0.17
Metal roofing	0.09
Horizontal air space	2.33
Extruded polystyrene board 1 inch	5.00
Inside surface	0.61
Total	8.20

Table 4.3. Mean black-globe temperature 7.6 cm (3 in.) below interior surface of roof unit.

	Mean daytime temperature (6 a.m. to 6 p.m.)	
Recording location	°F	°C
Outside shade under tree	90.0	32.2
Metal roof only (MR): over 27 years old	93.9	34.4
MR + double reflective air cell	90.9	32.7
MR + single reflective sheet: 1 year old	92.6	33.6
MR + 2.54 cm (1 in.) polystyrene board: new	92.2	33.4
MR + reflective coating 1: exterior	91.2	32.9
MR + reflective coating 2: exterior	91.4	33.0
MR + reflective coating 3: exterior	91.7	33.1
MR + reflective coating 4: exterior	91.6	33.1
MR + reflective coating 5: exterior	92.7	33.7

(Table 4.3). Most roof surfaces absorb large quantities of heat. An advantage of reflective roof coatings is that the material is on the exterior and therefore is not susceptible to insect or rodent damage. These coatings can be applied easily to existing buildings and will also extend the life of most roof materials. Some of these coatings will also seal small holes or leaks in older roofs. A disadvantage is that the reflective chacteristics of the material may decline with age and their effectiveness is reduced by dust accumulation, which often occurs in arid climates.

Reflective materials are also available for application below the roof. The effectiveness of two such materials is shown in Table 4.3. The single reflective sheet had been installed in this poultry house for over 1 year and had accumulated dust and rodent excreta on its upper surface. The double reflective air-cell material was clean at the initiation of the test. These results strongly indicate that reflective materials are most effective when the surface can be kept clean. Therefore, it is recommended that they be installed immediately below the roofing whenever possible, or below the support members, and in either case sealed with the special tape which is available to prevent dust and moisture penetration.

Low-conductance materials

Traditional insulation materials are those which are poor conductors of heat. These materials are available in the form of loose fill, batts or blankets, boards, spray-on foams or blown-in materials which dry to form a rigid fill. Boards, batts, foam or other rigid fill materials are preferred in vertical spaces because they do not settle. Any of these materials can be used above a ceiling. Spray-on foam is often used to retrofit insulation in existing buildings without ceilings because it is easy to apply.

Rodent and insect damage

Mice and rats often destroy insulation within poultry houses. Materials applied under the roof supports can provide an ideal rodent habitat. When these same materials are applied above the roof supports and immediately below the roof sheeting, there is a greatly reduced rodent problem. Soft materials such as polystyrene board or polyurethane foam are often damaged by darkling beetles, which are common pests of poultry houses. Soft materials are also often damaged by rodents and wild birds. Polystyrene or polyurethane board insulation can be purchased with an exterior film that insects will not penetrate. Batt or fill insulation should be protected by sheeting.

Amount of insulation

The benefit of insulation is greater in climates where greater temperature extremes occur (see Table 4.4 for general recommendations).

Vapour barriers

Moisture can penetrate some insulation materials and condense on a cool external surface. Condensation is a much greater concern in areas which have temperatures below 15.6°C (60°F). Since water is a poor insulator, this moisture can greatly reduce the effectiveness of the insulation. It may also cause the insulation to lose shape and sag (see Fig. 4.4).

Some insulation materials have a surface which is highly resistant to moisture penetration and do not require a vapour barrier for protection. However, to be effective the insulation material must be installed so that there are no gaps, tears or cracks. With blanket, batt or fill insulation, an additional vapour barrier is usually recommended, a polyethylene film is

Table 4.4. Suggested design R values for tropical housing.[a]

Expected high temperature		Recommended R value		Heat conductivity	
°F	°C	Roof	Walls[b]	(BTU ft^{-2} min^{-1} °F^{-1})	(kcal m^{-2} min^{-1} °C^{-1})
90	32.2	5	2	0.0033	0.01611
100	37.8	8	4	0.0021	0.01025
110	43.3	11	6	0.0015	0.00732
120	48.9	14[c]	8	0.0012	0.00586
130	54.4	17[c]	10	0.0009	0.00439

[a] If the house is located in an area which has winter temperatures below 5°C, insulation may be needed to protect birds from cold temperatures; this is not considered in this table.
[b] Applies to mechanically ventilated houses.
[c] Add reflective material for open housing.

Fig. 4.4. Damage to glass fibre insulation in wall of poultry house; note that no vapour barrier was used.

Fig. 4.5. Interior view of a naturally ventilated commercial broiler house showing insulated ceiling with vapour barrier.

often used for this purpose. Figure 4.5 shows a broiler house with a vapour barrier installed under ceiling insulation.

Design of naturally ventilated housing

Building orientation and spacing

Open-type poultry houses are often orientated in an east–west direction to take advantage of solar energy which enters during the cool season through openings in the side-wall. In climates where winter temperatures remain above 10°C this is not a significant advantage. Prevailing wind direction during the hot season is the predominate concern in most tropical locations. To achieve maximum air movement, houses should be orientated perpendicular to the prevailing wind direction.

Spacing between buildings is an important factor (Timmons, 1989b). Spacing buildings too close together can reduce air movement when natural ventilation is used. A building creates a zone on the downwind side where wind velocities are reduced. Air in this wake area also contains increased levels of dust and microorganisms emanating from the upwind house.

Recommended design spacing can be calculated from the following formula (Imperial units must be used):

$$D = 0.4 \times H \times (L)^{0.5}$$

where: D = separation distance, ridge to closest wall of next building in ft;
H = height of obstructing bulding (or other barrier) in ft;
L = length of obstructing building in ft.

Building width

Naturally ventilated buildings should not exceed 12 m (39.4 ft) in width, with the possible exception of very mild climates with consistent winds (e.g. some coastal areas). In general, it is best to use narrower houses in hotter climates. With floor housing, 10–11 m (32.8–36.1 ft) widths are popular in very hot climates.

Building length

Buildings can be of any convenient length. If mechanical feeding, egg collection belts or manure belts are used, there is a practical limit on length for most systems (check with manufacturer for recommendations). In long buildings doors are often placed in the side-walls at intervals of 15–30 m (50–100 ft) to provide access for service and bird removal.

Ventilation

CURTAINS

Most naturally ventilated poultry houses utilize poultry curtains to control air movement (see Fig. 4.6). These are available in a variety of materials. Some curtain materials allow light to enter the building. These are widely used in applications where the manager wants natural light to penetrate the building so that artificial light is not needed during daylight hours. Black (opaque) curtains are used in applications where it is desirable to exclude all external light (e.g. to provide blackout housing for broiler breeder pullets). Insulated curtains are also available for use in areas where it is desirable to reduce heat loss from buildings during the cool season (Timmons *et al.*, 1981).

In hot climates curtains should comprise as much of the side-wall as possible but never less than half the available opening. The side-wall height should be at least 2.1 m (7 ft) in hot climates. Experience has shown that higher side elevations (2.4–3.0 m (8–10 ft)) improve hot-weather ventilation.

Curtains can be easily raised and lowered by attaching a small steel

Fig. 4.6. Exterior view of a curtain-side broiler house showing ridge vents.

cable to the curtain and controlling its length with an electric or hand-operated winch.

Hot-weather protection

INSULATION

The amount of insulation recommended for best results is given in Table 4.4. The use of some reflective protection is highly recommended in very hot climates. This can be used in addition to materials that resist heat transfer. Some roofing materials, such as thatch, provide significant insulation value. Local materials are sometimes used to increase the insulation of a roof at a low cost (see Fig. 4.7).

ROOF COOLING

When an uninsulated metal roof is used, sprinkling during high-temperature periods is very effective in reducing heat load on confined poultry. Uniform wetting of the roof is highly desirable for optimum effectiveness. Any type of sprinkling system can be used, even a soaker hose. For economical use of water and energy, the system should be designed for minimum runoff. This can be accomplished by proper sizing of the sprinklers or by use of an intermittent timer.

INTERIOR COOLING SYSTEMS

Interior cooling systems are designed to increase heat loss by convection or evaporation. Increased convective and evaporative heat loss can be

Fig. 4.7. Use of grass cuttings on roof to improve insulation and reduce solar heat.

accomplished by using fans. One common approach is to suspend fans from the interior building structure so they are positioned approximately 1.2 m (4 ft) above the litter and in the centre of the building to move air down the length of the building. Fans may be spaced 6–15 m (20–50 ft) apart depending on their size and the air velocity desired. They are often tilted about 8° from vertical to direct air down over the birds. Fans should always be equipped with a protective grid to prevent injury to caretakers.

Vertical ceiling fans are also used to cool poultry. For greatest effectiveness, locate them about 3.7 m (12 ft) above the birds and 7.6–15 m (25–50 ft) apart.

Evaporative cooling systems can also be used and are required if temperatures consistently exceed 37.7°C (100°F). Interior fogging and sprinkling systems have been used effectively under practical conditions. In determining which cooling system to select, the housing, equipment and environmental conditions need to be considered.

Fogging systems vary greatly in pressure used and therefore in the droplet size obtained. High-pressure systems (20–80 atm) produce a small droplet size, resulting in greater surface area and rapid evaporation. High-pressure systems can achieve more cooling because more water can be evaporated. With high-pressure systems the moisture normally evaporates before it reaches the birds or the floor. This may be important if wet manure is a problem. Wet manure can result in excessive fly breeding in cage laying houses and can cause disease problems in litter floor houses.

Fig. 4.8. Direct application of water to cool poultry during high-temperature period.

Low-pressure systems or sprinklers usually wet the birds and the litter to some extent. In emergencies birds can be wet with a garden hose (see Fig. 4.8). This wetting does not harm the birds as long as fly breeding or disease problems associated with wet litter are not a problem.

To be most effective and to minimize costs, these internal cooling systems should be controlled by a thermostat which will turn the system on or off as the temperature fluctuates. A set-point of 29.5–32°C (85–90°F) might be used for the fans and 32–35°C (90–95°F) for the fogging system. Many growers control these systems manually. Success with a manual control system requires consistent attention by the caretaker. Fogging systems should be protected by a filter to prevent clogging of the nozzles.

ROOF VENTS

Various types of roof vents can be constructed or purchased. A roof can also be designed with a continuous ridge vent (see Fig. 4.6). Ridge vents were developed to allow hot air to exit so that cooler air will enter through the side openings of the house to provide some ventilation when there is little or no air movement. In practice there is usually some air movement and as a consequence the air from side openings far exceeds any ventilation which may occur as a result of the ridge ventilation. In controlled North

Carolina tests, a greatly enlarged ridge vent failed to have a significant impact on broiler house temperatures or bird performance (Timmons *et al.*, 1986).

PROTECTION FROM REFLECTED OR DIRECT SOLAR RADIATION

Shade is required at all times if poultry are confined in cages, because they cannot move to escape the sun. Direct solar radiation should be prevented by provision of adequate overhang on the roof or by providing shade of some sort with minimum obstruction of air movement. Many open houses in California have utilized lath slats on the side-walls to provide partial shade and some protection from high winds. Unfortunately, slats also cause some obstruction of air movement during periods of low wind speed. Side curtains which are lowered from the top can also be used effectively to block direct sunshine.

Indirect radiation or reflected radiation is also a source of heat load to poultry in some housing situations. This can be greatly reduced by planting and maintaining vegetative cover on the ground next to the poultry house. This vegetation should be mowed frequently so that it does not obstruct air movement.

Design of closed, fan-ventilated houses

Positive- or negative-pressure systems

Both of these types have been used to ventilate poultry houses with excellent success. With a positive-pressure ventilation system the air is forced into the house and allowed to exit through exhaust openings. With a negative-ventilation system the air is exhausted from the building by fans and enters the building through inlet openings.

With positive-pressure ventilation the incoming air must be distributed into the house in such a way that uniform air distribution is achieved. Therefore, construction costs are usually higher. With most positive-pressure systems a higher static pressure difference is maintained between the interior and exterior of the building, when compared with a negative-pressure house. This results in higher electrical costs for ventilation if the same volume of air is pumped through the building. Positive-pressure systems have the advantage that when a door is opened or if there are air leaks there is no change in the uniformity of air movement.

House orientation

The primary consideration for orientation of controlled-environment poultry houses is the direction of prevailing winds. Strong winds will often distort the pattern of air movement in houses when they blow into exhaust openings or against exhaust fan openings. It is desirable to orientate houses so that wind will have a minimum effect on the ventilation system to be used. Fan openings or exhaust openings can be protected from outside wind by use of hoods or overhanging structures.

Roof and side-wall construction

As with open houses, many building designs and types of construction materials can be used successfully. Consider the relative cost of materials in the area and how the structure will be ventilated. Interior surfaces should be durable and impervious to water.

Insulation

The recommended design R value for poultry houses in different hot climates is shown in Table 4.4.

Static pressure and air velocity

Since static pressure used in poultry houses is low, it is usually measured in centimetres, or inches, of water. If static pressure is too high, fan efficiency will decrease; if too low, air will not be delivered uniformly to all parts of the house. For uniform air distribution and good mixing of incoming air, an air speed of 152 m min^{-1} (500 ft min^{-1}) through the inlets has been found to be about optimum. This speed is achieved in a negative-pressure house at a negative static pressure of 0.3 cm ($\frac{1}{8}$ inch) of water pressure. The pad area or slot inlet area must be of appropriate size so that the static pressure does not exceed this at total fan capacity. For uniform ventilation it is desirable to have adjustable air inlet openings so that the optimum static pressure and air speed can be maintained regardless of the number of fans operating. Static pressure sensors are available which will control the inlet opening to maintain the desired static pressure. An economical alternative is an inlet designed so that the opening is adjusted by a weighted curtain.

Estimating design requirements for poultry house evaporatiave cooling systems

The system used in this chapter for estimation of heat load in poultry buildings and cooler capacity required was developed at the University of California (Hart *et al.*, 1964). These calculations are estimates because numerous assumptions are made in developing this simplified system. It is not necessary to know the exact cooling requirement to design a commercial poultry house. If a reasonable estimate can be made, then an appropriate safety factor can be engineered into the system. For example, if the weather records for expected high dry-bulb and wet-bulb temperatures are not reliable or are not known precisely for the location where the house will be constructed, an additional safety factor can be added.

Sources of Heat

Temperatures in evaporatively cooled poultry houses depend on: (i) heat entering through the walls and roof; (ii) heat given off by the birds; and (iii) the volume, temperature and humidity of air being moved through the house to remove the heat, moisture and gases produced by the birds and manure.

Estimating the Heat Entering through the Walls and Roof

The amount of heat which enters depends on the area of the walls and roof, the amount of insulation or the heat conductance of the structure and the difference between inside and outside temperature. This can be estimated using the following formula:

> Quantity of heat = area of heat transfer × heat conductivity of roof × temperature difference between inside and outside of house

Area of heat transfer is estimated by taking 1.25 × floor space in square metres (this allows for the sloping roof and wall surface). *Heat conductivity values* for various types of construction are given in Table 4.4. *Average temperature difference* between inside and outside temperature is obtained by subtracting 3.9°C (7°F) from the difference between the design dry-bulb and the design wet-bulb temperature.

If design temperatures are not available for the area where the poultry house will be located, appropriate design temperatures may be estimated from weather records. These records will be more reliable if they cover a long period of time (e.g. 10 years or more). Generally, a design temperature should never be exceeded by more than 3°C and should not be exceeded on more than 3–5% of daily temperature records. Remember that you are not designing for average high temperatures but rather for extremes. The design wet-bulb temperature should be estimated from the wet-bulb temperature which occurs at the time of the daily high dry-bulb temperature. If accurate temperature records are not available, it is advisable to select design temperatures with a considerable margin of safety.

Estimating Heat Produced by the Birds

This can be estimated using the following formula:

> Quantity of heat (kcal min^{-1}) = number of birds × average weight per bird × heat production per minute per kg

An adult chicken produces 0.0686 kcal min^{-1} kg^{-1} (0.124 BTU min^{-1} lb^{-1}) of body-weight. Young birds weighing 1.5 kg or less produce 0.0832 kcal min^{-1} kg^{-1} body-weight (0.15 BTU min^{-1} lb^{-1}). Large birds weighing 3 kg or more produce 0.0054 kcal min^{-1} kg^{-1} body-weight (0.10 BTU min^{-1} lb^{-1}).

Calculating Total Heat Entering the Building
Total heat load = heat from building + heat from birds

Calculating Cooler Capacity
The calculation is based on the objective of keeping the inside temperature low (about 3.9°C above the wet-bulb temperature).

Cooler capacity = total heat load/0.7652 (0.086 for BTU min^{-1})
A cubic metre of air absorbs 0.7652 kcal when it is warmed 2.8°C).

Calculating Pad Area Required
Pad area required (m^2) = cooler capacity (m^3 min^{-1})/design air speed for type of pad used; (ft^2) = cooler capacity (cubic feet per minute (c.f.m.))/design air speed for type of pad used. Design air speed for an aspen fibre pad 5 cm (2 in.) thick is 46 m min^{-1} (150 feet per minute (f.p.m.). Commercial pads are now available which can cool air effectively at higher air speed. The manufacturer should be consulted to obtain the appropriate design air speed for other types of pads.

Sample calculation (SI units)
A broiler house will be constructed 12 m × 150 m to house 20,000 broilers which will be raised to a maximum weight of 2.5 kg each. The house will have an average R value of 14; design dry-bulb temperature is 46°C, design wet-bulb temperature is 30°C.

Building heat = (12 × 150 × 1.25) × (0. 0059) × [(46 − 30) − 3.9] = 161 kcal min^{-1}

Bird heat = 20,000 × 2.5 × 0.0686 = 3430 kcal min^{-1}
Total heat load = 161 + 3430 = 3591 kcal min^{-1}
Cooler capacity = 3591/0.7652 = 4693 m^3 min^{-1}
Pad area required = 4693/46 = 102 m^2 (for an aspen fibre pad 5 cm thick)

Sample calculation British units
A broiler house will be constructed 40 ft × 500 ft to house 20,000 broilers, which will be raised to a maximum weight of 5 lb each. The house will have an average R value of 14; design dry-bulb temperature is 115°F; design wet-bulb temperature is 80°F.

Building heat = (40 × 500 × 1.25) 25,000 × 0.0012 × [(115 − 80) − 7] = 840 BTU min^{-1}

Bird heat = 20,000 × 5 × 0.124 = 12,400 BTU min^{-1}

Total heat to be removed = 840 + 12,400 = 13,240 BTU min^{-1}

Cooler capacity (c.f.m.) = 13,240/0.086 = 153,953 c.f.m.

Pad area required = 153,953/150 = 1026 sq. ft (for an aspen fibre pad 2 in. thick)

Pad system design

Several types of pad material can be used. Early pads were composed of wood fibre about 5 cm (2 in.) thick (see Fig. 4.9); these pads perform best with an air speed through the pad of 46 m min^{-1} (150 f.p.m.). Several commercial pads are now available which can cool air efficiently with higher air speeds. A higher air speed through the pad translates to a smaller pad area requirement. To ensure efficient cooling, be sure to use the appropriate design air speed for the type of pad installed.

Pad systems should be designed for uniform wetting and recovery of the water. A 5 cm diameter plastic pipe with 0.32–0.48 cm ($\frac{1}{8}$ to $\frac{3}{16}$ in.) holes drilled every 15 cm (6 in.) makes an acceptable water distribution system. The recirculation pump should have a capacity of about 4.1 l of water per minute per linear metre of pad ($\frac{1}{3}$ gal. min^{-1} ft^{-1} pad). The system should be equipped with a gutter to catch the overflow, which is returned to a sump. The sump should have a capacity of 15.5 l of water per metre of pad (1.25 gal. ft^{-1}). Water addition is conveniently controlled with a float valve. A filter should be installed in the return line to remove particles which might clog the holes in the distribution line. Some water should constantly be removed (bled) from the system to prevent build-up of minerals on the pad: 16 l h^{-1} bleed-off for every 100 m^3 min^{-1} cooling capacity is adequate.

Fan louvers

Louvers are used to prevent air backflow through fan openings when certain fans are not operating. If some fans are designed to operate constantly, they should not have louvers attached because louvers reduce fan output. Fans that will be operated intermittently should always have louvers attached. During very hot weather periods, when the fans are on continuously, it may be desirable to prop the louvers on these fans open to reduce back pressure and increase fan output. Mechanically operated louvers are available which result in less air resistance as the louver is not held open by the air pressure.

Fig. 4.9. Exterior view of a wood fibre evaporative cooling pad.

Fan safety guards

Fans that are within reach of personnel should always have safety guards attached. The author sustained a permanent injury from an unguarded fan on a poultry house and would like to emphasize that there should never be an exception to this rule. When the guard is installed within 10 cm (4 in.) of the moving parts, a woven wire mesh of at least 16 gauge with 1.3 cm ($\frac{1}{2}$ in.) openings should be used. If the guard can be installed further than 10 cm (4 in.) from the moving parts, a 12 gauge 5 cm (2 in.) mesh screen has less air resistance.

Pad placement, fan location and air inlet design

Pads can be located anywhere in the walls of the house or in a special structure attached to the building.

EARLY HOUSE DESIGNS

In early house designs for evaporatively cooled poultry houses, pads were often located in one wall with the fans in the opposing wall. This design worked best when all fans were in use but, with the slower air speed at the inlet that occurred during cooler weather (recommended design speed for wood fibre pads is 46 m min^{-1} (150 f.p.m.)), uniformity of air movement suffered.

Many of the early positive-pressure house designs utilized commercially available evaporative coolers spaced along the ridge of the building, with an exhaust slot low on the side-walls. This system worked but electrical efficiency was not as good as pad and fan systems because smaller fans and pump motors were used in these coolers. These systems were more difficult to maintain because each cooler used a fan and a water pump and their roof location made access more difficult. Most poultry farmers with this system installed a walkway on the roof connecting the coolers, a garden hose on the roof to clean the pads and a permanent ladder to reach the walkway.

HIGH-RISE DESIGN

Cage laying houses are often designed with a manure storage area below the cages (height 1.5–2.4 m (5–8 ft)). Walkways are installed below the lowest cage level for caretaker access to the birds (Fig. 4.10). With this design the fans are usually located in the manure storage area in order to ensure adequate air movement over the manure to enhance drying. Internal circulating fans are often needed with this system to assist manure drying (see Fig. 4.11).

A recent modification of this design, which is referred to as a Turbo™ house, uses a partial floor between the upper bird chamber and the lower manure storage chamber (Fig. 4.12). Manure falls through slots in this floor. This design prevents movement of ammonia and other gases from the manure storage to the bird chamber and increases manure drying as a result of directing air flow more uniformly over the manure.

SLOT-AND-FAN DESIGN

With this design air enters the house through an adjustable slot inlet which is located on both sides just below the eaves of the building. The cooling pad is often located vertically in a false wall which is attached to the side of the house. Fans should be located at least 1.8 m (6 ft) from the closest inlet. In conventional houses fans are usually placed in groups at one or more locations in the house and the air inlet starts 1.8–2.4 m (6–8 ft) on either side of the fan. In high-rise houses that use this design, fans should be equally spaced throughout the lower part of the wall in the manure storage area. This provides better air movement over the manure to increase drying.

This design has been successfully used with a high-pressure fogging system in place of the pad (see Fig. 4.13). The fogging system is easier to maintain and there are no pads to be cleaned and replaced. The water recycling system is also eliminated. The cooling efficiency of this system is somewhat less than a well-maintained pad system but it has proved

Fig. 4.10. A multiple-deck cage system in a high-rise laying house showing walkways (photo courtesy of Big Dutchman, Holland, Michigan).

Fig. 4.11. Interior view of a manure storage area in the lower portion of a high-rise cage layer house showing fans for manure drying.

Fig. 4.12. Turbo™ use design with multiple-deck cage system and manure storage area below cages. Air enters the house through the slots in the ceiling (photo courtesy of Chore-Time Equipment, Milford, Indiana).

adequate in many practical applications. It is common for growers to install two fogging lines in the chambers on each side of the house to provide a backup in the event one line needs to be shut down for maintenance or to be used for additional cooling during very hot conditions.

TUNNEL DESIGNS

Tunnel designs have been in use since evaporative pad systems were first developed. With this system the pad is put on one end of the building and the fans are located on the opposite end. This design results in high air velocity moving past the birds. The system works best in floor houses where the cool air falls to the floor where the birds are located. In long poultry houses there is considerable temperature difference between ends of the house. Bird migration toward the cooler end of the house can be a problem with this design. This can be controlled by placing cross fences in the building; gates are needed in the fences so that the caretaker can move throughout the house. This tunnel design has recently been installed in large numbers of broiler houses in the USA. The pads are usually closed during cold weather and a slot inlet is opened along the sides of the house so that the building is ventilated like a slot-and-fan design.

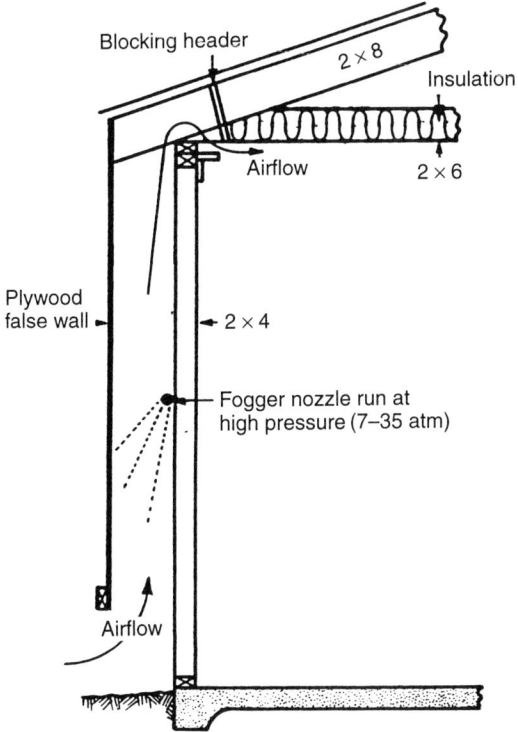

Fig. 4.13. Air inlet and cooling chamber design for a slot-and-fan poultry house (Parsons and Bell, 1968).

In single-tier cage laying houses this system is not recommended because the cool air moves below the cages without cooling the birds. This problem can be partially corrected by placing spoilers (a sheet of plastic from the floor up to the cage bottom) across the building at frequent intervals to force the air to mix with warmer air above; however, these spoilers interfere with caretaker movement and are unacceptable to most poultry caretakers. The design has been extensively used for multideck cage systems.

Attic Plena

The attic can be used as a plenum to distribute air throughout the building with either a negative or positive system. A flat ceiling is constructed, with air inlets which should be located to provide uniform air entry. For greatest effectiveness the area of the inlets should be adjustable so that adequate air speed (about 152 m min^{-1} (500 ft min^{-1})) can be maintained during periods when lower air volumes are needed to maintain good air conditions in the house (cooler periods).

Flex housing

This term is often applied to housing which can be operated either as open-type, naturally ventilated housing or as closed, fan-ventilated housing (Timmons and Baughman, 1986). Fans can be added to curtain side housing so that the house can be used with natural ventilation or with the fans exhausting air while the curtains are lowered to provide an appropriate inlet opening. Fans are usually arranged in banks and the curtains should not be closer than 1.8 m (6 ft) to an exhaust fan in order to avoid air short-circuiting. Flex housing offers economical natural ventilation when weather is favourable and forced ventilation when it is not. Flex houses can also be equipped with evaporative pad coolers or internal fogging systems to be used in hot weather.

Poultry House Maintenance

Direct-drive fans

Direct-drive fans require little maintenance except frequent cleaning. The blades, louvers and guard screens should be cleaned frequently. Fans can be most efficiently and safely cleaned with a jet of high-pressure air.

Belt-drive fans

Belt-drive fans need to have the belts checked at frequent intervals and adjusted when necessary. The belt adjustment can be tested by checking for a deflection of no more than 2.5 cm (1 in.); however, the use of a revolutions per minute (r.p.m.) counter on the fan shaft is a more reliable test (Parsons and Bell, 1968).

Pad maintenance

Pads collect dust, dirt and organic material, which should be cleaned off at regular intervals. A hose equipped with a nozzle that produces a concentrated spray is convenient for cleaning pads. The need for cleaning varies with local conditions. In some arid climates it may be desirable to protect pads from blowing sand (see Fig. 4.14). Algae build-up on pads can be prevented by addition of algicides (must be compatible with pad materials) and by running the fans for at least half an hour after the water is turned off to be sure the pads are dried quickly. The water distribution

Fig. 4.14. Sandstorm protection for an evaporative cooling pad.

pipe should be checked frequently for clogged holes and cleaned and flushed as necessary. The sump and distribution lines should be cleaned and flushed on a regular schedule. The supply line filter should be cleaned weekly or more often if needed.

Evaluating house and fan performance

Fan performance can be checked with a vane or 'pinwheel' anemometer (Parsons and Bell, 1968). The anemometer is moved uniformly over the fan opening for a 1-min period and the result is multiplied by the area of the opening to determine the air volume delivered by the fan. This result should approach the manufacturer's rating for the fan under comparable static pressure.

Cooling-pad efficiency can be checked by comparing the dry-bulb temperature inside the pad with the outside wet-bulb temperature. The inside temperature should not be more than 1–1.5°C above the outside wet-bulb temperature. If it is more than this, the system should be checked for the problems previously discussed.

Uniformity of ventilation is particularly important in cage layer housing because the birds can't move away from an area with poor air quality. Air movement in closed poultry houses can be visually checked by smoke from a bee smoker or smoke stick. Temperature uniformity, monitored in several locations, can be useful in determining how the system is working under a variety of weather conditions. Ammonia will often build up in dead spots and can be detected by smell or ammonia testing equipment.

Monitoring House Performance

An incline water manometer is often used to monitor static pressure in fan-ventilated poultry houses. Negative-pressure houses are usually designed to operate at 0.32 cm (0.125 in.) of static pressure. Positive-pressure houses usually perform best at slightly higher static pressure.

Emergency Systems

Several types of alarm systems can be used to alert management of a power outage or mechanical problem (Parsons and Bell, 1968). These systems can sound an audible alarm, dial one or more telephone numbers or activate an emergency power generator. Some of the types of alarms commonly used are: (i) high-temperature alarm; (ii) no electrical power; (iii) no air velocity from continuous fan(s); and (iv) smoke detector.

A standby power generator is essential for most fan-ventilated houses where the power supply is subject to interruption. There are designs where the 'flex' concept can be used with low-density housing so that birds can be ventilated adequately during outages without a standby generator. This is probably not a good idea in areas which have extremely high temperatures on a regular basis. If it is attempted, a gravity watering system should be considered to provide water during power outages. Simple emergency systems are available with electronic magnetic closures which open a poultry curtain or panels when the power fails.

References

Acme Engineering & Manufacturing Corporation (1983) *Environmental Control Handbook for Poultry Confinement Operations*. Acme Engineering & Manufacturing Corp., Muskogee, OK.
Albright, L.D. (1990) *Environment Control for Animals and Plants*. American Society of Agricultural Engineers, St Josepth, MI.
Esmay, M.L. (1978) *Principles of Animal Environment*. AVI Publishing, Co., Westport, CN.
Hart, S.A., Wilson, W.O. and Lett, P.J. (1964) *Light and Temperature*

Controlled Housing for Poultry. Circular 526, University of California, Agricultural Experiment Station, Davis, CA.

Leeson, S. and Summers, J.D. (1991) *Commercial Poultry Nutrition*. University Books, Guelph, Ontario, Canada.

Midwest Plan Service (1989) *Natural Ventilation Systems for Livestock Housing*. Iowa State University, Ames, IA.

Midwest Plan Service (1990a) *Heating, Cooling and Tempering Air for Livestock Housing*. Iowa State University, Ames, IA.

Midwest Plan Service (1990b) *Mechanical Ventilating Systems for Livestock Housing*. Iowa State University, Ames, IA.

North, M.O. and Bell, D.D. (1990) *Commercial Chicken Production Manual*. Van Nostrand Reinhold, New York, pp. 175–226.

Parsons, R.A. and Bell, D.D. (1968) *Summer Operation of Environmental Poultry Houses*. Leaflet AXT-239, University of California, Cooperative Extension Service, Davis, CA.

Siegel, H.S. and Drury, L.N. (1968) Physiological responses of chickens to variations in air temperature and velocity. *Poultry Science* 47, 1120–1127.

Sturkie, P.D. (ed) (1976) *Avian Physiology*. Springer-Verlag, New York.

Timmons, M.B. (1989a) Principles of cooling poultry houses. *Proceedings 1989 Poultry Symposium*. University of California, 1–16.

Timmons, M.B. (1989b) Improving ventilation in open-type poultry housing. *Proceedings 1989 Poultry Symposium*. University of California, 1–8.

Timmons, M.B. and Baughman, G.R. (1986) Flexible housing for broiler breeders. *Poultry Science* 65, 253–257.

Timmons, M.B., Baughman, G.R. and Parkhurst, C.R. (1981) Development and evaluation of an insulated curtain for poultry houses. *Poultry Science* 60, 2585–2592.

Timmons, M.B., Baughman, G.R. and Parkhurst, C.R. (1986) Effects of supplemental ridge ventilation on curtain-ventilated broiler housing. *Poultry Science* 65, 258–261.

5

Nutrient Requirements of Poultry at High Temperatures

N.J. Daghir
Faculty of Agricultural Sciences, UAE University, Al-Ain, UAE.

Introduction	101
General Temperature Effects	102
Energy Requirements	103
Protein and Amino Acid Requirements	106
Vitamins	109
Vitamin C	109
Vitamin A	111
Vitamins E and D_3	111
Thiamine	112
Mineral Requirements	112
Calcium	112
Phosphorus	114
Potassium	115
Turkey Nutrition at High Temperatures	116
Non-nutrient Feed Additives	116
Antibiotics	117
Aspirin	117
Coccidiostats	117
Reserpine	117
Flunixin	118
Conclusions and Recommendations	118
References	119

Introduction

Nutrition under temperature stress is one area of research that is beginning to receive a great deal more attention than it did in the past. There have been at least five reviews on this subject in the 1980s. The first was by Moreng (1980), in which he reviewed the effects of temperature on vitamin requirements of poultry. The second review was by the National Research Council (NRC, 1981), which appeared in the publication *Effect of Environment on Nutrient Requirements of Domestic Animals* and dealt mainly with feed intake as affected by temperature changes, efficiency of production, metabolizable energy (ME) requirements and water intake. The third review, by Austic (1985), was published as one of the chapters in *Stress Physiology in Livestock*, Vol. 3, *Poultry* and covered, in a concise and comprehensive way, energy, protein, amino acids, vitamins, minerals, essential fatty acids and water. The fourth review was by Leeson (1986), who dealt with various nutritional considerations during heat stress, and gave some specific dietary manipulations which could lead to improvement in performance of broilers and layers under heat-stress conditions. The fifth and last review was by Shane (1988), who discussed the interaction of heat and nutrition, with special emphasis on the immune system.

General Temperature Effects

Before getting into specific nutrient requirements, it should be stated that there is a great deal of disagreement as to what is the ideal temperature range for the different classes and age-groups of poultry. This is probably due to the fact that many factors influence the reaction of poultry to temperature changes. Humidity of the atmosphere, wind velocity and previous acclimatization of the bird are among the most important. Birds, in general, perform well within a relatively wide temperature range. This range, which extends between 10 and 27°C, is not too different for broilers, layers or turkeys (Milligan and Winn, 1964; de Albuquerque *et al.*, 1978; Mardsen and Morris, 1987).

Kampen (1984) found that the highest growth rate of broilers occurs in the range of 10–22°C. However, maximum feed efficiency is at about 27°C. In layers, he reported that, in the temperature range 10–30°C, the net energy available for egg production is almost constant and feed costs per egg are minimal at 30°C.

As for the optimum temperature range, what is ideal for growth is not ideal for feed efficiency, and what is ideal for feed efficiency is not ideal for egg weight. For example, we know that feed efficiency is always reduced at temperatures below 21°C (70°F). Egg production and growth rate are reduced at temperatures below 10°C (50°F). The overall optimum range is mainly dependent on the relative market value of the product produced, in proportion to feed cost. As the price ratio widens, the best temperature falls, and vice versa.

Another general consideration that is very relevant to this discussion is feed intake. There is no question that high and low environmental temperatures impose limitations on the performance of both broilers and laying hens that are unrelated to feed intake. Only part of the impairment in performance is due to reduced feed intake. NRC (1981) summarized several papers on laying hens, and concluded that the decrease in feed intake is about 1.5% per 1°C, over the range of 5–35°C, with a baseline of 20–21°C. Austic (1985) summarized work on growing chickens and concluded that it was 1.7% per 1°C over a base line of 18–22°C. Response of birds subjected to cyclic temperatures is not too different from that due to constant temperature. This decrease in feed intake is not linear, but becomes more severe as temperature rises. Table 5.1 shows the change in percentage decrease per 1°C rise in temperature as calculated and summarized from 12 references.

Some workers have tried to partition the detrimental effects on performance into those that are due to high temperature *per se* and those due to reduced feed intake, by conducting paired-feeding experiments. Smith and Oliver (1972), in work with laying hens subjected to 21°C and 38°C temperatures, showed that 40–50% reduction in egg production and egg weight at 38°C is due to reduced feed intake, while the reductions in shell thickness and shell strength are mainly due to high temperature. Dale and Fuller (1979), in work with broilers, showed that 63% of the reduction in growth rate is due to reduced feed intake. In contrast to paired feeding, another interesting approach to this area of research that needs to be pursued is testing the effect of force-feeding of birds in a hot environment on different performance parameters in layers and in broilers.

Energy Requirements

ME requirement decreases with increasing temperature above 21°C. This reduced requirement is mainly due to a reduction in energy requirements for maintenance and the requirement for production is not influenced by environmental temperature.

We have observed for many years that energy consumption during the

Table 5.1. Effect of temperature on feed intake of laying hens (values summarized from various references).

Temperature (°C)	% Decrease per 1°C rise
20	—
25	1.4
30	1.6
35	2.3
40	4.8

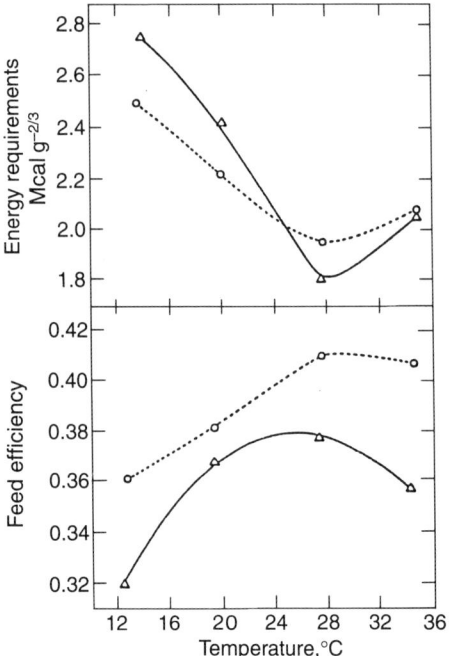

Fig. 5.1. Feed efficiency (lower graph) and energy requirements (upper graph) for maintenance of male (○) and female (△) broilers as functions of environmental temperature. (Hurwitz et al., 1980.)

summer drops significantly in contrast to winter or spring (Daghir, 1973). Energy intake during the summer was 10–15% lower than during the winter.

Energy requirement for maintenance decreases with environmental temperature to reach a low at 27°C, followed by an increase up to 34°C. This has been demonstrated by Hurwitz et al. (1980) in work with broilers. Figure 5.1 shows the effects of environmental temperature on both feed efficiency and energy requirements for maintenance of male and female broilers.

The use of high-energy rations for broilers has become quite common in warm regions. Some workers feel that this practice should be accompanied by raising the levels of the most critical amino acids. This is based on a few reports, one of which is McNaughton and Reece (1984), who found a significant weight response to additional energy only in the presence of relatively high levels of lysine (Fig. 5.2). They concluded that a diet energy response in warm weather is seen only when adequate amino acid levels are provided. This approach may increase performance, but it will also increase the heat load on the bird and its ability to survive.

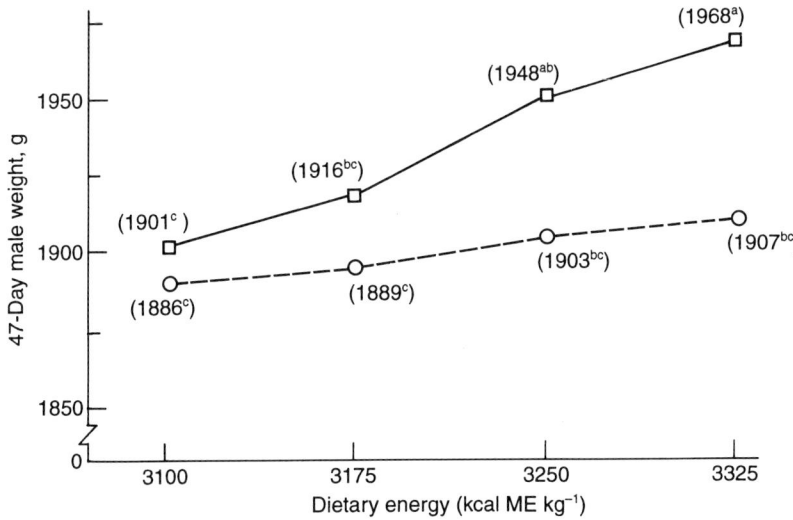

Fig. 5.2. Influence of dietary energy level on the lysine requirement of 23- to 47-day broiler males. Values in parentheses show mean of each treatment. Means followed by different small letters are significantly different ($P < 0.05$). Lysine at 0.308 (dashed line) or 0.322 (solid line) % Mcal^{-1} kg^{-1}. (McNaughton and Reece, 1984.)

The beneficial use of fats in hot-weather feeding programmes is well documented. We have known for some time now that the addition of fat stimulates feed and ME consumption at high temperatures. One of the very early reports on this was by Fuller and Rendon (1977), who showed that broilers fed a ration in which 33% of the ME was supplied by fat consumed 10% more ME and 10% more protein and gained 9% more weight than chicks fed a low-fat ration. Reid (1979) also observed improved performance in laying hens as a result of adding fat at high temperatures. We have observed that added fat at 31°C improved feed consumption in laying hens to a greater extent than at lower temperatures (Daghir, 1987). Table 5.2 shows that the addition of fat to laying rations increased feed intake by 17.2% at 31°C and only 4.5% at lower temperatures (10–18°C).

We have learned a great deal in recent years about the value of fat at

Table 5.2. Interaction of temperature and added fat on feed consumption (g hen^{-1} day^{-1}) (from Daghir, 1987).

Temperature (°C)	Added Fat (%)		% Increase
	0	5	
31	93	109	17.2
10-18	127	133	4.5

high temperatures. The higher fat content of the diet contributes to reduced heat production, since fat has a lower heat increment than either protein or carbohydrate. Energy intake is increased in both broiler and laying hens in a warm environment by the addition of fat. The addition of fat to the diet appears to increase the energy value of the other feed constituents (Mateos and Sell, 1981). Fat has also been shown to decrease the rate of food passage in the gastrointestinal (GI) tract (Mateos *et al.*, 1982) and thus increase nutrient utilization. This is interesting in light of an earlier observation by Wilson *et al.* (1980) who found that high environmental temperature caused an increase in food passage time in white Pekin ducks. Dietary fat can therefore help in counteracting this effect of high temperature.

Geraert *et al.* (1992) investigated the effect of high ambient temperature (32°C vs. 22°C) on dietary ME value in genetically lean and fat 8-week-old male broilers. Lean broilers exhibited higher apparent ME (AME) and true ME (TME) values than fat broilers. Hot climatic conditions significantly increased AME and TME values, particularly in leaner birds. Protein retention efficiency was enhanced by selection for leanness and increased with ambient temperature.

Protein and Amino Acid Requirements

Temperature changes neither increase nor decrease the requirement for protein. Bray and Gesell (1961) reported that egg production could be maintained at 30°C provided a daily protein intake of about 15 g was ensured by appropriate dietary formulation. Figure 5.3 shows that temperature has no effect on egg numbers as long as protein intake is maintained. Egg mass, however, reaches a plateau at 30°C earlier than at 24.4 or 5.6°C. Today we find responses from increasing protein intake to 18 g day^{-1}, and even 20 g day^{-1} in some strains, but this is independently of temperature.

Amino acid requirements as affected by temperature have been studied for many years. March and Biely (1972) demonstrated that high temperatures (31°C) do not affect the metabolic requirement for lysine. They carried out feeding experiments with chicks raised under two temperature conditions (20°C and 31°C). Chicks were fed diets with graded levels of lysine (0.73, 0.88, 1.03 and 1.33%). Two separate growth response lines were obtained when feed intake was related to body-weight gain at the two temperatures (Fig. 5.4). These results indicate that growth is reduced at the same rate as feed intake at the high temperature. March and Biely (1972) showed that, when these data are plotted on the basis of lysine intake versus body-weight gain, one response line can describe the relationship for both temperatures (Fig. 5.5). Balnave and Oliva (1990) reported that the

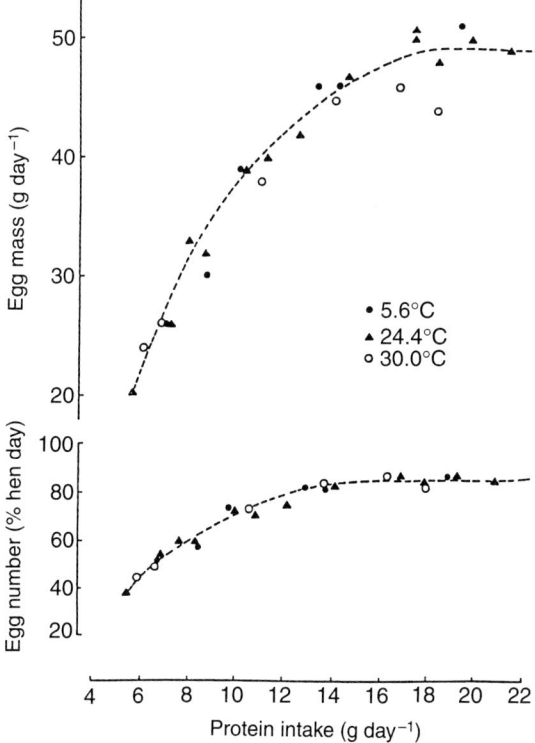

Fig. 5.3. Egg mass (upper curve) and number (lower curve) at different ambient temperatures, as affected by protein intake. Adapted from Bray and Gesell, 1961.

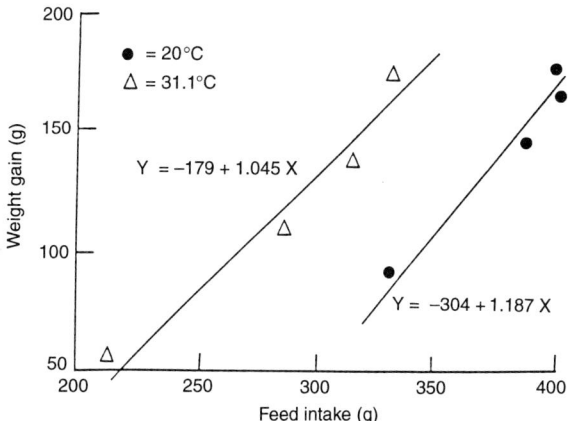

Fig. 5.4. Relationship between feed intake and weight gain of White Leghorn chicks fed for 15 days on diets with 0.73, 0.88, 1.03 or 1.33% lysine, at two ambient temperatures. (March and Biely, 1972.)

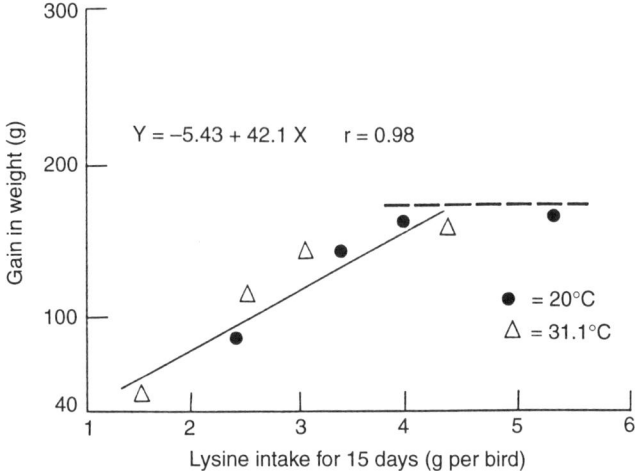

Fig. 5.5. Relationship between accumulative intake of lysine and accumulative growth of chicks fed for 15 days on diets with 0.73, 0.88, 1.03 or 1.33% lysine, at two ambient temperatures. (March and Biely, 1972.)

methionine requirement of 3–6-week-old broilers kept at a constant 30°C or at cycling temperatures of 25–30°C was reduced compared with broilers kept at 21°C.

The industry has followed the practice of adjusting the dietary levels of protein and amino acids in order to maintain a constant intake of these nutrients as house temperatures and thus feed intake levels vary. This is based on the assumption that temperature does not affect the efficiency with which amino acids are utilized for tissue growth or egg production. Hurwitz *et al.* (1980) suggested a method of estimating the protein and amino acid requirements which takes into account the reduced rate of production at high temperatures. They used a mathematical model which evaluates the amino acid requirements on the basis of the sum of the maintenance requirements per kcal or as percentage of diet increase as environmental temperature increases above the optimum range for growth. Then, as the temperature of 28–30°C approached, there was a decline in the amino acid requirements. This was demonstrated for arginine, leucine and the sulphur amino acid requirements. On this basis, Hurwitz *et al.* (1980) suggested that the use of a linear relationship between the amino acid requirements in feed formulation should be reconsidered. Austic (1985) recommended that we continue to increase amino acid levels as percentage of the diet up to 30°C. Beyond that temperature, further increases are not justified because both growth and egg production will be depressed.

Sinurat and Balnave (1985) reported that the food intake and growth rate of broilers in a cycling temperature (25–35 °C) were improved not only by increasing the dietary ME, but also by reducing the amino acid : ME ratio during the finishing period. The same workers (Sinurat and Balnave, 1986) observed that broilers at high temperatures (25–35 °C) on a free-choice system selected similar amino acid : ME ratios which were lower than the ratios contained in the complete diet. This is in agreement with work that goes back a number of years (Waldroup *et al.*, 1976), in which it was shown that growth rate and feed utilization of heat-stressed broilers was significantly improved when diets were formulated to minimize excess of amino acids. This is theoretically sound, since minimizing protein levels and improving the balance of amino acids should minimize the heat increment and thereby reduce the amount of heat which must be dissipated. Zuprizal *et al.* (1993) studied the effects of high temperature on total digestible protein (TDP) and total digestible amino acids (TDAA) of rape-seed meals and soyabean meals fed to broilers at 6 weeks of age. TDP and TDAA of both rape-seed and soyabean meals decreased as the ambient temperature increased from 21 to 32 °C. A 12 and 5% reduction in TDP value was observed with the rape-seed and soyabean meals respectively. Wallis and Balnave (1984) studied the influence of environmental temperature, age and sex on the digestibility of amino acids in growing broiler chickens. Although sex had no major effect on amino acid digestibility at 30 or 50 days of age, the influence of environmental temperature was found to be sex-related. There was a decrease in amino acid digestibility at higher temperature in female but not in male birds.

Scott and Balnave (1988) conducted studies to evaluate the use of diets varying in energy and nutrient density to overcome the nutritional stresses associated with the onset of lay and with periods of high temperature. Changes in dietary ME concentration had little influence on food and nutrient intake and egg mass output of hens in early lay kept at normal, cold or hot temperatures. At normal and cold temperatures, all dietary ME-nutrient density combinations allowed hens to meet the recommended daily protein intake, but only those fed the most concentrated diets were able to meet this recommendation at hot temperatures. Egg mass output of hens at hot temperature remained inferior even with the highest ME and protein intakes.

Vitamins

Vitamin C

Vitamin C is perhaps the most studied nutrient in relation to ambient temperature, and yet its effects are still not fully defined. There is some evidence which indicates that, under conditions of high environmental temperatures, some mammals and birds are not able to synthesize sufficient ascorbic acid to replace the severe losses of this vitamin that occur during stress. As early as 1961, Thornton showed that blood ascorbic acid decreased with an increase in environmental temperature from 21 to 31°C. This action was postulated to be a result of both partial exhaustion of the endogenous stores and a reduction in the amount of vitamin being synthesized. Ahmad *et al.* (1967) also showed that ascorbic acid limited the increase of body temperature during heat stress up to 35°C. Supplemental ascorbic acid has also been reported to improve heat resistance and reduce mortality associated with elevated ambient temperatures (Pardue *et al.*, 1984). Pardue *et al.* (1985a), in a study on the effects of high environmental temperatures on broilers, showed that ascorbic acid reduced mortality in both female and male broilers during a heating episode (38°C at bird level) in the production facility. Ascorbic acid supplementation stimulated growth in females during the early phase of growth, but not males. Kafri and Cherry (1984) also showed improved growth rate at 32°C in males, but not females. This suggests a need to investigate sex differences in ascorbic acid synthesis and/or metabolism.

Njoku (1984) showed improved growth performance of broilers reared in the tropics when their diet was supplemented with 200 mg kg^{-1} ascorbic acid. The same worker, in a second study, also showed improvement in gain to food ratios in broilers (Njoku, 1986).

A report by Thaxton (1986) showed that, following infection with infectious bursal disease (IBD) virus, vitamin C protected the immune biological tissues in growing birds and reduced their mortality to infection in a hot environment. Furthermore, Pardue *et al.* (1985b) showed that immunosuppression at high environmental temperatures could result from a reduction in thyroid activity. Therefore, it may be that the reduction of weight loss of an immunocompetent organ induced by feeding ascorbic acid may be associated with thyroid activity. Takahashi *et al.* (1991) studied the effects of supplemental ascorbic acid on broilers treated with propylthiouracil. Feeding ascorbic acid partly prevented the decreases in body-weight gain, feed conversion and weights of the bursa of Fabricius and thymus in chicks fed propylthiouracil. They suggested that ascorbic acid improves the performance of chicks with experimentally-induced hypothyroidism.

In work with laying hens, ascorbic acid supplementation was shown to improve egg weight, shell thickness and egg production (Perek and Kendler, 1962, 1963). Njoku and Nwazota (1989) found that the inclusion of ascorbic acid in the diet improved egg production, food intake and food utilization and decreased the cost of feed per kg egg. The addition of 400 mg ascorbic acid kg^{-1} diet gave the most efficient performance. Palm oil inclusion in the diet also reduced the effect of heat stress and increased egg production, egg weight, food intake and efficiency of utilization.

Ascorbic acid and palm oil when fed alone or in combination reduced the incidence of cracked eggs. These authors concluded that the addition of both ascorbic acid and palm oil ameliorated the effects of heat stress in a hot tropical environment.

With broiler breeders, dietary ascorbic acid supplementation has been shown to improve nutrient utilization as judged by the production of hatching eggs (Peebles and Brake, 1985). Wide variations are encountered in the recommendations for supplementing poultry diets with ascorbic acid. Maumlautnner *et al.* (1991) suggested that these variations might be due to high losses during storage at high temperature. They tested three forms of vitamin C (crystalline ascorbic acid, protected ascorbic acid and phosphate–ascorbic acid ester). These three were fed to hens from 21 to 30 weeks of age and the birds were kept at 20 or 34°C. Performance and eggshell quality of treated hens improved only at 34°C. The best results were with the protected ascorbic acid and the phosphate–ascorbic acid ester.

Vitamin A

Vitamin A requirements, as affected by temperature, have been investigated by several workers. As early as 1952, Heywang showed that hot weather caused a marked increase in vitamin A requirements of the laying and breeding chicken. Kurnick *et al.* (1964) reported more vitamin A storage in the liver in Leghorn pullets during cool periods than in those fed during hot weather. Smith and Borchers (1972) suggested that environmental temperature is of minor importance as a factor influencing the conversion of β-carotene to vitamin A. Elevated body temperatures may interfere with absorption. A review by Scott (1976) documented a threefold increase in vitamin A requirement of breeder hens at 38°C compared with those at normal room temperature.

Moreng (1980) reviewed the literature on the effects of high temperature on vitamin A requirements. His review shows that some evidence suggests improved bird performance in response to additional vitamin A. The differences in performance, however, are not always statistically significant.

Vitamins E and D_3

It is known that vitamin E requirements increase with increased stresses, particularly those that are related to high temperature (Cheville, 1979). Generally, vitamin E serves as a physiological antioxidant through inactivation of free radicals and thus contributes to the integrity of the endothelial cells of the circulatory system. High environmental temperature may exert an effect on health and performance by modifying the cellular and hence dietary requirement of vitamin E (Heinzerling *et al.*, 1974). Scott (1966) suggested that heat stress interferes with the conversion of vitamin D_3 to the active form, an important step for calcium metabolism. There is still no clear evidence that adding vitamin E or D_3 has any beneficial effect during heat stress.

Thiamine

Mills *et al.* (1947) found that the requirement for this vitamin was significantly increased for chicks grown at 32.5°C, as compared with 21°C. They could not detect a change in requirement for pyridoxine, nicotinic acid, folic acid or choline.

Mineral Requirements

Calcium

Egg weight and eggshell strength decline at high environmental temperatures. This appears to be partly due to reduced calcium intake, but several physiological mechanisms are involved: (i) reduced blood flow through shell gland due to peripheral vasodilatation; (ii) respiratory alkalosis; (iii) reduced blood ionic calcium content; (iv) reduced carbonic anhydrase in shell gland and kidneys; and reduced Ca mobilization from bone stores.

The methods that have been used to reduce environmental temperature effects on shell quality can be summarized as follows.

1. $NaHCO_3$ has been added to feed, but has not always given positive responses.
2. Carbon dioxide-enriched atmosphere has been very effective, but it is not practical if it involves reduced ventilation.
3. Introduction of oyster shell or hen-sized sources of Ca separately from other nutrients has been a very useful means of improving shell quality. The results of a study reported by Sauveur and Picard (1987) are shown in Table 5.3.
4. Night cooling has been very effective as a tool in maintaining shell quality at high temperatures.

There is very little work on the combined effects of temperature and humidity on performance. It is somewhat accepted that high humidity aggravates the detrimental effects of high temperature. Picard *et al.* (1987) studied the effects of high temperature and relative humidity on egg composition. They found that, when high temperatures are combined with high humidity, there is a further decrease in egg weight and shell weight. There is also a decrease in egg components (Table 5.4).

Table 5.3. Effects of temperature and Ca source on egg weight and shell quality (from Sauveur and Picard, 1987).

Diet		End of control period (20°C)	2nd day at 33°C	24-28th day at 33°C
Pulverized limestone	Egg wt (g)	61.1	60.0	57.7
	Eggshell wt (g)	5.84	4.47	5.22
	SWUSA*	8.04	6.26	7.48
Oyster shell	Egg wt (g)	62.2	60.0	58.9
	Eggshell wt (g)	5.82	4.96	5.42
	SWUSA	7.93	6.90	7.67

*SWUSA = shell weight per unit surface area (g per 100 cm^2).

Table 5.4. Effects of high temperature and relative humidity on egg composition (from Picard et al., 1987).

Temperature/relative humidity	20°/50%	33°/30%	33°/85%
Egg weight (g)	58.1	56.2	54.2
Eggshell weight (g)	5.72	5.47	4.89
Yolk weight (g)	15.0	14.3	13.9
Albumen weight (g)	37.5	36.4	35.3
Yolk dry matter (%)	52.7	51.0	50.9
Albumen dry matter (%)	12.8	12.4	12.0

Another dietary manipulation, tested by Odom *et al.* (1985), was to study the effect of chronic drinking of carbonated water on production parameters of laying hens exposed to high ambient temperature (Table 5.5). Their data indicate that the use of carbonated drinking water during periods of hot weather can help relieve the associated problem of high environmental temperature-induced eggshell thinning. This has recently been confirmed by Koelkebeck *et al.* (1992), who reported that a carbonated drinking water system can be operated efficiently in a commercial cage layer facility and can help improve eggshell quality of flocks experiencing shell quality problems during the summer. The test was performed in a cage layer facility on both 46- and 86-weeks-old commercial layers during a 12-week period in the summer.

One of the problems of the fowl processing industry is that processing yield from spent hens is low due to a high incidence of bone breakage. After one year of production, the hens' bones weaken, causing them to

Table 5.5. Effect of carbonated water on shell quality (from Odom et al., 1985).

		Temperature (°C)	
	Water treatment	23	35
% Eggshell	Tap	9.89	8.44
	Carbonated	9.84	8.88
Egg specific gravity	Tap	1.086	1.072
	Carbonated	1.088	1.074

break easily, and this increases the incidence of bone fragments in the processed meat. Koelkebeck *et al.* (1993), in an experiment on heat-stressed laying hens, found that providing these birds with carbonated drinking water improves tibia breaking strength. They suggested that carbonated drinking water during heat stress may reduce bone breakage during the processing of spent hens. This has also been previously reported in male chickens by Kreider *et al.* (1990), who showed that carbonated drinking water improved tibia bone breaking strength of cockerels exposed to a 37°C environment.

Phosphorus

Garlich and McCormick (1981) reported that Ca and P balance seems to have an effect on survival time during periods of acute heat stress. They showed a direct relationship between plasma phosphorus and survival time, and an inverse relationship with plasma calcium. Survival time in fasted chicks was greater when their previous diet contained low levels of Ca and high levels of P. This may be important to consider where feed withdrawal is practised to reduce mortality during high-temperature spells.

The detrimental effects of high P levels on shell quality during high temperature conditions have been well documented (Miles and Harms, 1982; Miles *et al.*, 1983). We have shown (Daghir, 1987) that the decrease in shell thickness at high temperatures is greater with higher P levels at high temperatures, than with low P levels (Table 5.6). Egg weight is affected similarly (Table 5.7).

Teeter *et al.* (1985) suggested that alkalosis and weight gain depression attributed to heat stress can be alleviated by dietary measures. They showed that including 0.5% $NaHCO_3$ in the diet enhanced body-weight gain by 9% (Table 5.8). Adding 0.3% or 1% NH_4Cl to diets increased body-weight gains by 9.5 and 25% respectively. Supplementing the 1% NH_4Cl diet with 0.5% $NaHCO_3$ increased weight gain an additional 9%. $CaCl_2$ addition had very little effect.

Table 5.6. Interaction of temperature and dietary phosphorus on shell thickness (mm) (from Daghir et al., 1985).

Available phosphorus (%)	Temperature (°C)		% Difference
	10-18	31	
0.45	0.362	0.336	−7.7
0.35	0.367	0.344	−6.7
0.25	0.351	0.348	−0.9

Table 5.7. Interaction of temperature and dietary phosphorus on egg weight (from Daghir et al., 1985).

Available phosphorus	Temperature (°C)		% Difference
	10-18	31	
0.45	60.7	47.4	−28.2
0.35	60.0	53.9	−11.4
0.25	56.9	52.1	−9.4

Table 5.8. Effect of NH_4Cl, $NaHCO_3$ and $CaCl_2$ on broilers (from Teeter et al., 1985).

Treatment	Body-weight gain (g)	
	Thermoneutral*	Hot*
Basal	933^A	442^C
Basal + 0.3% NH_4Cl		484^{BC}
Basal + 1% NH_4Cl		553^B
Basal + 3% NH_4Cl		464^C
Basal + 1% NH_4Cl + 0.5% $NaHCO_3$		594^B
Basal + 0.5% $CaCl_2$		481^{BC}
Basal + 1% $CaCl_2$		474^{BC}

* Means with different superscripts are significantly different ($P < 0.05$).

Potassium

It has been known for some time that the K requirement of growing chickens increases with increased temperature. Huston (1978) observed that blood K concentrations were reduced in growing chickens by high environmental temperature. The same has been reported for laying hens (Deetz and Ringrose, 1976). The K requirement was reported to increase from 0.4% of the diet at 25.7°C to 0.6% at 37.8°C.

Teeter and Smith (1986) and Smith and Teeter (1987) conducted several experiments to study the impact of ambient temperature and relative humidity (35°C and 70%) upon K excretion and the effects of KCl supplementation of broilers exposed to chronic heat and cycling temperature stress. They concluded that dietary K levels should be increased for birds reared in heat-stressed environments. A level of 1.5–2% total or 1.8–2.3 g K daily is needed to maximize gain in 5–8-week-old broilers. They also suggested that this could be added to drinking water at 0.24–0.3% K in the form of KCl. This may be preferred since birds are more likely to drink water than consume feed under heat stress.

Turkey Nutrition at High Temperatures

Hellickson et al. (1966) were among the earliest to report on the effects of temperature on the performance of broad-breasted turkeys. They observed optimum performance (weight gain) between 15.6 and 21.1°C. Above 26.1°C, there was an appreciable decrease in fat deposition.

Auckland and Liddle (1972) suggested that a higher dietary protein level might be required for early finishing of turkeys during summer because of the reduced feed intake at high temperatures. This was confirmed by Ward and Brewer (1984), who observed a 450 g extra gain in male turkeys in a four-ration regime, with 10% higher amino acid levels during hot weather.

de Albuquerque et al. (1978) observed not only reduced body-weight gain at high temperatures, but also changes in eviscerated yields, with the highest yields obtained at 18.3°C and the lowest at 35.0°C.

Hurwitz et al. (1980) compared the chicken and the turkey and concluded that the turkey was less resistant to high temperatures than the chicken. Feeding of a high-protein diet to turkeys did not prevent the growth depression resulting from high temperatures.

Rose and Michie (1987) compared the effects of four temperatures (14, 17, 20 and 23°C) on growing turkeys from 10 to 15 weeks of age, fed three protein levels. Although there were no significant differences in growth, breast meat yields were decreased at the higher temperatures. This could not be counteracted by high dietary protein concentrations.

Non-nutrient Feed Additives

Several non-nutrient feed additives have been tested to reduce the harmful effects of high-temperature stress. Examples of these follow.

Antibiotics

The incorporation of antibiotics in so-called 'stress feeds' has been widely practised all over the world. Extremes in temperature are among the stresses that have at times been handled by antibiotic feeding. The literature on this subject is very skimpy and reports are often contradictory (Freeman *et al.*, 1975).

Aspirin

Aspirin has been used as a tranquillizing drug in order to help birds subjected to stress, particularly heat stress. The effects of aspirin on growth and egg production have been studied by few researchers. The results of feeding this drug to broilers during heat stress have been variable. An early report by Glick (1963) showed that supplementation of acetylsalicylic acid (ASA) at 0.3% of the diet significantly improved growth. Later studies, however, using ASA levels ranging from 0.005% to 0.9%, found no improvement or even negative effects on growth rate (Reid *et al.*, 1964; Nakaue *et al.*, 1967; Adams and Rogler, 1968).

Few reports are available on the effects of dietary aspirin on laying chickens. Balog and Hester (1989) fed aged layer breeders 0.05% ASA for a period of 4 weeks. Aspirin reduced production of shell-less eggs in these birds, but had no effect on soft-shelled eggs.

The bases of heat stress reduction by ASA is not known. It may be through cyclooxygenase inhibition, since Edens and Campbell (1985) reported reduced heat stress in broilers given flunixin, a non-steroidal cyclooxygenase inhibitor.

Coccidiostats

McDougal and McQuistion (1980) studied the effects of anticoccidial drugs on heat-stress mortality in broilers. Overall mortality during the 8-week study averaged 6% in unmedicated or monensin-medicated birds, 10% in arprinocid-medicated birds and 36% in nicarbazin-medicated birds.

Reserpine

Several reports have shown improved heat tolerance occurring following administration of reserpine, an alkaloid that is extracted from the *Rauwolfia* plant. Edens and Siegel (1974) evaluated the role of reserpine in three experiments with young chicks. They showed that pretreatment with the alkaloid prevented the rapid loss of CO_2 which normally occurs when birds are subjected to acute high temperature and thereby stabilize acid–base blood status.

Flunixin

This anti-inflammatory analgesic has been studied by Edens (1986), who added it to the drinking water at levels ranging from 0.28 to 2.20 mg kg^{-1} body-weight day^{-1}. It was given to 5-week-old broilers that were exposed to 37°C for 5 hours per day. Flunixin treatment increased water consumption by 100 ml before heat treatment and during heat stress. Water consumption was increased by 150–300 ml per bird per day.

Conclusions and Recommendations

1. Nutritional manipulations can reduce the detrimental effects of high environmental temperatures, but cannot fully correct them, for only part of the impairment in performance is due to poor nutrition.
2. The addition of fat stimulates feed and ME consumption at high temperature. This beneficial effect of fat is well established in broiler feeding programmes and has been shown to be of value in pullet rearing and laying rations in hot climates.
3. Temperature changes neither decrease nor increase the requirements for proteins and amino acids. The only adjustments needed for temperature changes are those that ensure adequate daily intakes for the various amino acids. Furthermore, avoiding excesses of amino acids and giving more attention to amino acid balance is important in helping to overcome some of the adverse effects associated with high-temperature stress because nutritional imbalances are more detrimental in hot than in temperate climates.
4. Research conducted so far on vitamin requirements as affected by temperature changes does not indicate any change in the absolute requirements. There is some evidence in the literature to suggest improved bird performance at high temperatures in response to additional vitamins A and C.
5. Besides the effect on calcium and electrolyte balance, there is little work on the effect of high temperature on mineral metabolism. Calcium and phosphorus balance has some effect on survival time during acute heat stress. The practice of separate feeding of oyster shell or hen-sized sources of calcium continues to be the most effective dietary approach to reducing the effect of environmental temperature on shell quality.
6. Several reports indicate the need for increased dietary K in heat-stressed conditions. In doing this, care should be taken to prevent the negative effect that this may have on the acid–base balance.
7. High environmental temperatures not only decrease growth rate in turkeys, but also reduce eviscerated yields and breast meat yields. This decrease cannot be counteracted by high dietary protein concentration.

8. Several non-nutrient feed additives have been tested to reduce the harmful effects of high-temperature stress. Examples of these are antibiotics, aspirin, reserpine, flunixin, etc.

References

Adams, R.L. and Rogler, J.C. (1968) The effects of dietary aspirin and humidity on the performance of light and heavy breed chicks. *Poultry Science* 47, 1344–1348.

Ahmad, M.M., Moreng, R.E. and Muller, H.D. (1967) Breed responses in body temperature to elevated environmental temperature and ascorbic acid. *Poultry Science* 46, 6–15.

Auckland, J.N. and Liddle, P.C. (1972) Protein requirements for early finishing of male and female turkeys. *British Poultry Science* 13, 273–278.

Austic, R.E. (1985) Feeding poultry in hot and cold climates. In: Youssef, M. (ed.) *Stress Physiology In Livestock*, Vol. 3, *Poultry*, CRC Press, Boca Raton, Florida pp. 123–136.

Balnave, D. and Oliva, A. (1990) Responses of finishing broilers at high temperature to dietary methionine source and supplementation levels. *Australian Journal of Agricultural Research* 41, 557–564.

Balog, J.M. and Hester, P.Y. (1989) The effect of dietary ASA on egg shell quality. *Poultry Science* 68 (Suppl. 4), 9 (Abstract).

Bray, D.J. and Gesell, J.A. (1961) Studies with corn–soya laying diets. IV. Environmental temperature. *Poultry Science* 40, 1328–1335.

Cheville, N.F. (1979) Environmental factors affecting the immune response of birds – a review. *Avian Diseases* 23, 166–170.

Daghir, N.J. (1973) Energy requirements of laying hens in a semi-arid continental climate. *British Poultry Science* 14, 451–461.

Daghir, N.J. (1987) Nutrient requirements of laying hens under high temperature conditions. *Zootecnica International* 5, 36–39.

Daghir, N.J., Farran, M.T. and Kaysi, S.A. (1985) Phosphorus requirements of laying hens in a semi-arid continental climate. *Poultry Science* 64, 1382–1384.

Dale, N.M. and Fuller, H.L. (1979) Effects of diet composition on feed intake and growth of chicks under heat stress. I. Dietary fat levels. *Poultry Science* 58, 1529–1534.

de Albuquerque, K., Leighton, A.T., Mason, J.P. and Potter, L.M. (1978) The effects of environmental temperature, sex and dietary energy levels on growth performance of large white turkeys. *Poultry Science* 57, 353–362.

Deetz, L.E. and Ringrose, R.C. (1976) Effect of heat stress on the potassium requirement of the hen. *Poultry Science* 55, 1765–1769.

Edens, F.W. (1986) Flunixin-induced increased water consumption in broiler chickens before and during heat stress. *Poultry Science* 65, 166–168.

Edens, F.W. and Campbell, D.G. (1985) Reduced heat stress in broilers given flunixin, a non-steroidal cyclooxygenase inhibitor. *Poultry Science* 64 (Suppl. 4), 93 (Abstract).

Edens, F.W. and Siegel, H.S. (1974) Reserpine modification of the blood pH,

pCO_2 and pO_2 of chickens in high environmental temperature. *Poultry Science* 53, 279–284.

Freeman, B.M., Manning, A.C.C., Harrison, G.F. and Coates, M.E. (1975) Dietary aureomycin and the responses of the fowl to stressors. *British Poultry Science* 16, 395–404.

Fuller, H.L. and Rendon, M. (1977) Energetic efficiency of different dietary fats for growth of young chicks. *Poultry Science* 56, 549–557.

Garlich, J.D. and McCormick, J.D. (1981) Interrelationships between environmental temperature and nutritional status of chicks. *Federation Proceedings* 40, 73–76.

Geraert, P.A., Guillaumin, S. and Leclercq, B. (1992) Effect of high ambient temperature on growth, body composition and energy metabolism of genetically lean and fat male chickens. *Proceedings of the 19th World's Poultry Congress* 2, 109–110.

Glick, B. (1963) Research reports. *Feedstuffs* 35, 14.

Heinzerling, R.H., Nockls, C.F., Quarles, C.L. and Tengardy, R.P. (1974) Protection of chicks against *E. coli* infection by dietary supplementation with vitamin E. *Proceedings of the Society of Experimental Biology and Medicine* 146, 279–282.

Hellickson, M.A., Butchbaker, A.F., Witz, R.L. and Bryant, R.L. (1966) Performance of growing turkeys as affected by environmental temperature. *Proceedings of the American Society of Agricultural Engineers*, Chicago, Illinois, Vol. 10, pp. 793–795.

Heywang, B.W. (1952) The level of vitamin A in the diet of laying and breeding chickens during hot weather. *Poultry Science* 31, 294–301.

Hurwitz, S., Weiselberg, M., Eisner, U., Bartov, I., Riesenfield, G., Shareit, M., Nir, A. and Bornstein, S. (1980) The energy requirements and performance of growing chickens and turkeys as affected by environmental temperature. *Poultry Science* 59, 2290–2299.

Huston, T.M. (1978) The effect of environmental temperature on potassium concentrations in the blood of the domestic fowl. *Poultry Science* 57, 54–56.

Kafri, I. and Cherry, J.A. (1984) Supplemental ascorbic acid and heat stress in broiler chicks. *Poultry Science* 63, 125–126.

Kampen, M.V. (1984) Physiological responses of poultry to ambient temperature. *Archiv fur Experimentelle Veterinar Medizin* 38, 384–391.

Koelkebeck, K.W., Harrison, P.C. and Parsons, C.M. (1992) Carbonated drinking water for improvement of egg shell quality of laying hens during summer time months. *Journal of Applied Poultry Research* 1, 194–199.

Koelkebeck, K.W., Harrison, P.C. and Mandindou, T. (1993) Effect of carbonated drinking water on production performance and bone characteristics of laying hens exposed to high environmental temperature. *Poultry Science* 72, 1800–1803.

Kreider, E.M., Nelson, S.M. and Harrison, P.C. (1990) Influence of carbonated drinking water on tibia strength of domestic cockerels reared in hot environments. *American Journal of Veterinary Research* 51, 1948–1949.

Kurnick, A.A., Heywang, B.W., Hulett, B.J., Vavich, M.G. and Reid, B.L. (1964) The effect of dietary vitamin A, ambient temperature and rearing location on growth, feed conversion and vitamin liver storage of white Leghorn pullets. *Poultry Science* 43, 1582–1586.

Leeson, S. (1986) Nutritional considerations of poultry during heat stress. *World Poultry Science Journal* 42, 69–81.

McDougal, L.R. and McQuistion, T.E. (1980) Mortality from heat stress in broiler chickens influenced by anticoccidial drugs. *Poultry Science* 59, 2421–2423.

McNaughton, J.L. and Reece, F.N. (1984) Response of broiler chickens to dietary energy and lysine levels in a warm environment. *Poultry Science* 63, 1170–1174.

March, B.E. and Biely, J. (1972) The effect of energy supplied from the diet and from environment heat on the response of chicks to different levels of dietary lysine. *Poultry Science* 51, 665–668.

Mardsen, A. and Morris, T.R. (1987) Quantitative review of the effects of environmental temperature on food intake, egg output, and energy balance in laying pullets. *British Poultry Science* 28, 693–704.

Mateos, G.G. and Sell, J.L. (1981) Influence of fat and carbohydrate source on rate of food passage of semi-purified diets for laying hens. *Poultry Science* 60, 2114–2119.

Mateos, G.G., Sell, J.L. and Eastwood, J.A. (1982) Rate of food passage as influenced by level of supplemental fat. *Poultry Science* 61, 94–100.

Maumlautnner, K., Singh, R.A. and Kamphues, J. (1991) Influence of varying vitamin C sources on performance and egg shell quality of layers at varying environmental temperature. *Vitamine und Weitere Zusatzstofft bei Mensch und Tin. Symposium Proceedings*, pp. 266–269.

Miles, R.D. and Harms, R.H. (1982) Relationship between egg specific gravity and plasma phosphorus from hens fed different dietary calcium, phosphorus and sodium levels. *Poultry Science* 61, 175–177.

Miles, R.D., Costa, P.T. and Harms, R.H. (1983) The influence of dietary phosphorus level on laying hen performance, egg shell quality and various blood parameters. *Poultry Science* 62, 1033–1037.

Milligan, J.L. and Winn, P.N. (1964) The influence of temperature and humidity on broiler performance in environmental chambers. *Poultry Science* 43, 817–824.

Mills, C.A., Cottingham, E. and Taylor, E. (1947) The influence of environmental temperature on dietary requirement for thiamine, pyridoxine, nicotinic acid, folic acid and choline in chicks. *American Journal of Physiology* 149, 376–379.

Moreng, R.E. (1980) Temperature and vitamin requirements of the domestic fowl. *Poultry Science* 59, 782–785.

Nakaue, H.S., Weber, C.W. and Reid, B.L. (1967) The influence of ASA on growth and some reparatory enzymes in broiler chicks. *Proceedings of the Society of Experimental Biology and Medicine* 125, 663–664.

National Research Council (NRC) (1981) *Effect of Environment on Nutrient Requirements of Domestic Animals*. National Academy Press, Washington, DC, pp. 109–133.

Njoku, P.C. (1984) The effect of ascorbic acid supplementation on broiler performance in a tropical environment. *Poultry Science* 63 (Suppl.) 156.

Njoku, P.C. (1986) Effect of dietary ascorbic acid supplementation on broiler chickens in a tropical environment. *Animal Feed Science and Technology* 16, 17–24.

Njoku, P.C. and Nwazota, A.O.U. (1989) Effect of dietary inclusion of ascorbic acid and palm oil on the performance of laying hens in a hot tropical environment. *British Poultry Science* 30, 831–840.

Odom, T.W., Harrrison, P.C. and Darre, M.J. (1985) The effects of drinking carbonated water on the egg shell quality of single comb white Leghorn hens exposed to high environmental temperature. *Poultry Science* 64, 594–596.

Pardue, S.L., Thaxton, J.P. and Brake, J. (1984) Effects of dietary ascorbic acid in chicks exposed to high environmental temperature. *Journal of Applied Physiology* 58, 1511–1516.

Pardue, S.L., Thaxton, J.P. and Brake, J. (1985a) Influence of supplemental ascorbic acid on broiler performance following exposure to high environmental temperature. *Poultry Science* 64, 1334–1338.

Pardue, S.L., Thaxton, J.P. and Brake, J. (1985b) Role of ascorbic acid in chicks exposed to high environmental temperature. *Journal of Applied Physiology* 58, 1511–1520.

Peebles, E.D. and Brake, J. (1985) Relationships of dietary ascorbic acid to broiler breeder performance. *Poultry Science* 64, 2041–2048.

Perek, M. and Kendler, J. (1962) Vitamin C supplementation to hen's diet in a hot climate. *Poultry Science* 41, 677–678.

Perek, M. and Kendler, J. (1963) Ascorbic acid as a dietary supplement for white Leghorn hens under conditions of climatic stress. *British Poultry Science* 4, 196–200.

Picard, M., Angulo, I., Antoine, H., Bouchot, C. and Sauveur, B. (1987) Some feeding strategies for poultry in hot and humid environments. *Proceedings of the 10th Annual Conference of the Malaysian Society of Animal Production*, pp. 110–116.

Reid, B.L. (1979) Nutrition of laying hens. *Proceedings, Georgia Nutrition Conference*, University of Georgia, Athens, pp. 15–18.

Reid, B.L., Kurnick, A.A., Thomas, J.M. and Hulett, B.J. (1964) Effect of ASA and oxytetracycline on the performance of white Leghorn breeders and broiler chicks. *Poultry Science* 43, 880–884.

Rose, S.P. and Michie, W. (1987) Environmental temperature and dietary protein concentrations for growing turkeys. *British Poultry Science* 28, 213–218.

Sauveur, B. and Picard, M. (1987) Environmental effects on egg quality. In: Wells, R.G. and Belyavin, C.G. (eds) *Egg Quality: Current Problems and Recent Advances*. Butterworths, London, pp. 219–234.

Scott, M.L. (1966) Factors in modifying the practical vitamin requirements of poultry. *Proceedings, Cornell Nutrition Conference*, pp. 34–35.

Scott, M.L. (1976) Effects of heat on vitamin metabolism. In: Tromps, S.W. (ed.) *Progress in Biometeorology*. Swets and Zeitinger, Amsterdam, 275–282.

Scott, T.A. and Balnave, D. (1988) Influence of dietary energy, nutrient density and environmental temperature on pullet performance in early lay. *British Poultry Science* 29, 155–165.

Shane, S.M. (1988) Factors influencing health and performance of poultry in hot climates. *Critical Reviews in Poultry Biology* 1, 247–267.

Sinurat, A.P. and Balnave, D. (1985) Effect of dietary amino acids and ME on

the performance of broilers kept at high temperatures. *British Poultry Science* 26, 117–128.
Sinurat, A.P. and Balnave, D. (1986) Free-choice feeding of broilers at high temperatures. *British Poultry Science* 27, 577–584.
Smith, A.J. and Oliver, J. (1972) Some nutritional problems associated with egg production at high environmental temperatures: the effect of environmental temperature and rationing treatments on the productivity of pullets fed on diets of different energy content. *Rhodesian Journal of Agricultural Research* 10, 3–21.
Smith, J.E. and Borchers, R. (1972) Environmental temperature and the utilization of β-carotene by the rat. *Journal of Nutrition* 102, 1017–1024.
Smith, M.O. and Teeter, R.G. (1987) Potassium balance of the 5–8-week-old broiler exposed to constant heat or cycling high temperature stress and the effects of supplemental potassium chloride on body weight gain and feed efficiency. *Poultry Science* 66, 487–492.
Takahashi, K., Akiba, Y. and Horiguctti, M. (1991) Effects of supplemental ascorbic acid on performance, organ weight and plasma cholesterol concentration in broilers treated with propylthiouracil. *British Poultry Science* 32, 545–554.
Teeter, R.G. and Smith M.O. (1986) High chronic ambient temperature stress effects on broiler acid–base balance and chloride and potassium carbonate. *Poultry Science* 65, 1777–1781.
Teeter, R.G., Smith, M.O., Owens, F.N. and Arp, S.C. (1985) Chronic heat stress and respiratory alkalosis: occurrence and treatment in broiler chicks. *Poultry Science* 64, 1060–1064.
Thaxton, J.P. (1986) Role of ascorbic acid for relief of stress effects. *Proceedings, Colorado State University 2nd Poultry Symposium on the Impact of Stress*, pp. 53–65.
Thornton, P.A. (1961) Increased environmental temperature influences on ascorbic acid activity in the domestic fowl. *Federation Proceedings* 20, 210A.
Waldroup, P.W., Mitchell, R.J., Payne, J.R. and Hazen, K.R. (1976) Performance of chicks fed diets formulated to minimize excess levels of essential amino acids. *Poultry Science* 55, 243–253.
Wallis, I.R. and Balnave, D. (1984) The influence of environmental temperature, age, and sex on the digestibility of amino acids in growing broiler chickens. *British Poultry Science* 25, 401–407.
Ward, J.B. and Brewer, C.E. (1984) Effect of lasalocid, a coccidiostat, potassium, and sodium on the performance of broilers. *Poultry Science* 63, 200–201 (Abstract).
Wilson, E.K., Pierson, F.W., Hester, P.V., Adams, R.L. and Stadelman, W.J. (1980) The effects of high environmental temperature on feed passage time and performance traits of white Pekin ducks. *Poultry Science* 59, 2322–2330.
Zuprizal, Larbier, M., Chagneau, A.M. and Geraert, P.A. (1993) Influence of ambient temperature on true digestibility of protein and amino acids of rapeseed and soybean meals in broilers. *Poultry Science* 72, 289–295.

6

Feedstuffs Used in Hot Regions

N.J. Daghir
Faculty of Agricultural Sciences UAE University, Al-Ain, UAE

Introduction	126
Cereals and Their By-products	126
Barley (*Hordeum sativum*)	126
Millet (*Setoris* spp.)	128
Rice by-products (*Oryza sativa*)	129
Sorghum (*Sorghum bicolor*)	131
Triticale	132
Protein Supplements	134
Coconut meal (copra)	134
Cottonseed meal (*Gossypium* spp.)	134
Peanut meal (groundnut)	136
Sunflower seed and meal (*Helianthus annuus*)	137
Sesame meal (*Sesamum indicum*)	138
Linseed meal	139
Mustard seed meal	140
Single-cell protein	140
Novel Feedstuffs	141
Cassava root meal	141
Palm kernel meal	142
Ipil-ipil leaf meal (*Leucaena leucocephala*)	142
Salseed (*Shorea robusta*)	143
Dried poultry waste	143
Buffalo gourd meal (*Cucurbita foetidisima*)	144
Guar meal (*Cyamopsis tetragonaloba*)	144
Bambara groundnut meal (*Voandizeia subterrenea*)	145
Jojoba meal (*Simmoensia chinensis*)	145
Conclusions	145
References	147

Introduction

The future development of the poultry industry in many regions of the world depends to a large extent on the availability of feedstuffs in those areas that are suitable or can be made suitable for use in poultry feeds. This is because feed costs constitute about 50–70% of the total cost of producing eggs and poultry meat and the less a country can depend on imported feeds, the lower feed costs can be. In North America, yellow maize and soyabean meal are the two major ingredients used in poultry feed. With the exception of some Latin American countries, such as Brazil and Argentina, these two feed ingredients are not plentiful in most of the hot regions of the world. This has stimulated poultry nutritionists in the hot regions to search for cheaper locally available feedstuffs and investigate the composition and nutritional value of these feedstuffs. Furthermore, the use of cereal grains for poultry feeds is questionable because cereal grains still form the staple diet of the people in those areas. In many of these countries, the feed industry is owned and directed by multinational corporations, who prefer to rely on known technology rather than investing in the development of new technology appropriate to a developing region.

The value of this chapter, therefore, is to point out the ingredients that can be used and are being used for poultry feeding in tropical and hot regions and present research findings on chemical composition and nutritional value of these ingredients. The major sources of nutrients in poultry feeds are the grains and their by-products, oil-seed meals, meat-packing by-products, fish meal, several other agricultural and industrial by-products and a host of feed additives. This chapter covers three groups of feedstuffs: the cereals and their by-products, protein supplements and novel feedstuffs. The first two groups are those that are commonly used in hot regions, and the novel feedstuffs group consists of those that are less commonly used but have potential for greater use in these regions. It is hoped that collecting these data on feedstuffs composition and nutritional value will help countries in these regions to take full advantage of available knowledge and will stimulate further research in this field.

Cereals and Their By-products

Barley (Hordeum sativum)

Barley is one of the most extensively cultivated cereal grains in the world. It is of significance in hot dry regions where annual precipitation is not sufficient for the production of other grains, such as maize and wheat. It is the most commonly raised feed grain in Europe, where it is grown on relatively poorer soil than that used for other grains. According to FAO

(1985), the annual production of barley is over 75 thousand metric tons compared with maize production of about 62 thousand metric tons. Barley types include spring and winter varieties, with two groups predominating: the six-rowed and the two-rowed barleys. The mature grain contains a hull which encloses the kernel. In most varieties, the hulls are cemented to the kernel and are carried through the threshing process. In the naked varieties, the kernel threshes free. The hull forms about 10–14% of the weight of the grain.

Barley is considered a low-energy feed grain due to its high fibre content, and the metabolizable energy (ME) value is about 12.5 MJ per kg dry matter. Therefore, in diets requiring high energy levels, such as those for broilers, the amount of barley used is minimal. The crude protein content of barley grain ranges from about 6 to 16% on a dry matter basis. The lipid content of this grain is low, usually less than 2.5%. Crude fibre content of barley is about 5–6%, with higher levels occurring if grown in arid regions. Barley contains antinutrient compounds called β-glucans. These are non-starch polysaccharides made up of glucose units and found in the endosperm cell wall. They are considered to be responsible for the sticky droppings and reduced feed utilization and growth rate in growing chickens fed high levels of barley. Acen and Varum (1987) compared the β-glucan levels in ten Scandinavian varieties of barley and found a variation in mean values for 2 years from 3.08 to 4.83%. There were more β-glucans in two-rowed barley than in six-rowed barley. The β-glucan content of barley increases with stage of ripeness (Hasselman et al., 1981).

Research workers have for some time been investigating not only factors that influence the value of this grain as a feed, but also the possible methods of improving its nutritional value. Both water treatment and enzyme supplementation have been shown to improve the nutritional value of barley. The effectiveness of such treatment was shown to be dependent on geophysical area and weather conditions in which barley was grown. Daghir and Rottensten (1966) reported on experiments conducted to study the influence of variety and enzyme supplementation on the nutritive value of barleys grown in the same location and receiving similar cultural treatments. Enzyme supplementation significantly improved growth of chicks in all varieties tested and had a more marked effect on growth than the variety used. In general, barley grown under arid conditions responds more positively to the enzyme and water treatments than that grown in more humid areas. Figures of ME content given by Allen (1990) show that barley from the dry Pacific coast of the USA is lower than that of Midwestern barley. Herstad (1987) in a study on broilers and laying hens showed that supplementation of broiler diets with β-glucanase improved litter conditions and overall performance. In laying hens, using barley at 73% of the diet did not reduce egg production or egg weight, but had a slight negative effect on feed efficiency. Coon et al. (1988) fed two barley cultivars (Morex and Glenn) and a high-protein feed-grade barley (12.6% crude protein (CP)) to laying hens at different dietary levels. They concluded that the use of barley can be beneficial for regulating egg size and minimizing body-weight gains in post-peak layers if barley is priced low enough to offset resulting increased feed consumption and lower feed utilization.

Barley varieties vary considerably in proportion of hull. Hull-less barley has an energy value similar to that of wheat, but yield of this variety is much lower per acre than the hulled varieties. Some researchers reported no difference in performance of chicks fed hull-less or hulled barley (Anderson *et al.*, 1961; Coon *et al.*, 1979). Newman and Newman (1988) evaluated the hull-less barley mutant Franubet in broiler feeding trials and found that this variety supported chick growth equal to that of maize and without the problem of wet sticky droppings. The authors suggested that the superior value of this barley cultivar may be due to absence of antinutritional factors, the presence of a different form of β-glucans and the character of its unique starch, which could affect energy availability.

Rotter *et al.* (1990) studied the influence of enzyme supplementation on the bioavailable energy of a hull-less (Scout) barley. Barley replaced up to 75% of the wheat in the diet. The available energy for young chicks increased significantly due to enzyme supplementation as the barley component of the diet increased. No significant increase in ME was observed when the same diets were fed to adult roosters. They suggested that the mature bird has a more developed gastrointestinal tract, which can neutralize the negative effects of the β-glucans.

Classen *et al.* (1991) in a review on the use of feed enzymes concluded that the efficacy of these enzyme products has been well demonstrated in both university and on-farm trials. These workers believe that enzyme supplementation will become an invaluable tool within the feed industry. They cautioned, however, that feed manufacturers should be aware that enzyme products not designed for use in animal feeds should not be used.

Barley has to be checked for moisture, which should not exceed 13%, and crude protein, which can vary from 7 to 12%. Crude fibre in barley analyses at about 5 to 6%. The texture of a mash feed containing more than 20% barley may adversely affect palatability and feed consumption. Pelleting will improve palatability and will enable use of higher levels of this grain.

Millet (Setoris *spp.*)

The name millet is applied to several species of cereals which are widely cultivated in the tropics and warm temperate regions of the world. Millet has a chemical composition and ME content very similar to those of sorghum. Its feeding value, however, is more like that of barley and oats.

The composition of millet is very variable: the protein content is generally within the range of 10–12%, the ether extract 2–5% and the crude fibre 2–9%. Few studies have been published on the feeding value of millet for chickens. Sharma *et al.* (1979) found that broilers fed on millet competed favourably with those on maize, wheat and sorghum. Abate and Gomez (1983) also reported similar results. Karmajeewa and Than (1984) fed pullets from 8 to 20 weeks a protein concentrate plus either wheat, millet or paddy rice as crushed or whole grain. Pullets fed on millet had a higher linoleic acid content in their livers and laid larger eggs than those reared on wheat. They concluded that replacement pullets can be grown successfully on whole wheat or millet and a protein concentrate offered on a free-choice basis with no adverse effects on their subsequent laying performance. Millet has a higher linoleic acid content than wheat and, when included in growing diets, would improve egg weight in the subsequent laying phase.

Pearl millet (*Pennisetum typhoides*), also known in India and Pakistan as bajra, has been investigated for use in laying hen diets by several workers. Chawla *et al.* (1987) reported that it can be used at 20% of the diet, although Reddy and Reddy (1970) were able to incorporate as much as 32% into the diet without detrimental effects on production. Mohan Ravi Kumar *et al.* (1991) found that the inclusion of pearl millet at 60% in laying hen diets made isocaloric and isonitrogenous with maize-based diets did not influence hen-day egg production, feed intake, feed efficiency or body weight. Such high levels of pearl millet in laying hen diets were possible because these workers had incorporated fairly high levels of fish meal in those diets.

Rice by-products *(Oryza sativa)*

Rice is second to wheat in worldwide production. Only when it is produced in abundance is rice incorporated in sizeable amounts in poultry feed. In the processing of whole rice, the first step is removal of the husks. Once the husks have been removed, the rice becomes 'brown rice'. The second step is removal of the bran, which yields white rice. White rice is usually further processed or polished and the residues are called rice polishings. Basically, there are two rice by-products used in poultry feeding. These are rice bran and rice polishings. Because many of the hot and tropical regions of the world are rice-producing areas, huge quantities of these two by-products, as well as other rice by-products, are produced in those regions.

Rice bran is recognized as being variable in its composition, depending on the severity with which the rice is threshed and the extent to which the oil is extracted. McNab (1987) looked at 14 samples of rice bran and found them to vary from 2 to 20% in oil and from 7.17 to 12.91 kJ g^{-1} in energy. Protein content did not vary much and was in the range of

13–14%. Panda (1970) reported on rice brans produced in India and varying from 1 to 14% in oil content and 13 to 14% in crude protein. The oil is highly unsaturated and may become rancid very quickly. It is therefore removed to produce a product with better keeping quality.

Feltwell and Fox (1978) reported that the feeding value of rice polishings depends upon the degree of polishing to which the grain has been subjected. It is unlikely to contain much of the rice flour. Panda (1970) gives a figure of 11–12% crude protein and 12–13% oil for rice polishings produced in India, which is close to Scott *et al.* (1982), who list this product as having 12% crude protein and 12% oil.

Both rice bran and rice polishings can be used in poultry rations at fairly high levels if they are low in rice hulls and if the high oil level in them can be stabilized by an antioxidant so that much of the energy value will not be lost through oxidative degradation. Hussein and Kratzer (1982) demonstrated the detrimental effects of rancid rice bran on the performance of chickens. Cabel and Waldroup (1989) tested the effects of adding an antioxidant or a metal chelator or both on the nutritive quality of rice bran stored at high temperature (35–38°C) and high humidity (80–90%). The addition of 250 p.p.m. ethoxyquin proved effective in significantly reducing the initial peroxide value and 1000 p.p.m. elthylenediamine tetra-acetic acid (EDTA) significantly lowered the 20 h active oxygen method value. They therefore concluded that rancidity development in stored rice bran can be slowed by the addition of ethoxyquin or EDTA.

Because the production of edible oil from rice bran can provide a much needed source of energy for people in rice-producing areas, Randall *et al.* (1985) developed a process to stabilize the bran and prevent enzymatic hydrolysis of its oil. This stabilization process heats the freshly milled bran in an extrusion cooker to 130°C. The hot extruded bran is maintained at near 100°C on an insulated holding belt for 3 min prior to cooling in an ambient air stream. The processed rice bran is in the physical form of small, free-flowing flakes with a low microbiological load. Lipolytic enzymes are permanently inactivated and the extracted bran residue retains the flake form, thereby reducing dust and fines. Sayre *et al.* (1987) conducted feeding trials with broilers to compare this stabilized rice bran with the raw bran. Free fatty acid (FFA) content in oil from raw rice bran stored at elevated temperatures (32°C) reached 81% whereas FFA in stabilized bran oil remained at about 3%. Chickens fed stabilized rice bran made significantly greater gains than those fed raw bran diets.

Paddy rice

In the process of dehulling and milling of paddy rice, a huge quantity of broken grains, popularly known as rice kani, is obtained as a by-product. Verma *et al.* (1992) examined the usefulness of this by-product as a substitute for maize. Graded amounts of rice kani were substituted for yellow maize up to 40% in isonitrogenous diets for broilers. No significant differences were observed in body-weight gain, feed intake or feed efficiency. The ME value obtained for rice kani was reported by these workers to be 13.3 MJ kg^{-1}. The authors concluded that rice kani is a good substitute for maize in broiler diets.

Sorghum *(*Sorghum bicolor*)*

Sorghum is one of the most important feed and food crops in the arid and semiarid tropics (Hulse *et al.*, 1980). It is the main food grain in Africa and parts of India and China. Sorghum is extensively used in poultry feeding in many countries of the world. One of the major limitations for its use in poultry feeds is its relatively high tannin content. The detrimental effects of high-tannin sorghums on the growth and feed efficiency of the growing chicken are well documented (Chang and Fuller, 1964; Armstrong *et al.*, 1973). Vohra *et al.* (1966) showed that as little as 0.5% tannic acid will depress growth. The question has always been raised as to whether naturally occurring sorghum tannins are as toxic as tannic acid. Dale *et al.* (1980) observed a growth depression regardless of the source of the tannin, but a higher sorghum tannin content was necessary to cause growth depression equivalent to commercial tannic acid. Supplementation of high-tannin sorghum diets with methionine or choline and methionine alleviated the growth depression. Elkin *et al.* (1978) found that addition of 0.15% methionine to a high-tannin sorghum–soyabean meal diet brought the growth of broilers up to that obtained with a similar diet containing a low-tannin sorghum grain. Other methods for improving the nutritional value of high-tannin sorghum are the addition of fats and the adequate grinding of the grain (Douglas *et al.*, 1990a). Nir *et al.* (1990) concluded that the main effect of grinding is to improve feed utilization, which is accomplished by increasing the surface area of the grain relative to its reduced particle size.

There are several causes for the growth-depressing and toxic effects of tannins. Tannins affect the palatability of diets and thus reduce feed intake, but this is not a major factor in growth depression caused by high-tannin diets (Vohra *et al.*, 1966). Nelson *et al.* (1975) reported a high variability in the metabolizable energy values of sorghum grains. These workers observed that both metabolizable energy and amino acid availability increased as the tannic acid content of the sorghum grains decreased. Douglas *et al.* (1990a) also showed that the ME content of high-tannin sorghum was lower than that of the low-tannin sorghum varieties. Furthermore, the high-tannin variety contained higher levels of both acid detergent fibre and neutral detergent fibre than the low-tannin variety.

Besides the negative effects of tannins in sorghum grains, the presence of phytates has also been shown to reduce growth and increase incidence of locomotor disorders (Luis *et al.*, 1982; El-Alaily *et al.*, 1985; Mohammadain *et al.*, 1986). Ibrahim *et al.* (1988) were able to improve the nutritional quality of Egyptian and Sudanese sorghum grains by the addition of phosphates. These authors suggested that treatment of sorghum by dry-mixing with dicalcium phosphate could extend the use of high-tannin sorghum in poultry feeds.

Few studies have been conducted to determine the effects of sorghum tannin on the performance of the laying hen. Supplementation of tannic acid to a laying hen diet reduced feed consumption, egg production and egg weight and resulted in abnormal colouring and mottling of the yolk (Armanious *et al.*, 1973). Bonino *et al.* (1980) conducted a 32-week study in which high-tannin sorghum was found to cause a small but significant decrease in egg production in comparison with low-tannin sorghum-fed controls. Sell *et al.* (1983) observed that tannin significantly reduced egg production and feed efficiency but had no effect on egg weight or bodyweight. At the end of the experiment, all hens were placed on a commercial laying ration for a 31-day period and recovery was complete by the end of this period.

In general, low-tannin sorghums are nearly equal to maize when fed to broilers. High-tannin sorghums, however, are lower in energy than maize and therefore, when sorghum grains are used in place of maize in broiler and layer diets, xanthophyll supplement and an additional amount of fat are needed in the ration. Finally, sorghum grains must be adequately ground to ensure maximum utilization.

Triticale

Triticale is a synthetic small grain crop resulting from the intergeneric cross between durum wheat and rye. Its name is derived from a combination of the two generic terms of the parent cereals (*Triticum* and *Secale*). The objectives of crossing these two cereals was to produce a grain having the high quality and productivity of wheat and the hardiness of rye. Triticale is now grown commercially in many countries of the world. Its composition is variable, with crude protein ranging from 11 to 18%. It has been grown in several hot regions of the world and found to be as high-yielding as wheat.

The use of triticale for broilers has been tested by several workers. Sell *et al.* (1962) fed chicks increasing levels of triticale at the expense of wheat in a basal diet and concluded that methionine and lysine were limiting for chick growth when the diets containing the triticale were isonitrogenous with the basal wheat diet. Fernandez and McGinnis (1974) compared triticale with maize and provided evidence that chicks grown on diets with 53% or 73% triticale were lighter in weight than those on a maize diet.

Wilson and McNab (1975) also found that broiler weights at 56 days were depressed when triticale was substituted for maize at a 50% level. Proudfoot and Hulan (1988) concluded that triticale should not be used over 15% in broiler diets. Bragg and Sharby (1970), on the other hand, found that triticale could partially or totally replace wheat in broiler diets without adverse effects on growth or feed efficiency. Rao et al. (1976) were also able to replace 75% of the maize in broiler diets with no adverse effect on weight gain or feed efficiency. If included in the diet in finely ground form, triticale impairs feed intake. Therefore, part of the negative effect observed by the above researchers may have been due to differences in the grinding of the grain.

Ruiz et al. (1987), in studying the nutritive value of triticale (Beagle 82) as compared with wheat, found that broiler growth response was inconsistent with triticale fed at levels of 25% and 50% of the diet. These workers indicated that variations in chemical composition of different triticale cultivars may account for the variation in results from studies conducted to evaluate the nutritive value of triticale. Daghir and Nathanael (1974) studied the composition of triticale (Armadillo 107) as compared with wheat (Mexipak) and found that the proximate analysis of these two grains was practically identical. In mineral composition, they were also similar, except in iron, which was lower in triticale than in wheat. ME values as determined by these authors with both chicks and laying hens were found to be about 5–6% lower in triticale than in wheat.

In studies with broilers and laying hens, Jerock (1987) concluded that, because of the great variability among varieties of triticale, it is necessary to limit the level of usage to 30 and 50% in the rations of broilers and laying hens, respectively. Flores et al. (1992) confirmed these varietal differences in triticale and added that enzyme supplementation could be a tool to improve some of these poor varieties.

Leeson and Summers (1987) observed a decrease in laying hen performance when triticale was used at 70% level in isocaloric diets, compared with maize. Cuca and Avila (1973) were able to use 50% triticale in a milo-based diet with no detrimental effect on laying hen performance. Luckbert et al. (1988) showed that triticale can be used up to 40% in diets for laying hens.

Castanon et al. (1990) studied the effect of high inclusion levels of triticale in diets for laying hens containing 30% field beans. Partial replacement of maize by triticale did not affect feed consumption, laying rate or feed-to-gain ratio. Mean egg weight increased slightly with increasing levels of triticale, while yolk colour decreased with increases in triticale.

The above studies indicate that triticale can partially replace maize and/or wheat in broiler and layer rations. The level of replacement depends on the cultivar being used and the chemical composition of that cultivar.

Furthermore, higher levels can be used in laying than in broiler rations. When used at high levels, it may be advisable to supplement the ration with synthetic lysine.

Protein Supplements

Coconut meal (copra)

The use of coconut meal in poultry diets is somewhat limited because of its high fibre content, averaging about 12%. The average protein content of this meal (solvent processed) is about 22%. Light-coloured meals are usually better in quality than dark-coloured ones. The meal protein is low in the amino acids lysine and histidine. The oil content of this meal varies from 2.5 to 7.5% and, in areas where energy sources are scarce, it can be useful for the preparation of high-energy diets. Such diets, however, have the disadvantage of being susceptible to becoming rancid during storage unless they are properly treated with an antioxidant. Panigrahi *et al*. (1987) reported that coconut meals from small-scale screw-press expellers generally have a high lipid content and produce good growth in broiler chicks when fed at high levels of the ration. These workers, however, found that these high-coconut meal diets induced inquisitive and excited behaviour and increased feed spillage and water intake. Increasing the energy content of the diet by the addition of maize oil improved efficiency of food utilization, food intake and growth rate and reduced abnormal behaviour (Panigrahi, 1992). The maximum level of coconut meal used in poultry diets should not exceed 10% of the diet. When used at such levels, amino acid supplementation may be necessary.

Cottonseed meal (Gossypium *spp.*)

Early work on cottonseed meal (CSM) composition and nutritive value has been reviewed by Phelps (1966). Since that date, a great deal has been written on CSM meal and on methods of improving its nutritional value for poultry. Decorticated good-quality CSM has become a very useful source of supplementary protein for poultry in many regions of the world, particularly the cotton-producing areas.

Although CSM is a rich source of protein, its use in poultry rations has been limited because of certain undesirable characteristics. It contains gossypol, a polyphenolic pigment found in the pigment glands of most varieties of CSM. Gossypol has been known for many years to depress growth, feed intake, feed efficiency and hatchability. Additional deleterious effects that have been reported include enlarged gall-bladders and blood and bone marrow changes, including a reduction of haemoglobin, red cell count and serum proteins. Other symptoms reported were oedema in body cavities, degenerative changes in liver and spleen, haemorrhages of liver, hypoprothrombinaemia and diarrhoea. The anaemia resulting from gossypol toxicity is due to its property of binding iron. Gossypol is present either in the bound or free state. Our concern is mainly with free gossypol because the bound form is non-toxic to animals. The amount of free gossypol in CSM depends to a large extent on the type of processing, i.e. screw-press, prepress or solvent extraction (Jones, 1981). Although screw-pressing produces a meal that is lowest in free gossypol, prepress solvent extraction is the method in common use. Waldroup (1981) suggested that 100 p.p.m. of free gossypol in broiler diets is considered acceptable on the basis of growth and feed efficiency. The presence of gossypol in poultry diets may be counteracted by the addition of iron salts, which bind the gossypol. Scott *et al.* (1982) suggested that gossypol was inactivated or chelated by iron in a 1:1 molar ratio. This has been confirmed by El-Boushy and Raternick (1989), who indicated that free gossypol can be chelated by added iron or iron available naturally in the feed. These workers also cautioned against using high levels of iron because this can have a negative effect on body-weight gains of broilers.

Another method of improving CSM was reported by Rojas and Scott (1969) and consists of treatment of the meal with phytase produced by *Aspergillus ficcum*. The hydrolysis of phytin in the CSM released the phosphorus for utilization and may have freed some proteins from protein–phytate complexes according to these workers.

The hull-less 50% protein CSM available in some areas of the world has given excellent results when fed to broilers. This is particularly true if the gossypol content is minimized. This type of CSM has an energy value close to that of soyabean meal. Plant breeders have developed cultivars that are practically free of gossypol. Unfortunately, however, these cultivars are more susceptible to insect infestation and do not yield as much as the gossypol-rich varieties.

Another problem with CSM is the cyclopropene fatty acids. The cyclopropene fatty acids, malvalic and sterculic acids, existing in cottonseed oils have been shown to cause a pink discoloration of egg white (Phelps *et al.*, 1965). The level of these acids in the crude oil ranges from 0.6 to 1.2% (Bailey *et al.*, 1966). The level of these acids in the meal depends upon the amount of the residual oil but usually averages about 0.01% in meals obtained from commercial processors (Levi *et al.*, 1967). These fatty acids also cause a greater deposition of stearic and palmitic acid in depot fat. Thus egg and body fat of hens consuming cottonseed oil has a higher proportion of stearic acid than that found when other fats are fed. Besides the effects of the cyclopropene fatty acids on pink discoloration of egg whites, free gossypol also causes yolk discoloration. Heywang *et al.* (1955) reported that as little as 10 p.p.m. free gossypol in the laying hen diet produced discoloured yolks. Panigrahi and Hammonds (1990) reviewed the literature on the effects of screw-press CSM on eggs produced.

The appearance of aflatoxin in CSM has prompted some workers to use ammonia treatment, which effectively alters the aflatoxins without affecting the performance of laying hens (Vohra et al., 1975). Isopropanol extraction, which has been used for aflatoxin removal, also reduces the free gossypol and residual oil levels in the extracted CSM. Reid et al. (1987) reported that isopropanol-treated CSM can be safely utilized in laying hen diets at levels up to 15% without any detrimental effects on either performance or yolk colour and quality.

Panigrahi et al. (1989) studied the effect of feeding a screw-press-expelled CSM to laying hens at dietary concentrations of up to 30% of the diet. The overall performance of hens fed 7.5% CSM was not significantly different from that of controls, but a 30% CSM diet, containing 255 g free gossypol kg^{-1} and 87 mg cyclopropenoid fatty acids, significantly reduced feed intake and egg production. The 15% CSM diet did not produce adverse effects initially, but egg production was depressed towards the end of the production cycle. Treatment of 30% CSM diet with a solution of ferrous sulphate heptahydrate (100 mg supplemented dietary iron per kg) further reduced feed intake and egg production. Storage of eggs at room temperature for up to 1 month did not lead to discoloration of any kind in the CSM diet groups, but resulted in yolk mottling. Storage of eggs at cold temperatures for 3 months resulted in brown yolk discoloration and the initial stages of pink albumen discoloration when the 30% CSM diet was fed. The brown yolk discoloration was reduced by treatment of the CSM with iron.

Panigrahi et al. (1989) showed variability in the responses of different flocks of hens to dietary CSM. In a later study, Panigrahi and Morris (1991) examined the effects of genotype of the hen on the depression of egg production and discolorations in eggs resulting from dietary screw-pressed CSM. They reported that the strongest interaction between breed and diet occurred with food intake. The susceptibility of eggs to the cyclopropenoid fatty acid-related cold storage effects also depended on the genotype of the hen.

Before use in poultry feeds, CSM should be checked for protein, fat, fibre and gossypol content.

Peanut meal (groundnut)

Peanut meal is also known as groundnut, *Arachis* nut, earth-nut and monkey-nut meal. It is used extensively in poultry feeds in many hot and tropical regions. Generally, peanut meal should not be used as the major source of protein unless the diet is supplemented with the essential amino acids, lysine and methionine. Furthermore, it is one of those feedstuffs that is very susceptible to contamination by *Aspergillus flavus*, which produces the group of toxins known as aflatoxins.

The early literature on the use of peanut meal as a feedstuff for poultry was reviewed by Rosen (1958). Daghir *et al.* (1969) evaluated peanut meals produced in the Middle East by using animal assays as well as several types of chemical assays. In all meals, methionine and lysine were the first and second limiting amino acids, respectively. In an earlier study, Daghir *et al.* (1966) reported that in all-plant diets, in which peanut meal served as the principal source of protein, lysine was more limiting than methionine. This amino acid limitation of peanut meal has been confirmed by several workers (Anderson and Warnick, 1965; Mezoui and Bird, 1984; El-Boushy and Raternick, 1989).

Sunflower seed and meal (Helianthus annuus)

Sunflower seed production for the primary purpose of producing oil is significant in many countries in the hot regions of the world as well as in temperate regions. In fact, sunflower seed production in the USA is second only to soyabeans among the oil seeds. Sunflower meal (SFM) is, therefore, available in large quantities for use in animal feeds.

Morrison *et al.* (1953) showed that one reason for the variability in nutrient value of SFM found by earlier workers was the relatively high temperature used in their processing. An excellent review of the value of SFM for poultry was published by McNaughton and Deaton (1981). Klain *et al.* (1956) were the first to report that the protein of SFM was deficient in lysine for chicks. Rad and Keshavarz (1976) and Valdivie *et al.* (1982) used isocaloric diets and showed that SFM could be used in lysine-supplemented diets at about 15–20% without adversely affecting chick performance. Gippert *et al.* (1988) reported that Hungarian SFM can replace 25%, 50% and 75% of the soyabean meal in a starter, grower and finisher broiler ration, respectively, without any detrimental effect on body-weight or feed conversion. Zatari and Sell (1990) reported that up to 10% SFM could be used in diets adequate in lysine and containing 6% animal–vegetable fat without adversely affecting growth or feed efficiency of broilers to 7 weeks of age.

There is very little information on the feeding value of full-fat raw or full-fat treated sunflower seed for poultry. A small amount of sunflower seed (1–2%) has been reported to have been included in commercial mixed scratch grains for poultry in the past because of the distinct and attractive appearance of the seeds (Morrison, 1956). The seeds contain 25–32% oil, about 16% protein and 12–28% fibre. The reason for this wide variation in oil is variety, soil and climatic conditions, and fibre content variation depends on variety and the extent to which the seed was cleaned prior to its analysis. Daghir *et al.* (1980b) tested the use of sunflower seed as full-fat raw or full-fat steamed or heated as a source of energy and protein for broilers. They concluded that, in practical-type broiler rations, sunflower seed fed in the full-fat raw form can constitute at least 10% of the ration without any adverse effect on performance. Cheva-Isarakul and Tagtaweewipat (1991) studied the utilization of sunflower seed fed at levels of 0–50% of the diet to broilers up to 7 weeks of age. These authors concluded that sunflower seed is a good source of crude protein and ME in broiler diets. Lysine supplementation improved diets containing high levels of sunflower seed. They recommended that, to avoid plugging due to high fat content, sunflower seed should be milled in two steps, the first through a small mesh-size sieve and the second through a sieve with a larger mesh. Also, to avoid rancidity, seeds should be ground just before feed mixing and the mixed diet kept no longer than 2 weeks.

Few studies have dealt with the value of sunflower seed or meal for laying hens. The early literature was reviewed by Rose *et al.* (1972), who evaluated the replacement of soyabean meal with sunflower seed meal in maize–soyabean meal rations for laying hens. Sunflower seed meal replaced 50% of the soyabean meal protein without adversely affecting hen performance. Lysine supplementation did not consistently improve utilization of the diets used. These workers reported that characteristic eggshell stains, which develop after egg laying when sunflower seed meal is used in mash rations, were markedly reduced when similar rations were fed in crumble form. They suggested that chlorogenic acid present in sunflower seeds is responsible for these stains. ME values of 2205 and 2139 kcal kg^{-1} were reported by these workers for the two sunflower seed meals used in their study. Uwayjan *et al.* (1983) evaluated unprocessed whole sunflower seed as partial replacement for soyabean meal and yellow maize in laying rations. Sunflower seed used at 10, 20, or 30% in the ration had no detrimental effect on performance. Neither lysine nor lysine plus methionine supplementation of sunflower seed-containing rations improved hen performance. Rations containing sunflower seed gave a significantly lower yolk colour score and a significant rise in yolk cholesterol content. Lee and Moss (1989), in two feeding trials with laying hens, observed that unhulled confectionery-type sunflower seeds depressed egg production when fed at levels exceeding 20% of the diet. Feed efficiency, egg weight, albumen height and eggshell thickness were not significantly different among treatments.

Sesame meal (Sesamum indicum)

Sesame is also known as benne, gingili and teel or til. The leading countries in the production of sesame are India, Iraq, Egypt and Pakistan.

Sesame is known to be high in methionine but deficient in lysine; hence it cannot be fed as the major protein supplement in broiler rations

(Daghir and Kevorkian, 1970). The very early work on the nutritive value of sesame meal and the extent to which it can replace soyabean meal in broiler rations was reviewed by Daghir et al. (1967). These workers showed that sesame meal may replace 50% of the soyabean meal in broiler rations supplemented with 2% fish meal without significant change in weight gains or feed efficiency. Chicks receiving a combination of soyabean meal and sesame meal in the proportion of 10:20 supplemented with 0.32% lysine had significantly greater body-weight than those receiving an all-soyabean meal diet.

Cuca and Sunde (1967) reported that sesame meal binds dietary calcium due to its high phytic acid content. Using maize–soyabean meal diets, they found normal bone ash values with calcium levels as low as 0.8%. With California sesame meal, 1.05% calcium was required for normal bone ash values and with Mexican sesame meal 1.5% calcium was needed. Lease et al. (1960) reported that sesame meal interferes with the biological availability of zinc. In later work, Lease (1966) showed that autoclaving sesame meal for 2 h caused an increase in chick growth, a decrease in leg abnormalities and increased bone ash as compared with chicks receiving similar diets containing unautoclaved sesame meal. More recently, Bell et al. (1990) reported that sesame seed meal may provide an acceptable alternative to soyabean meal in broiler rations when the substitution level is 15% or less.

Linseed meal

Linseed meal contains high levels of mucilage, which is almost completely indigestible by poultry. It also contains a small amount of a cyanogenetic glycoside called linamarin and an enzyme (linase) that is capable of hydrolysing this glycoside, producing hydrogen cyanide. The meal has a poor-quality protein that is low in both methionine and lysine.

Because of the above limitations, this meal is not considered a satisfactory protein supplement for poultry. Depressed growth has been reported when chicks were fed diets containing as little as 5% linseed meal. The adverse effects of this meal can be partially reduced by autoclaving and increasing the level of vitamin B_6 in the ration. McDonald et al. (1988) did not recommend using it at levels exceeding 3% of the diet.

In light of the importance of dietary n-3 fatty acids for human health, Ajuyah et al. (1993) investigated the effect of dietary full-fat linseed on the fatty acid composition of chicken meat. The white meat of birds fed this seed had elevated levels of n-3 fatty acids.

Mustard seed meal

Wild mustard is widely spread all over the world and grows heavily in wheat and barley fields. It is mainly the presence of sinigrin, a sulphur-containing glucoside, which has prevented the use of mustard seed meal as a protein source for poultry. The glucoside is also present in rapeseed, crambe seed and other seeds of the Brassica family. This glucoside when digested by an endogenous enzyme (myrosinase) is hydrolysed to volatile allylisothiocyanate, an antinutritional goitrogenic factor (Fainan et al., 1967). Attempts to detoxify sinigrin in mustard seed by soaking, boiling and/or roasting failed (Daghir and Mian, 1976). Detoxification of allylisothiocyanate in rapeseed has been achieved by chemical additives, particularly $FeSO_4 \cdot 7H_2O$ (Bell et al., 1971). Daghir and Charalambus (1978) tested this $FeSO_4$ treatment on mustard seed meal and found it to be an effective method of detoxification of the meal for use in broiler rations.

Single-cell protein

Microbial fermentation for the production of protein has been known for a long time. Single-cell organisms such as yeast and bacteria grow very rapidly and can double their cell mass in 3–4 h. A range of nutrient substrates from plant sources can be used, as well as unconventional materials such as methanol, ethanol, alkanes, aldehydes and organic acids. The protein content of bacteria is higher than that of yeasts and contains higher levels of the sulphur-containing amino acids but a lower level of lysine. Single-cell proteins contain high levels of nucleic acids, ranging from 5 to 12% in yeasts and 8 to 16% in bacteria. Yeast single-cell protein (YSCP) has been studied for use as a protein supplement in poultry rations since the early part of this century. Clark (1931) was the first to report problems with feeding yeast and claimed that it was unsuitable for growing chicks. The early literature on feeding yeast to chickens has been reviewed by Saoud and Daghir (1980). These workers reported that molasses-grown yeast results in growth depression and lower feed efficiencies as compared with a ration not containing single-cell protein. The detrimental effects of single-cell protein were mainly due to changes in blood nitrogenous constituents. Daghir and Abdul-Baki (1977) found that yeast protein produced from molasses can be used at 5 and 10% levels of well-balanced broiler rations with adequate supplementation of lysine and methionine. The addition of 15% yeast protein to the ration depresses growth and feed efficiency in broilers even with methionine and lysine supplementation. Daghir and Sell (1982) studied the amino acid limitations of YSCP when fed as the only source of protein in semipurified diets for broilers. They showed that methionine and arginine were the first and second most limiting amino acids in the YSCP tested. Their data suggested that palatability of the diet containing YSCP was also an important factor in reducing feed intake of those diets. Based on these studies as well as others, single-cell protein is not recommended for use at levels exceeding 5% for broilers and 10% for layers.

Novel Feedstuffs

Cassava root meal

Cassava root meal, also known as tapioca, mandioca, manioca, yucca and manioc meal, is a readily digestible carbohydrate food that is made from the roots of the cassava plant. It is a perennial root crop native to the humid tropics. The cortical layers of roots contain linamarin, a cyanogenic glycoside. These compounds yield hydrogen cyanide upon treatment with acid or appropriate hydrolytic enzymes. Because of this compound, many of the early studies showed growth depression in chicks fed increasing amounts of cassava. Sweet cassava is low in hydrocyanic acid and can be fed after drying. Bitter cassava, however, needs to be grated and pressed to remove the juice, which contains most of the glucoside. Therefore, effective processing of cassava reduces its cyanide content, thereby reducing its toxic effects (Reddy, 1991).

Studies on cassava root meal have shown a wide variation in response of broilers to this feedstuff. Recommended levels of the meal for broilers have ranged from 10% (Vogt, 1966) to 50% (Enriquez and Ross, 1967; Olson et al., 1969). This variability in results may be due to differences in experimental conditions, as well as in cassava-processing method. Cassava, if properly detoxified by sun-drying whole-root chips on a concrete floor, can be satisfactorily used in broiler diets (Gómez et al., 1983). Gómez et al. (1987) in Colombia studied the effects of increasing the ME of diets containing 0, 20 and 30% cassava root meal by adding either vegetable oil or animal tallow. Diets containing 20% cassava root meal with either vegetable oil or animal tallow produced the greatest body-weight of broilers. Eruvbetine and Afolami (1992) were able to feed as much as 40% cassava in place of maize to broilers in Nigeria without any significant effect on body-weight, feed conversion or mortality. Fuentes et al. (1992) evaluated sun-dried cassava meal in Brazil as a replacement for maize in broiler diets. They concluded that cassava meal can replace up to 75% of the maize without a significant effect on weight gain or feed conversion.

According to Scott et al. (1982), the dried cassava root meal contains about 1.8% crude protein, 1.3% fat, 85% nitrogen-free extract and 1.8% fibre. The mineral content includes 0.35% phosphorus, 0.3% Ca and 16 p.p.m. manganese.

Palm kernel meal

This meal has a fairly good-quality protein and a good balance of calcium and phosphorus. It has not been widely used for poultry because of its unpalatability and high fibre content (15%). A number of studies conducted in Nigeria (Onwudike, 1986a,b,c) have shown that palm kernel meal can be a valuable source of protein for poultry. Onwudike (1988) reported that the meal could be fed to laying hens at high levels if the diet contained fish meal. The same worker (Onwudike, 1992) was able to replace fish meal in the high-palm kernel meal diet by blood meal, along with the addition of DL-methionine, without any significant drop in egg production. Panigrahi (1991) observed behaviour changes in broiler chicks fed on diets containing high levels of palm kernel meal. This altered behaviour was associated with lower feed intake and reduced weight gains. When these diets containing high levels of palm kernel meal were supplemented with high levels of maize oil, behaviour was normal and weight gains were only 3.5% less than those of controls. Although Onwudike (1992) suggested that in the presence of an animal protein supplement palm kernel meal can be used at high levels in poultry rations, because of the palatability problem it probably should not be used at levels exceeding 20% of the ration.

Ipil-ipil leaf meal (Leucaena leucocephala)

Ipil-ipil is a leguminous tree that grows abundantly in the Philippines, Hawaii, Thailand and other tropical countries. The leaves of this tree, which are called *koa haole* in Hawaii, are used in animal feeds. This meal contains approximately 24% protein, 3.25% fat, 14% fibre and 530 mg β-carotene activity per kg (Scott *et al.*, 1982). The major problem of this meal is the presence of a toxic alkaloid mimosine. This alkaloid constitutes about 97% of the acid fraction extracted from the plant (Alejandrino *et al.*, 1976). In studies with chickens (Librijo and Hathcock, 1974), feed intake declined 33% and egg production declined 49% when this meal was included at 30% of the diet. Berry and D'Mello (1981) showed that egg production and live weight gain of chickens on diets containing 20% of this meal were significantly reduced. Iji and Okonkwo (1991) reported that regional differences exist in the composition of ipil-ipil meals. These differences should be better known before levels higher than 3% can be used in poultry diets. Scott *et al.* (1982) reported that levels of ipil-ipil meal above 5% cause reduced growth of broilers and reduced egg production in laying hens. This is probably due to poor amino acid digestibility as well as the toxic alkaloid mimosine, since Picard *et al.* (1987) presented very low amino acid digestibility values for this meal. These workers concluded that ipil-ipil meals are not suitable for poultry protein nutrition and should be regarded only as a pigment source to be used at low levels of the diet.

*Salseed (*Shorea robusta*)*

Salseed is produced in large quantities in many tropical and subtropical regions, primarily for its oil. After removal of the oil, about 85% of the seed is available as meal. Panda (1970) reviewed the early work on salseed and salseed meal and pointed out the economic importance of using this meal in poultry feeds as a source of energy.

Zombade *et al.* (1979) reported that salseed meal contains 9.8% crude protein, 2.2% ether extract, 45.0% available carbohydrates and 11.7% tannins. These workers observed that more than 5.0% salseed meal in the diet of White Leghorn male chickens resulted in poor growth and food conversion. Verma and Panda (1972) also observed that 5.0% salseed caused growth depression in chicks. Most of this growth depression is due to the high tannin content of the meal since about 0.5% tannin in the diet has been found to affect chicks adversely (Vohra *et al.*, 1966).

Dried poultry waste

Dried poultry waste (DPW) has been successfully used for feeding ruminants for many years past. Its use in poultry diets, however, has not been widely accepted. Interest in the use of DPW has developed because the energy-yielding components of high-energy poultry rations were digested and metabolized only to the extent of 70–80% (Young and Nesheim, 1972). Most of the early work on the use of DPW was conducted on laying hens at Michigan State University (Flegal and Zindel, 1969). Most of these studies indicated that including DPW in layer rations at relatively low levels had no adverse effects on performance as long as the low energy content of the product is considered in formulating diets. Nesheim (1972) and Young (1972) conducted several studies on DPW collected from cage layer operations and found that the apparent utilization of this material is not more than 30%. Several workers were able to show that laying hens can utilize some of the essential amino acids found in DPW (Blair and Lee, 1973; Rinehart *et al.*, 1973). Daghir and Amirullah (1978) did not observe any differences in egg production or egg weight between birds receiving 10% DPW and those receiving a standard maize–soyabean diet. Furthermore, they did not detect any differences in overall acceptability of boiled eggs produced by hens fed DPW at a 10% level. Vogt (1973), on the other hand, reported that incorporation of 10% in a laying ration significantly decreased egg production and feed conversion without affecting egg weight. In areas where feedstuffs are scarce and expensive, this product may contribute to reducing feed costs if it is properly handled and dried.

Buffalo gourd meal (Cucurbita foetidisima)

The buffalo gourd is a desert plant that has been recognized since the middle of this century as a potential source of oil and protein. The oil portion has been found to be edible and rich in the essential fatty acids. The protein is similar to other oil-seed proteins and is especially rich in arginine, aspartic acid and glutamic acid, but low in lysine, threonine and methionine (Daghir and Zaatari, 1983).

The main problem limiting the use of buffalo gourd meal in practical poultry rations is the presence of naturally occurring toxins in the seed. Daghir *et al*. (1980a) reported that buffalo gourd meal, prepared by hexane extraction and fed at levels sufficient to contribute 8% protein to the diet, is toxic to growing chicks. It produces a neuromuscular condition that is characterized by an abnormal position of the neck and inability of the bird to keep its head up. Further work showed that the toxin is mainly located in the embryo of the seed since feeding the hulls or the oil produced zero incidence of this abnormality (Daghir and Sell, 1980). Several bitter principles have been isolated from the family Cucurbitaceae and have been given the name cucurbitacins. Some of these have been chemically identified as triterpenoid glycosides. Daghir and Flaifel (1984) evaluated several detoxification procedures on hexane-extracted buffalo gourd meal and tried to verify the adequacy of these methods by feeding the differently treated meals to growing chicks. The treatments, which consisted of water soaking, alcohol extraction, chloroform–methanol extraction or chloroform–methanol–alcohol extraction, had no positive effect on the meal. The detoxification of desert plant seeds like the buffalo gourd should have a tremendous economic impact on the poultry industry since all these plants are potential crops for the arid lands of the hot regions.

Guar meal (Cyamopsis tetragonaloba)

Guar is a drought-resistant legume that is widely cultivated in India and Pakistan. The endosperm of the guar seed is a rich source of a galactomannan polysacharide called guar gum. Guar meal is obtained after mechanical separation of the endosperm from the hulls and germ of the ground seeds. The meal contains 4% fat, 45% crude protein, 6% fibre and 4.5% ash. It is rich in lysine and methionine. Vohra and Kratzer (1964a) showed that as little as 7.5% guar meal in the ration caused a significant depression in chick growth. Furthermore, they showed that a major part of this depression is attributed to the guar gum in the meal. Their results indicated that guar meal can be slightly improved by autoclaving. The same workers (Vohra and Kratzer, 1964b) found growth of chickens to be inhibited about 25–30% by the inclusion of guar gum at levels that contributed 2% or 4% pectin to the diet. This depression was overcome by treatment with enzymes capable of hydrolysing this gum, namely pectinase and cellulase or a preparation from the sprouted guar beans. Patel et al. (1980) studied the effects of γ-irradiation, penicillin and pectic enzyme on chick growth depression and faecal stickiness caused by guar gum and found that the faecal condition of birds fed guar gum was significantly improved by a combination of γ-irradiation and pectic enzyme supplement.

Bambara groundnut meal (Voandizeia subterrenea)

Bambara groundnut (BGN) is an indigenous African leguminous crop that has been described as the most resistant pulse (National Academy of Science, 1979). BGN seeds contain a trypsin inhibitor (Poulter, 1981), which would limit its use at high levels in poultry diets. Onwudike and Eguakum (1992) evaluated meals from raw seeds and seeds boiled for 30 and 60 min. Raw BGN had trypsin inhibitory activity of 20.8 units mg^{-1} and boiling for 30 min completely eliminated all that activity. Raw BGN meal contained 7.5% ether extract, 2.03% fibre and 20.6% protein. The ME value of the raw BGN was 2.65 kcal g^{-1} and the ME value of the heat-treated meal was 3.88 kcal g^{-1}. Phosphorous was the element with the highest concentration in BGN (0.11%) and zinc was the highest microelement (36 p.p.m.).

Jojoba meal (Simmoensia chinensis)

This meal is a by-product of the oil produced from the jojoba seed. It contains 25–30% protein (Verbiscar and Banigan, 1978). Feeding it to chickens resulted in impaired body-weight, and reduced feed intake and feed efficiency (Ngou Ngoupayou et al., 1982). This effect may be due to several compounds present in the meal, such as glycosides, polyphenols, phytic acids and trypsin inhibitors (Wiseman and Price, 1987). Arnouts et al. (1993) studied the possibility of using jojoba meal to inhibit feed intake of broiler breeder pullets and thus limit body-weight gain as recommended by breeder companies. They found that the growth retardation required was obtained with a 4% level of jojoba meal in the diet and that reduced growth was the result of decreased feed intake linked with the simondsin content of the meal, as well as other antinutritional compounds affecting digestibility.

Conclusions

This chapter summarizes research work on the composition and nutritional value of feedstuffs that are available in the hot regions of the world.

Table 6.1. Recommended levels of inclusion of selected cereals and protein supplements in poultry feeds as % of diet.

Feedstuff	Broiler starter	Broiler finisher	Chick starter	Chick grower	Chick developer	Layer and breeder
Barley	0-5	0-5	0-10	0-10	0-20	0-60
Millet	0-10	0-10	0-15	0-30	0-50	0-40
Rice bran	0-5	0-5	0-10	0-15	0-20	0-15
Rice polishings	0-5	0-5	0-10	0-15	0-20	0-15
Rice paddy	0-5	0-5	0-5	0-10	0-15	0-15
Sorghum	0-20	0-40	0-20	0-30	0-40	0-50
Triticale	0-20	0-30	0-20	0-30	0-30	0-60
Cottonseed meal	0-5	0-15	0-10	0-15	0-15	0-10
Coconut meal	0-2	0-3	0-2	0-5	0-5	0-3
Peanut meal	0-5	0-10	0-5	0-10	0-15	0-10
Sesame meal	0-5	0-10	0-5	0-10	0-15	0-10
Sunflower seed	0-10	0-15	0-10	0-15	0-15	0-20
Sunflower meal	0-5	0-10	0-5	0-10	0-15	0-10
Mustard seed meal	0-3	0-5	0-3	0-5	0-5	0-3
Linseed meal	0-2	0-3	0-2	0-3	0-3	0-3
Single-cell protein	0-3	0-5	0-3	0-5	0-10	0-10

Although a great deal of work has been done on the composition of agricultural and industrial by-products and several feeding tests have been conducted, more work is needed, particularly on methods to render these more palatable and free of antinutritional factors. Table 6.1 presents recommended levels of inclusion of cereals and protein supplements, while Table 6.2 presents recommended levels of inclusion of novel feedstuffs covered in this chapter. It is observed that, in general, recommended levels of inclusion for the novel feedstuffs are low because of problems either in palatability of these feedstuffs or the presence of antinutritional factors. Therefore, research on methods of improving the palatability and reducing antinutritional factors in those feedstuffs is badly needed. This type of research can lead to more extensive use of these novel feedstuffs, which would enhance further expansion of the poultry industry in the hot regions of the world without competing with humans for scarce and costly cereal grains. It should be pointed out that, until more information is obtained on these novel feedstuffs and further improvements are made on their nutritional value, the levels of their inclusion in poultry feedstuffs will continue to be relatively low. Furthermore, even with these low levels of inclusion, greater care should be taken in the balancing and formulation of poultry rations containing these ingredients than in those containing commonly used feedstuffs.

Table 6.2. Recommended levels of inclusion of selected novel feedstuffs in poultry feeds as % of diet.

Feedstuff	Broiler starter	Broiler finisher	Chick starter	Chick grower	Chick developer	Layer and breeder
Cassava root meal	0-10	0-20	0-10	0-15	0-20	0-20
Salseed	0-3	0-5	0-3	0-5	0-5	0-3
Ipil-ipil leaf meal	0-2	0-3	0-2	0-3	0-5	0-3
Buffalo gourd meal	0-2	0-3	0-2	0-3	0-5	0-3
Guar meal	0-2	0-3	0-2	0-3	0-5	0-3
Palm kernel meal	0-2	0-3	0-2	0-3	0-5	0-5
Bambara groundnut meal	0-2	0-3	0-2	0-3	0-5	0-3
Dried poultry waste	0-2	0-3	0-2	0-3	0-5	0-5
Jojoba meal	0-2	0-3	0-2	0-3	0-3	0-3

References

Abate, A.N. and Gomez, M. (1983) Substitution of finger millet (*Eleusine coracana*) and bulvish millet (*Pennisetun typhoides*) for maize in broiler feeds. *Animal Feed Science and Technology* 10, 291-299.

Acen, A. and Varum, K. (1987) Beta-glucaner i bygg. NLVF's fagutvalg for Korn fordling seminar. Norges Landbrukshoegskole, Norway, pp. 16-17.

Ajuyah, A.O., Hardin, R.T. and Sim, J.S. (1993) Effect of dietary full fat flax seed with and without antioxidant on the fatty acid composition of major lipid classes of chicken meats. *Poultry Science* 72, 125-136.

Alejandrino, A.L., Concepcion, F. and Belone, B. (1976) A modified method of isolating and determining mimosine from ipil-ipil leaf meal. *Philippine Agriculture* (Suppl.) 9, 10.

Allen, D. (1990) Feedstuffs ingredient analysis table. *Feedstuffs* 62.

Anderson, J.O. and Warnick, R.E. (1965) Amino acid deficiencies in peanut meal and in corn and peanut meal rations. *Poultry Science* 44, 1066-1072.

Anderson, J.O., Wagstaff, R.K. and Dobson, D.C. (1961) Studies on the value of hull-less barley in chick diets and means of increasing this value. *Poultry Science* 40, 1571-1584.

Armanious, M.W., Britton, W.M. and Fuller, H.L. (1973) Effect of methionine and choline on tannic acid and toxicity in the laying hen. *Poultry Science* 52, 2160-2168.

Armstrong, W.D., Featherston, W.R. and Rogler, J.C. (1973) Influence of methionine and other dietary additions on the performance of chicks fed bird resistant sorghum grain diets. *Poultry Science* 52, 1592-1599.

Arnouts, S., Bugse, J., Cokelaere, M.U. and Decuypere, E. (1993) Jojoba meal in the diet of broiler breeder pullets. *Poultry Science* 72, 1714-1721.

Bailey, A.V., Harris, J.A., Skau, E.L. and Kerr, T. (1966) Cyclopropenoid fatty acid content and fatty acid composition of crude oils from twenty-five varieties of cottonseed. *Journal of the American Oil Chemists Society* 43, 107-110.

Bell, D.E., Ibrahim, A.A., Denton, G.W., Long, G.G. and Bradley, G.L. (1990) An evaluation of sesame seed meal as a possible substitute for soyabean oil meal for feeding broilers. *Poultry Science* 69, 157 (Abstract).

Bell, J.M., Young, G.G. and Downey, R.K. (1971) A nutritional comparison of various rapeseed and mustard seed solvent-extracted meals of different glucocinolate composition. *Canadian Journal of Animal Science* 51, 259–269.

Berry, S. and D'Mello, J.P.F. (1981) A comparison of *Leucaena leucocephala* and grass meals as sources of pigments in diets for laying hens. *Tropical Animal Production* 6, 167–173.

Blair, R. and Lee, D.J.W. (1973) Supplementation of low protein layer diets. *British Poultry Science* 14, 9–16.

Bonino, M.F., Azcona, J.O. and Sceglio, O. (1980) The effect of ammoniation, methionine and gluten meal supplementation on the performance of laying hens fed high tannin sorghum grain diets. *Proceedings 6th European Poultry Conference* 3, 481–484.

Bragg, D.B. and Sharby, T.F. (1970) Nutritive value of triticale for broiler chick diets. *Poultry Science* 49, 1022–1027.

Cabel, M.C. and Waldroup, P.W. (1989) Ethoxyquin and ethylenediamine tetra acetic acid for the prevention of rancidity in rice bran stored at elevated temperature and humidity for various lengths of time. *Poultry Science* 68, 438–442.

Castanon, J.I., Ortiz, R.V. and Perez-Lansac, J. (1990) Effect of high inclusion levels of triticale in diets for laying hens containing 30% field beans. *Animal Feed Science and Technology* 31, 349–353.

Chang, S.I. and Fuller, H.L. (1964) Effect of tannin content of grain sorghums on their feeding value for growing chicks. *Poultry Science* 43, 30–36.

Chawla, J.S., Nagra, S.S. and Pannu, M.S. (1987) Different cereals for laying hens. *Indian Journal of Poultry Science* 22, 95–100.

Cheva-Isarakul, B. and Tagtaweewipat, S. (1991) Effect of different levels of sunflower seed in broiler rations. *Poultry Science* 70, 2284–2294.

Clark, M. (1931) Practical food evaluation. *Food Technology (London)* 1, 98–99.

Classen, H.L., Graham, H., Inborr, J. and Bedford, M.R. (1991) Growing interest in feed enzymes to lead to new products. *Feedstuffs* 63(4), 22–25.

Coon, C.N., Shepler, R., McFarland, D. and Nordhein, J. (1979) The nutritional evaluation of barley selection and cultivars from Washington State. *Poultry Science* 58, 913–918.

Coon, C.N., Obi, I. and Hamre, M.L. (1988) Use of barley in laying hen diets. *Poultry Science* 67, 1306–1313.

Cuca, M.G. and Avila, E.G. (1973) Preliminary studies on triticale in diets for laying hens. *Poultry Science* 52, 1973–1974.

Cuca, M. and Sunde, M.L. (1967) The availability of calcium from Mexican and Californian sesame meals. *Poultry Science* 46, 994–1002.

Daghir, N.J. and Abdul-Baki, T.K. (1977) Yeast protein in broiler rations. *Poultry Science* 56, 1836–1841.

Daghir, N.J. and Amirullah, I. (1978) Dehydrated poultry waste and urea as feed supplements in layer rations. *Iran Journal of Agricultural Research* 6, 91–97.

Daghir, N.J. and Charalambus, K. (1978) Metabolizable energy of $FeSO_4$-treated and untreated mustard seed meal. *Poultry Science* 57, 1081–1083.

Daghir, N.J. and Flaifel, F.A. (1984) Buffalo gourd protein for growing chickens. *Proceedings 17th World's Poultry Congress*, pp. 295-297.

Daghir, N.J. and Kevorkian, K.A. (1970) Limiting amino acids in sesame meal chick diets. *Proceedings of the 14th World's Poultry Congress*, Madrid, pp. 711-715.

Daghir, N.J. and Mian, W.A. (1976) Mustard seed meal as a protein source for chickens. *Poultry Science* 55, 1699-1703.

Daghir, N.J. and Nathanael, A.S. (1974) An assessment of the nutritional value of triticale for poultry. *Proceedings of the 15th World Poultry Congress*, pp. 612-614.

Daghir, N.J. and Rottensten, K. (1966) The influence of variety and enzyme supplementation on the nutritional value of barley for chicks. *British Poultry Science* 7, 159-163.

Daghir, N.J. and Sell, J.L. (1980) Buffalo gourd seed and seed components for growing chickens. *Nutrition Reports International* 22, 445-452.

Daghir, N.J. and Sell, J.L. (1982) Amino acid limitations of yeast single cell protein for growing chickens. *Poultry Science* 61, 337-344.

Daghir, N.J. and Zaatari, I.M. (1983) Detoxification and protein quality of buffalo gourd meal for growing chickens. *Nutrition Reports International* 27, 339-346.

Daghir, N.J., Hajj, R. and Akrabawi, S.S. (1966) Studies on peanut meal for broilers. *Proceedings of the 13th World's Poultry Congress*, Kiev, pp. 238-246.

Daghir, N.J., Ullah, M.F. and Rottensten, K. (1967) Lysine supplementation of sesame meal broiler rations. *Tropical Agriculture, Trinidad* 44, 235-242.

Daghir, N.J., Ayyash, B. and Pellet, P.L. (1969) Evaluation of groundnut meal protein for poultry. *Journal of the Science of Food and Agriculture* 20, 349-354.

Daghir, N.J., Mahmoud, H.K. and El-Zein, A. (1980a) Buffalo gourd meal: nutritive value and detoxification. *Nutrition Reports International* 21, 837-847.

Daghir, N.J., Raz, M.A. and Uwayjan, M. (1980b) Studies on the utilization of full fat sunflower seed in broiler rations. *Poultry Science* 59, 2272-2278.

Dale, N.M., Wyatt, R.D. and Fuller, H.L. (1980) Additive toxicity of aflatoxins and dietary tannins in broiler chicks. *Poultry Science* 59, 2417-2420.

Douglas, J.H., Sullivan, T.W., Bond, P.L. and Steuwe, F.J. (1990a) Nutrient composition and metabolizable energy values of selected grain sorghum varieties and yellow corn. *Poultry Science* 69, 1147-1155.

Douglas, J.H., Sullivan, T.W., Bond, P.L., Struwe, F.J., Baier, J.G. and Robesone, L.G. (1990b) Influence of grinding, rolling, and pelleting on the nutritional value of grain sorghums and yellow corn for broilers. *Poultry Science* 69, 2150-2156.

El-Alaily, H., Soliman, H., Anwar, A., El-Zeiny, M. and Ibrahim, S. (1985) Sorghum grains as an energy source for chicks. *Egyptian Poultry Science* 5.

El-Boushy, A.R. and Raternick, R. (1989) Replacement of soyabean meal by cottonseed meal and peanut meal or both in low energy diets for broilers. *Poultry Science* 68, 799-804.

Elkin, R.G., Featherston, W.R. and Rogler, J.C. (1978) Investigation of leg abnormalities in chicks consuming high tannin sorghum grain diets. *Poultry Science* 57, 757–762.

Enriquez, F.Q. and Ross, E. (1967) The value of cassava root meal for chicks. *Poultry Science* 46, 622–626.

Eruvbetine, D. and Afolami, C.A. (1992) Economic evaluation of cassava (*Manihot esculenta*) as a feed ingredient for broilers. *Proceedings of the 19th World's Poultry Congress*, Vol. 3, pp. 532–535.

Fainan, C., Ryan, R.J. and Eickel, H.T. (1967) *Endocrinology* 81, 279.

FAO (1985) *Production Yearbook*, Vol. 39. Food and Agriculture Organization of the United Nations, Rome, Italy.

Feltwell, R. and Fox, S. (1978) *Practical Poultry Feeding*. Faber and Faber, London, 267 pp.

Fernandez, R. and McGinnis, J. (1974) Nutritive value of triticale for young chicks and effect of different amino acid supplements on growth. *Poultry Science* 53, 47–53.

Flegal, C.A. and Zindel, H.C. (1969) The utilization of dehydrated poultry waste by laying hens. *Poultry Science* 48, 1807–1810.

Flores, M.P., Castanon, J.I.R. and McNab, J.M. (1992) Use of enzymes to improve the nutritive value of triticale in poultry diets. *Proceedings of the 19th World's Poultry Congress*, Vol. 2, pp. 253–254.

Fuentes, M.F.F., Coelho, M.G.R., Souza, F.M., Lopes, I.R.V. and Pereira, L.I. (1992) Sun-dried cassava meal in tropical broiler diets. *Proceedings of the 19th World's Poultry Congress*, Vol. 3, p. 551.

Gippert, T., Halmagyi-Valler, T. and Gati, L. (1988) Utilization of differently treated, extracted sunflower coarse meal in the nutrition of broiler chickens. *Proceedings of the 18th World's Poultry Congress*, pp. 927–928.

Gómez, G., Valdirieso, M., Santos, J. and Hoyos, C. (1983) Evaluation of cassava root meal prepared from low- or high-cyanide containing cultivars in pig and broiler diets. *Nutrition Reports International* 28, 693–704.

Gómez, G., Tellez, G. and Caicedo, J. (1987) Effects of the addition of vegetable oil or animal tallow to broiler diets containing cassava root meal. *Poultry Science* 66, 725–731.

Hasselman, K., Elwinger, K., Nilsson, M. and Thomke, S. (1981) The effect of beta-glucanase supplementation, stage of ripeness, and storage treatment of barley in diets fed to broiler chickens. *Poultry Science* 60, 2664–2671.

Herstad, O. (1987) Cereals with higher fiber content (barley, oats, millet). *Proceedings of the Sixth European Symposium on Poultry Nutrition*, pp. A15–A23.

Heywang, B.W., Bird, H.R. and Altschul, A.M. (1955) Relationship between discolorations in eggs and dietary free gossypol supplied by different cottonseed products. *Poultry Science* 34, 81–90.

Hulse, J.H., Laing, E.M. and Pearson, O.E. (1980) *Sorghum and Millets: Their Composition and Nutritive Value*. Academic Press, London.

Hussein, A.S. and Kratzer, F.H. (1982) Effect of rancidity on the feeding value of rice bran for chickens. *Poultry Science* 61, 2450–2455.

Ibrahim, S., Fisher, C., El-Alaily, H., Soliman, H. and Anwar, A. (1988) Improvement of the nutritional quality of Egyptian and Sudanese sorghum grains by the addition of phosphates. *British Poultry Science* 29, 721–728.

Iji, P.A. and Okonkwo, A.C. (1991) Leucaena and neem leaf meals: potential protein sources for layers? *Feed International* December, 29–32.

Jerock, H. (1987) Nutritional value of wheat, rye, and triticale in broiler chickens and laying hens. *Proceedings of the Sixth European Symposium on Poultry Nutrition*, pp. A4–A14.

Jones, L.A. (1981) Special cottonseed products report. *Feedstuffs* 53(52), 19–21.

Karmajeewa, H. and Than, S.H. (1984) Choice feeding of the replacement pullet on whole grains and subsequent performance on laying diets. *British Poultry Science* 25, 99–109.

Klain, G.J., Hill, D.C., Branion, H.D. and Gray, J.A. (1956) The value of rapeseed oil meal and sunflower seed oil meal in chicks starter rations. *Poultry Science* 32, 542–547.

Lease, J.G. (1966) The effect of autoclaving sesame meal on its phytic acid content and on the availability of its zinc to the chick. *Poultry Science* 45, 237–241.

Lease, J.G., Barnett, B.D., Lease, E.J. and Turk, D.E. (1960) The biological unavailability to the chick of zinc in a sesame meal ration. *Journal of Nutrition* 72, 66–71.

Lee, K. and Moss, C.W. (1989) Performance of laying chickens fed diets containing confectionery-type sunflower seeds. *Poultry Science* 68 (Suppl. 1), 84 (Abstract).

Leeson, S. and Summers, J.D. (1987) Response of white Leghorns to diets containing ground or whole triticale. *Canadian Journal of Animal Science* 67, 583–585.

Levi, R.S., Reilich, H.G., O'Neill, H.J., Cucullu, A.F. and Skau, E.L. (1967) Quantitative determination of cyclopropenoid fatty acids in cottonseed meal. *Journal of the American Oil Chemists Society* 44, 249–252.

Librijo, N.T. and Hathcock, J.N. (1974) Metabolism of mimosine and other compounds from *Leucaena lucocephala* by the chicken. *Nutrition Reports International* 9, 217–222.

Luckbert, J., Maitre, I. and Castaing, J. (1988) Using triticale in laying hen diets. *Proceedings of the 18th World's Poultry Congress*, pp. 797–798.

Luis, E.S., Sullivan, T.W. and Nelson, L.A. (1982) Nutrient composition and feeding value of proso millet, sorghum grain and corn in broiler diets. *Poultry Science* 61, 311–320.

McDonald, P., Edwards, R.A. and Greenbagh, J.F.D. (1988) *Animal Nutrition*, 4th edn. Wiley, New York, pp. 463–464.

McNab, J.M. (1987) The energy value of roots and mill by-products. *Proceedings of the Sixth European Symposium on Poultry Nutrition*, pp. A26–A34.

McNaughton, J.L. and Deaton, J.W. (1981) Sunflowers poultry applications. *Feed Management* 32(6), 27.

Mezoui, C.M. and Bird, F.H. (1984) Peanut as a protein source for growing broilers. *Revue Science et Technique, série Sciences de la Santé* $1(\frac{1}{2})$, 45–51 (Abstract).

Mohammadain, G.M., Babiker, S.A. and Mohammad, T.A. (1986) Effect of feeding millet, maize and sorghum grains on performance, carcass yield and chemical composition of broiler meat. *Tropical Agriculture (Trinidad)* 63, 173–176.

Mohan Ravi Kumar, A., Reddy, V.R., Reddy, P.V.V.S. and Reddy, P.S. (1991) Utilization of pearl millet for egg production. *British Poultry Science* 32, 463–469.

Morrison, A.B., Clandinin, D.R. and Robblee, A.R. (1953) The effects of processing variables on the nutritive value of sunflower seed oil meal. *Poultry Science* 32, 492–496.

Morrison, F.B. (1956) *Feeds and Feeding*, 21st edn. Morrison, Ithaca, NY.

National Academy of Science (1979) *Tropical Legumes, Resources for the Future*. Washington, DC.

Nelson, T.S., Stephenson, E.L., Burgos, A., Floyd, J. and York, J.O. (1975) Effect of tannin content and dry matter digestion on energy utilization and average amino acid availability of hybrid sorghum grains. *Poultry Science* 54, 1620–1623.

Nesheim, M.C. (1972) Evaluation of dehydrated poultry manure as a potential poultry feed ingredient. *Animal Waste Conference Proceedings*, Cornell University, Ithaca, NY.

Newman, R.K. and Mewman, C.W. (1988) Nutritive value of a new hull-less barley cultivar in broiler chick diets. *Poultry Science* 67, 1573–1579.

Ngou Ngoupayou, J.D., Maiorind, P.M. and Reid, B.L. (1982) Jojoba meal in poultry diets. *Poultry Science* 61, 1692–1696.

Nir, I., Melcion, J.P. and Picard, M. (1990) Effect of particle size of sorghum grains on feed intake and performance of young broilers. *Poultry Science* 69, 2177–2184.

Olson, D.W., Sunde, M.L. and Bird, H.R. (1969) Amino acid supplementation of mandioca meal in chick diets. *Poultry Science* 48, 1949–1953.

Onwudike, O.C. (1986a) Palm kernel meal as a feed for poultry: 1. Composition and availability of its amino acids to chicks. *Animal Feed Science and Technology* 16, 179–186.

Onwudike, O.C. (1986b) Palm kernel meal as a feed for poultry: 2. Diets containing palm kernel meal for starter and grower pullets. *Animal Feed Science and Technology* 16, 187–194.

Onwudike, O.C. (1986c) Palm kernel meal as feed for poultry: 3. Replacement of groundnut cake by palm kernel meal in broiler diets. *Animal Feed Science and Technology* 16, 195–202.

Onwudike, O.C. (1988) Palm kernel meal as a feed for poultry: 4. Use of palm kernel meal by laying hens. *Animal Feed Science and Technology* 20, 279–286.

Onwudike, O.C. (1992) Effect of blood meal on the use of palm kernel meal by laying birds. *Proceedings 19th World's Poultry Congress*, Vol. 3, pp. 514–519.

Onwudike, O.C. and Eguakum, A. (1992) Effect of heat treatment on the composition, trypsin inhibitory activity, ME level and mineral bioavailability of Bambara groundnut meal with poultry. *Proceedings 19th World's Poultry Congress*, Vol. 3, pp. 508–511.

Panda, B. (1970) Processing and utilization of agro-industrial by-products as livestock and poultry feed. *Indian Poultry Gazette* 54, 39–71.

Panigrahi, S. (1991) Behaviour changes in broiler chicks fed on diets containing palm kernel meal. *Applied Animal Behaviour Science* 3–4, 277–281.

Panigrahi, S. (1992) Energy-deficit induced behaviour changes in broiler chicks fed on copra meal based diets. *Proceedings 19th World's Poultry Congress*, Vol. 3, pp. 503-507.

Panigrahi, S. and Hammonds, T.W. (1990) Egg discoloration effects of including screw-press cottonseed meal in laying hen diets and their prevention. *British Poultry Science* 31, 107-120.

Panigrahi, S. and Morris, T.R. (1991) Effects of dietary cottonseed meal and iron treated cottonseed meal in different laying hen genotypes. *British Poultry Science* 32, 167-184.

Panigrahi, S., Machin, D.H., Parr, W.H. and Bainton, J. (1987) Responses of broiler chicks to dietary copra cake of high lipid content. *British Poultry Science* 28, 589-600.

Panigrahi, S., Plumb, V.E. and Machin, D.H. (1989) Effects of dietary cottonseed meal, with and without iron treatment on laying hens. *British Poultry Science* 30, 641-651.

Patel, M.B., Jami, M.S. and McGinnis, J. (1980) Effect of gamma irradiation, pennicillin and/or pectic enzyme on chick growth depression and fecal stickiness caused by rye, citrus pectin and guar gum. *Poultry Science* 59, 2105-2110.

Phelps, R.A. (1966) Cottonseed meal for poultry: from research to practical application. *World's Poultry Science Journal* 22, 86-112.

Phelps, R.A., Shenstone, F.S., Kemmerer, A.R. and Evans, R.J. (1965) A review of cyclopropenoid compounds: biological effects of some derivatives. *Poultry Science* 44, 358-394.

Picard, M., Angulo, I., Antoine, H., Bouchot, C. and Sauveur, B. (1987) Some feeding strategies for poultry in hot and humid environments. *Proceedings 10th Annual Conference of the Malaysian Society of Animal Production*, pp. 110-116.

Poulter, N.H. (1981) Properties of some protein fractions from bambara groundnut. *Journal of the Science of Food and Agriculure* 32, 44-50.

Proudfoot, F.G. and Hulan, H.W. (1988) Nutritive value of triticale as a feed ingredient for broiler chickens. *Poultry Science* 67, 1743-1749.

Rad, F.H. and Keshavarz, K. (1976) Evaluation of the nutritional value of sunflower meal and the possibility of substitution of sunflower meal for soyabean meal in poultry diets. *Poultry Science* 55, 1757-1760.

Randall, J.M., Sayre, R.N., Schultz, W.G., Fong, R.Y., Mossman, A.P., Triberlhorne, R.E. and Saunders, R.M. (1985) Rice bran stabilization by extrusion cooking for extraction of edible oil. *Journal of Food Science* 50, 361-364.

Rao, D.R., Johnson, W.M. and Sunki, G.R. (1976) Replacement of maize by triticale in broiler diets. *British Poultry Science* 17, 269-274.

Reddy, C.V. (1991) Tapioca instead of maize. *Poultry International*, October, 42-44.

Reddy, D.R. and Reddy, C.V. (1970) Influence of source of grain on the performance of laying stock. *Indian Veterinary Journal* 47, 157-163.

Reid, B.L., Galaviz-Moreno, S. and Maiorino, P.M. (1987) Evaluation of isopropanol-extracted cottonseed meal for laying hens. *Poultry Science* 66, 82-89.

Rinehart, K.E., Snetsinger, D.C., Rogland, W.W. and Zimmerman, R.A. (1973)

Feeding value of dehydrated poultry waste. *Poultry Science* 52, 2079-2083.

Rojas, S.W. and Scott, M.L. (1969) Factors affecting the nutritive value of cottonseed meal as a protein source in chick diets. *Poultry Science* 48, 819-835.

Rose, R.J., Coit, R.N. and Sell, J.L. (1972) Sunflower seed meal as a replacement for soyabean meal protein in laying hen rations. *Poultry Science* 51, 960-967.

Rosen, G.D. (1958) Groundnuts (peanuts) and groundnut meal. In: Altshul, A.M. (ed.) *Processed Plant Protein Foodstuffs*. Academic Press, New York, pp. 419-448.

Rotter, B.A., Friesen, O.D., Guenter, W. and Marawardt, R.R. (1990) Influence of enzyme supplementation on the bioavailable energy of barley. *Poultry Science* 69, 1174-1181.

Ruiz, N., Marion, J.E., Miles, R.D. and Barnett, R.B. (1987) Nutritive value of new cultivars of triticale and wheat for broiler chicken diets. *Poultry Science* 66, 90-97.

Saoud, N.B. and Daghir, N.J. (1980) Blood constituents of yeast fed chicks. *Poultry Science* 59, 1807-1811.

Sayre, R.N., Earl, L., Kratzer, F.H. and Saunders, R.M. (1987) Nutritional qualities of stabilized and raw rice bran for chicks. *Poultry Science* 66, 493-499.

Scott, M.L., Nesheim, M.C. and Young, R.J. (1982) *Nutrition of the Chicken*. M.L. Scott, Ithaca, NY.

Sell, D.R., Rogler, J.C. and Featherston, W.R. (1983) The effects of sorghum tanin and protein level on the performance of laying hens maintained in two temperature environments. *Poultry Science* 62, 2420-2428.

Sell, J.L., Hodgson, G.C. and Shebeski, L.H. (1962) Triticale as a potential component of chick rations. *Canadian Journal of Animal Science* 42, 158-166.

Sharma, B.D., Sadagopan, V.R. and Reddy, V.R. (1979) Utilization of different cereals in broiler diets. *British Poultry Science* 20, 371-378.

Uwayjan, M.G., Azar, E.J. and Daghir, N.J. (1983) Sunflower seed in laying hen rations. *Poultry Science* 62, 1247-1253.

Valdivie, M.L., Sardinas, O. and Garcia, J.A. (1982) The utilization of 20% sunflower seed meal in broiler diets. *Cuban Journal of Agricultural Sciences* 16, 167-171.

Verbiscar, A.J. and Banigan, T.F. (1978) Composition of jojoba seeds and foliage. *Journal of Agricultural Food Chemistry* 26, 1456-1459.

Verma, S.V.S. and Panda, B. (1972) Studies on the metabolizable energy values of salseed (*Shorea robusta*) and salseed cake by chemical and biological evaluation in chicks. *Indian Journal of Poultry Science* 7, 5-12.

Verma, S.V.S., Tyagi, P.K. and Singh, B.P. (1992) Nutritional value of broken rice (rice kani) for chicks. *Proceedings of the 19th World's Poultry Congress*, pp. 524-527.

Vogt, H. (1966) The use of tapioca meal in poultry rations. *World's Poultry Science Journal* 22, 113-125.

Vogt, H. (1973) The utilization of production and processing waste from egg production. *World's Poultry Science Journal* 29, 157-158.

Vohra, P. and Kratzer, F.H. (1964a) The use of guar meal in chick rations. *Poultry Science* 43, 502-503.

Vohra, P. and Kratzer, F.H. (1964b) Growth inhibitory effect of certain polysaccharides for chickens. *Poultry Science* 43, 1164–1170.

Vohra, P., Kratzer, F.H. and Joslyn, M.A. (1966) The growth depressing and toxic effects of tannins to chicks. *Poultry Science* 45, 135–142.

Vohra, P., Hafez, Y., Earl, L. and Kratzer, F.H. (1975) The effect of ammonia treatment of cottonseed meal on its gossypol-induced discoloration of egg yolks. *Poultry Science* 54, 441–447.

Waldroup, P.W. (1981) Cottonseed meal in poultry diets. *Feedstuffs* 53(52), 21–24.

Wilson, B.J. and McNab, J.M. (1975) The nutritive value of triticale and rye in broiler diets containing field beans (*Vicia faba* L.). *British Poultry Science* 16, 17–22.

Wiseman, M.O. and Price, R.L. (1987) Characterization of protein concentrates of jojoba meal. *Cereal Chemistry* 64, 91–93.

Young, G.J. and Nesheim, M.C. (1972) Dehydrated poultry waste as feed ingredient. *Proceedings Cornell Nutrition Conference*, p. 46.

Young, R.J. (1972) Evaluation of poultry waste as a feed ingredient and recycling waste as a method of waste disposal. *Proceedings Texas Nutrition Conference*, p. 1.

Zatari, I. and Sell, J.L. (1990) Sunflower meal as a component of fat-supplemented diets for broiler chickens. *Poultry Science* 69, 1503–1507.

Zombade, S.S., Lodhi, G.N. and Ichhponani, J.S. (1979) The nutritional value of salseed (*Shorea robusta*) meal for growing chicks. *British Poultry Science* 20, 433–438.

7

Mycotoxins in Poultry Feeds

N.J. Daghir
Faculty of Agricultural Sciences, UAE University, Al-Ain, UAE.

Introduction	157
Aflatoxins	158
Occurrence	159
Aflatoxicosis in broilers	160
Aflatoxicosis in layers and breeders	162
Aflatoxin residues in eggs and poultry meat	162
Influence of aflatoxin on resistance and immunity	163
Aflatoxin and vitamin nutrition	165
Aflatoxin and zinc in feeds and feedstuffs	166
Citrinin	166
Fumonisins	167
Ochratoxins	168
Oosporein	169
T-2 Toxin	170
Vomitoxin (Deoxynivalenol – DON)	171
Zearalenone	172
Detection and Control of Mycotoxins	172
Conclusions	176
References	176

Introduction

The term mycotoxin is used to refer to all toxins derived from fungi. The names of most mycotoxins are based on the names of the fungi that produce them. In some cases, the name is the actual chemical name, or it may be based on a toxic manifestation of the toxin. Information on the existence and spread of these toxins in the hot regions of the world has been accumulating very rapidly in recent years. Methods of harvesting, storage, processing and handling of feedstuffs and complete feeds in these areas are conducive for fungal growth and therefore for the existence of these toxins in them. Mycotoxin contamination of feedstuffs was first recognized in the early 1960s in the UK with the discovery of aflatoxin in imported peanut meals as the cause of 'turkey X-disease'. The author detected aflatoxin poisoning in Lebanon as far back as 1965 in birds receiving contaminated peanut meals (Daghir *et al.*, 1966). In a small survey conducted on peanut meals used by feed manufacturers during that year, samples were found to vary from 0.5 to 3.0 p.p.m. of aflatoxin. Since that date, aflatoxins have been reported to occur in poultry feeds in several countries of the hot regions. The early literature on the degree of contamination of feeds in various countries has been reviewed by Jelinek *et al.* (1989). Table 7.1 shows that the incidence of aflatoxins in poultry feeds in some selected countries of the hot region in which it was studied varies from 18.9 to 94.4%. The level of aflatoxin reported by these countries exceeds the Food and Drug Administration and European Community permissible level of 20 p.p.b. A survey was conducted in the USA (Anonymous, 1988) on seven large poultry companies in the southeast (Georgia, North Carolina and Virginia). Approximately 1000 samples were gathered from feed troughs at regular intervals and checked for levels of aflatoxin, vomitoxin, zearalenone and T-2 toxin. Table 7.2 shows the results of this survey. Although these companies were all located in the southeast, they all stated that they were buying Midwestern maize for their rations. It is interesting that the vomitoxins were more of a problem on these farms than the aflatoxins. Blaney and Williams (1991) reported that, in Australia, the mycotoxins that produce substantial losses are the aflatoxins, zearalenone and deoxynivalenol, occurring mainly in grains.

By far the most important mycotoxins to the poultry industry today are the aflatoxins, ochratoxins, vomitoxins, zearalenone, oosporein and T-2 toxins, and therefore this chapter will cover these toxins, as well as citrinin, a mycotoxin produced by *Aspergillus* and *Penicillium*, and fumonisins, produced by *Fusarium moniliforme*. These latter two have also been shown to be toxic to poultry. The biological effects of mycotoxins, other than aflatoxin, on poultry have been reviewed by Wyatt (1979).

Aflatoxins

Aflatoxin is the common name for a group of structurally related compounds that include B_1, B_2, G_1, G_2 produced by fungi of the flavus parasiticus group of the genus *Aspergillus*. These toxins have received a great deal of attention since their discovery in the UK in 1960. They are by far the most widespread in poultry feeds. Besides their occurrence in peanut meals, they have been found in cottonseed meals, soyabean meals, maize, barley, wheat and oats, as well as other feedstuffs (Goldblatt, 1969).

Table 7.1. Incidence of aflatoxins in poultry feeds in some countries of the hot regions.

Country	No. of samples tested	% Incidence	Reference
Egypt			
Assiut region	70	85.7	Shaaban et al., 1988
India			
Haryana	70	63.0	Mahipal and Kaushik, 1983
Uttar Pradesh	208	47.0	Johri et al., 1986
Uttar Pradesh	266	46.0	Johri et al., 1987
Madras	189	40.0	Selvasubramanian et al., 1987
Punjab	77	45.4	Raina and Singh, 1991
Indonesia			
Jakarta	8	85.0	Ginting, 1984
Pontianak	31	64.0	Ginting, 1984
Bagor	290	94.4	Widiastuti et al., 1988
Jakarta	56	91.0	Purwoko et al., 1991
Mexico	290	18.9	Rosiles, 1987
Nigeria	122	57.4	Oyejide et al., 1987
Pakistan			
Lahore	300	42.0	Rizvi et al., 1990
Saudi Arabia			
Riyadh	71	77.0	Ewaidah, 1988
Riyadh	371	30.2	Shaaban et al., 1991
South Africa			
Natal	142	18.5	Westlake and Dutton, 1985
Turkey			
Istanbul	170	64.5	Babila and Acktay, 1991
Marmara	74	60.8	O'zpnar et al., 1988
USA			
Southern states	1018	38.0	Muirhead, 1989

Occurrence

Aflatoxicosis has been reported to occur in several avian species besides the chicken. Ducks, geese, pheasants, quail and turkey have all been shown to be affected by aflatoxin. It is believed that all domestic birds are susceptible, but some are more sensitive than others. Chickens are least and ducks most susceptible. There are apparently even breed and strain differences in susceptibility to aflatoxin. The New Hampshire breed is susceptible to diets containing as little as 0.5 p.p.m., which normally does not seriously affect other chicken breeds (Gumbman et al., 1970).

Table 7.2. Tests for mycotoxin contamination of feed on poultry farms in the southeastern USA (from Anonymous, 1988).

Toxin	Levels considered positive (p.p.b.)	% Positive
Aflatoxin	> 5	35
Vomitoxin	> 250	64
Zearalenone	> 100	11
T-2 toxin	> 50	5

Aflatoxicosis in broilers

Most studies on the effects of aflatoxin in broiler chickens have been designed as subacute studies in which aflatoxin was fed at moderate to high levels (0.625–10 p.p.m.) from 0 to 3 weeks of age (Smith and Hamilton, 1970; Tung *et al.*, 1975; Huff, 1980). In these studies, about 2.5 p.p.m. aflatoxin was usually required to decrease body-weight significantly. Even in turkey poults, which are considered more sensitive than broilers, 2 p.p.m. are required to decrease body-weight significantly (see Fig. 7.1). Besides reduction in body-weight and feed conversion, birds show a variety of symptoms, such as enlarged livers, spleen and pancreas, repressed bursa and pale combs, shank and bone marrow. Aflatoxin has also been shown by Osborne *et al.* (1975) to inhibit fat digestion in broilers by decreasing the enzymes and bile acids required for fat digestion. A high-fat diet made aflatoxicosis less severe in broilers, and so did high-protein diets.

Several studies have shown that aflatoxin is a hepatotoxin in broilers. The hepatotoxicity is reflected in elevated liver lipid levels (Tung *et al.*, 1973), as well as disruption of hepatic protein synthesis (Tung *et al.*, 1975). This hepatotoxicity also induces severe coagulopathies (Doerr *et al.*, 1974; Doerr and Hamilton, 1981) and anaemia (Tung *et al.*, 1975). Aflatoxin has also been reported to increase the susceptibility of young broiler chickens to bruising (Tung *et al.*, 1971). Huff *et al.* (1983) observed elevated prothrombin times and increased incidence and severity of thigh and breast bruises in broilers fed 2.5 p.p.m. aflatoxin up to 6 weeks of age. These authors suggested that aflatoxin is implicated in increased susceptibility of broilers to bruising during live-haul and processing. It is now well established that this relationship of aflatoxin to bruising in broilers is very prevalent, and normal events, such as catching of birds prior to processing, physical insults in the broiler house and removal from the coops at the processing plant, may cause rupture of fragile capillaries. This is referred to as haemmorhages of the musculature and skin and often leads to condemnations or downgrading of these birds. Wyatt (1988a) concluded that this bruising effect will occur with aflatoxin concentrations of less than 100 p.p.b. in the feed. Hamilton (1990) confirmed this and showed that

Fig. 7.1. Effect of aflatoxin on body-weight in turkey poults. Control is on far left.

Table 7.3. Aflatoxin and bruising in chickens (from Hamilton, 1990).

Aflatoxin (ng g^{-1})	Body-weight (g)	Minimal bruising energy (joules)	Capillary fragility (mmHg)
0	444^a	0.43^a	350^a
625	451^a	0.37^b	260^b
1250	438^a	0.31^c	185^c
2500	380^b	0.23^d	—

$^{a-d}$ Values in a column with different superscripts differ significantly ($P < 0.05$).

the level of aflatoxin in the feed that can cause bruising is much lower than that causing growth inhibition (Table 7.3).

In another study, by Doerr *et al.* (1983), it was shown that the abnormalities normally encountered in broilers fed moderate to high levels of aflatoxin can be produced with much lower levels of the toxin (0.075–0.675 p.p.m.) if fed from day-old to market. These workers showed reduced growth, poor pigmentation and fatty livers resulting from this chronic low-level aflatoxicosis. Their data demonstrate that fatty livers may result in broiler chickens consuming a very low dose of aflatoxin and thus may explain the occasional idiopathic occurrence of fatty livers in broilers at processing. Their work also demonstrates the difficulty in establishing a 'safe' level of aflatoxin-contaminated feed, one which can be fed to broiler chickens with no adverse effects on performance, because toxicity in their experiments varied from one experiment to the other, depending on the stress factors present in each case.

Aflatoxicosis in layers and breeders

The effects of aflatoxins on laying birds have been studied by Garlich *et al.* (1973). Aflatoxin decreased egg production about 2–4 weeks after administration of the toxin. The toxin decreased egg weight, but had no significant effect on shell thickness or on percentage of shell (Hamilton and Garlich, 1971). These workers also suggested that dietary aflatoxin can cause a fatty liver syndrome in laying hens. This was confirmed later by Pettersson (1991), who reported that more than 2 p.p.m. aflatoxin in the feed decreases egg production and egg weight and increases the incidence of fatty livers in laying hens.

The effects of aflatoxin in the diet were also investigated with broiler breeder hens. Hens exhibited typical symptoms of aflatoxicosis, including enlarged livers and spleens. Although fertility was not affected, both egg production and hatchability of fertile eggs decreased (Dalvi, 1986).

Aflatoxin residues in eggs and poultry meat

Aflatoxin B_1 is the mycotoxin that has been most studied for possible residues in tissues. Practical analytical methods have been developed for aflatoxin B_1 in tissues and for only one of the aflatoxin B_1 transformation products, aflatoxin M_1. Residues of aflatoxin B_1 have been found in eggs and in tissues from hens and broilers fed aflatoxin-contaminated rations (Jacobson and Wiseman, 1974).

Some workers have been concerned about the potential of finding aflatoxicol (R_0), which is the most toxic of the known metabolites, in eggs or meat. Trucksess *et al.* (1983) demonstrated that aflatoxin B_1 and its metabolite aflatoxicol can be detected in eggs and edible tissues from hens given feed contaminated with aflatoxin B_1 at a level of 8 p.p.m. The levels of aflatoxin residues increased steadily for 4–5 days to a plateau and decreased after toxin withdrawal at about the same rate that it increased. At 7 days after withdrawal, only trace amounts could be found in eggs.

Kan *et al.* (1989) did not detect any residues of aflatoxin B_1 in breast muscle of broilers and laying hens or in eggs or liver of broilers after feeding these birds diets with maize contaminated with 150 or 750 p.p.b. aflatoxin B_1 for 6 and 3 weeks, respectively. These authors concluded that aflatoxin B_1 levels in feed of about 100 p.p.b., which is much higher than the current EC tolerance of 20 p.p.b., do not impair performance of broilers and laying hens.

Influence of aflatoxin on resistance and immunity

There is now enough experimental evidence that aflatoxins can diminish innate resistance and immunogenesis in birds. It is also fairly well established that aflatoxin affects the production of certain non-specific humoral substances, the activity of thymus-derived lymphocytes and the formation of antibodies. The early literature on this subject has been reviewed by Richard *et al.* (1978). Susceptibility to several infectious diseases, including salmonellosis, candidiasis and coccidiosis, has been shown to be enhanced by aflatoxin consumption.

Studies on the effects of aflatoxin on susceptibility to infection in the chicken are summarized in Table 7.4. It is seen that, in all diseases tested, aflatoxins enhanced susceptibility to the parasite except in the case of *Salmonella gallinarum* and *Candida albicans* (Pier, 1986).

Elzanaty *et al.* (1989) studied the effect of aflatoxins on Newcastle disease (ND) vaccination in broiler chicks. Different concentrations of aflatoxin (0, 8, 16, 32 and 48 μg bird^{-1}) were given daily for 2 weeks. The birds were vaccinated against ND at 4 and 18 days of age with Hitchner B$_1$ and Lasota strains, respectively, and at 35 days with Komarov strain. Adverse effects on humoral immunity to ND virus were observed, as measured by a haemagglutination inhibition test at 10, 17 and 24 days

Table 7.4. Enhancement of pathogenicity of certain infectious agents in chicken by aflatoxin.

Agent	Dose	Enhanced susceptibility*	Reference
Salmonella gallinarum	5 p.p.m.	−	Pier, 1986
Candida albicans	0.625-10 p.p.m.	=	Pier, 1986
Eimeria tenella	0.2 p.p.m. B$_1$	+	Edds, 1973
Eimeria tenella	2.5 p.p.m. B$_1$	+	Wyatt *et al.*, 1975
Salmonella spp. S. *worthington*, S. *derby* and S. *thompson*	10 p.p.m.	+	Boonchuvit and Hamilton, 1975
S. *typhimurium*	0.625-10 p.p.m.	+	Boonchuvit and Hamilton, 1975
Marek's disease virus	0.2 p.p.m. B$_1$	+	Edds, 1973
Infectious bursal disease	0.625-10 p.p.m.	+	Chang and Hamilton, 1982

* Increased incidence of infection or severity of disease.

post-treatment, and no effect on protection, as indicated by a challenge test at the end of the experiment.

Under experimental conditions, the effects of aflatoxin on immune responsiveness in the chicken have been studied by several workers (Pier, 1986). Data show that several cell products participating in non-specific defence mechanisms in the chicken are affected. The effects of aflatoxin on γ-globulin levels and antibody titres are less constant than the effects on non-specific humoral substances. Consumption of moderate levels of aflatoxin does not decrease the level of immunoglobulin. However, a lag or decrease in antibody titre to antigens tested has been reported (Thaxton et al., 1974; Boonchuvit and Hamilton, 1975). Decreased levels of certain immunoglobulins (IgG and IgA) have also been reported (Tung et al., 1975; Giambrone et al., 1978). These decreases in IgG and IgA have been observed only when relatively high doses of aflatoxin were administered (2.5–10 p.p.m.). Mohiuddin (1992) reported that the feeding of aflatoxin to poultry resulted in a decrease in antibody and cell-mediated immune responses, resulting in severe disease outbreak even after vaccination. The decrease in humoral immunity was established by determining the antibody titres in the serum. The decrease in cell-mediated immune response was established by decreases in phagocytic activity and in T lymphocytes as demonstrated by the α-naphthyl esterase (ANAE) reaction and adenosine deaminase esterase (ADA) reaction in separated lymphocytes. This author suggested that several vaccination failures in the chicken might be the effect of aflatoxin on the immune system.

The effects of aflatoxin on the different lymphatic tissues in poultry seem to depend on the dose given. Moderate levels of intake (0.5 p.p.m. B_1) caused thymic involution, but the bursa was not affected (Pier, 1986). Higher doses (2.5–10 p.p.m.), however, caused both thymic and bursal hypoplasia (Thaxton et al., 1974).

Aflatoxin has been shown to affect other aspects of the immune system in poultry. Complement activity has been shown to be suppressed by this toxin in a study by Campbell et al. (1983). These workers evaluated the combined effects of aflatoxin and ochratoxin on immunity in young broilers. They observed anaemia, hypoproteinaemia, lymphocytopaenia, heterophilia and decreased bursa weight when both toxins were present. They concluded, however, that the most pronounced effect of aflatoxin on the immune system was depression of complement activity.

Vasenko and Karteeva (1987) studied the influence of feeds containing fungi on the immune system in hens housed in laying cages. Hens showed symptoms of mycotoxicosis. Neutropenia and lymphopenia were significant in blood samples. Serum lysozyme activity decreased by $1.2\,\mu\text{g ml}^{-1}$ compared with controls. The complement activity decreased twofold. The phagocytosis index decreased by 0.54–1.2. Total serum immunoglobulins decreased by 3.03 mg ml^{-1} and IgM by 1.11 mg ml^{-1} compared with controls.

Several field observations have been made relating aflatoxicosis with increased incidence and severity of several poultry diseases. Samvanshi et al. (1992) reported on enhanced susceptibility to aflatoxicosis in infectious bursal disease on broiler farms in Uttar Pradesh, India. Rao et al. (1985) reported on a severe outbreak of aspergillosis in Khaki Campbell ducklings receiving maize containing 0.28 p.p.m. aflatoxin. They suggested that the presence of aflatoxin in the feed may have predisposed the ducklings to *Aspergillus* infection.

Hegazy et al. (1991) studied the interaction of naturally occurring aflatoxins in poultry feed and immunization against fowl cholera. Out of 1175 poultry feed samples examined, 30.7% proved positive for aflatoxin. Outbreaks of fowl cholera were diagnosed on two farms where aflatoxin was detected in the rations used. The impact of aflatoxins in the feed on the efficacy of immunization against fowl cholera was monitored by a haemagglutination test and found in chickens of the involved farms, which were compared with groups of chickens and ducks fed on aflatoxin-free rations and vaccinated with the same polyvalent fowl cholera bacterin. The antibody titre of the chickens fed aflatoxin-free diets was 4–15 times higher than in those of the involved farms.

In summary, aflatoxicosis reduces the ability of the chicken to synthesize proteins, and thus the ability to synthesize antibodies is also reduced. This will result in very low antibody titres if aflatoxin has been consumed either prior to, during or after antigen administration or exposure. Aflatoxin ingestion has been shown to cause atrophy of the bursa and the thymus. Primary lymphatic tissue, such as the bursa and thymus, is required for development of immunity. Thus, because of the effect on these tissues, chickens will show deficiencies in both humoral and cell-mediated immunity. Aflatoxicosis has been demonstrated to cause an impairment in the development of immunity to infectious bursal disease (gumboro), Newcastle disease, Marek's disease, salmonellosis and coccidiosis. Furthermore, aflatoxicosis can result in vaccine failures even though the quality of the vaccine may be excellent and the vaccination techniques adequate.

Aflatoxin and vitamin nutrition

Aflatoxins interact with the fat-soluble vitamins. They have been shown to depress hepatic storage of vitamin A and to increase the dietary requirement for vitamin D_3 by 6.6 IU kg^{-1} of diet for each 1 p.p.m. of aflatoxin B_1 in the ration (Bird, 1978). We have previously mentioned that aflatoxin-fed chicks show increased prothrombin times, capillary fragility and bruise susceptibility, which indicates a relationship to vitamin K.

Bababunmi and Bassir (1982) reported a delay in blood clotting of chickens and ducks induced by aflatoxin treatment.

These studies all indicate that a marginal deficiency of these vitamins may be aggravated in the presence of aflatoxins. One exception seems to be thiamine, whose deficiency had a protective effect on aflatoxicosis (Wyatt *et al.*, 1972).

Vitamin requirements of the chicken are therefore increased during aflatoxicosis, as well as certain amino acid requirements and protein in general. Fortification of poultry rations with synthetic methionine, over and above National Research Council (NRC, 1984) requirements, has been shown to alleviate the growth depression usually seen during aflatoxicosis.

Aflatoxin and zinc in feeds and feedstuffs

Ginting and Barleau (1987) studied 48 samples of poultry feeds and 53 maize samples in Bogor, India. Data showed a significant correlation between aflatoxin and zinc content in both maize and poultry feed. The zinc content of the feed samples was about ten times the reported nutrient requirements for chickens. Jones and Hamilton (1986) studied factors influencing fungal activity in feeds. As feed moved from the feed mill to the feeder pans, there was an increase in fungal activity and this increase was associated with an increase in zinc concentration. Jones *et al.* (1984) analysed feed samples from five commercial chicken operations and found that aflatoxin content correlated with zinc content. They suggested that stricter control of zinc levels during manufacture could reduce aflatoxin contamination of chicken feed.

Citrinin

Citrinin is a mycotoxin produced by various species of the genera *Penicillium* and *Aspergillus*. It is a nephrotoxin whose toxicity is also characterized by reduced growth, decreased feed consumption and increased water consumption (Wyatt, 1988a). Increased water intake is followed by acute diarrhoea and, within 6 hours after removal of citrinin from diet, water consumption is normal (Ames *et al.*, 1976).

Citrinin at dietary levels of 250 p.p.m. was shown by Ames *et al.* (1976) to cause significant increases in the size of the liver and kidney, by 11 and 22% respectively. There were also alterations in serum sodium levels. Necropsy of birds with citrinin toxicity revealed the presence of pale and swollen kidneys. Roberts and Mora (1978) confirmed the effect of citrinin on water consumption and the resulting diarrhoea and observed haemorrhagic jejunums and mottled livers with dietary levels of 130 and 160 p.p.m. These workers also reported that *P. citrinum*-contaminated

maize free of citrinin was also toxic to young chicks. This demonstrates that
P. citrinum may be producing toxic principles other than citrinin.

Dietary citrinin at levels as high as 250 p.p.m. fed to laying hens for
3 weeks had no effect on body-weight, feed consumption or egg production
(Ames *et al.*, 1976). Laying hens, like broiler chicks, however, developed
acute diarrhoea after initiation of citrinin feeding.

Smith *et al.* (1983) studied the effects of temperature, moisture and
propionic acid on mould growth on whole maize and subsequent citrinin
production. They found that temperature was the predominant factor in
determining the degree of mould on maize. At each moisture level studied
(10, 15 and 20% added H_2O), mould growth was greater on samples
incubated at room temperature than that of samples incubated at a lower
temperature. The effectiveness of propionic acid depended on the moisture
content of the maize. Higher levels of propionic acid were needed to
prevent mould growth and toxin production as the moisture content of the
maize increased.

Fumonisins

Fusarium contamination of poultry feeds came to the forefront in the USA
in the late 1980s when it was affecting several poultry operations and
causing severe growth depression, decreased feed intake, high mortality,
oral lesions and proventricular and gizzard erosion in young broilers
(Wyatt, 1988a). The predominant species found in those contaminated
feeds was *Fusarium moniliforme* (Wu *et al.*, 1991). Now we know that this
species produces a group of toxins known as fumonisins.

Fumonisins are water-soluble toxins that include fumonisin A_1, A_2,
B_1, B_2, B_3 and B_4 (Gelderblom *et al.*, 1992). Studies with broilers have
shown that a high concentration of fumonisin B_1 present in *Fusarium
moniliforme* cultures caused poor performance and increased organ
weights, diarrhoea, multifocal hepatic necrosis, biliary hyperplasia and
rickets (Brown *et al.*, 1992; Ledoux *et al.*, 1992). Weibking *et al.* (1993)
showed that broiler chicks fed 450 mg fumonisin B_1 or over per kg diet
had significantly lower feed intakes and body-weight gains, increased liver
and kidney weights and increased mean cell haemoglobin concentration.
These workers reported that levels of fumonisin B_1 as low as 75 mg per
diet may have an effect on the physiology of chicks because of its inhibition
of sphingolipid biosynthesis. Wu *et al.* (1991) found that some isolates of
Fusarium moniliforme when fed to broiler chicks resulted in decreased
antibody responses to sheep red blood cells.

Reports about the occurrence of fumonisins in poultry feeds are starting
to appear in countries other than the USA. Sydenham *et al.* (1992) tested
21 samples of *Fusarium moniliforme*-contaminated feeds associated with

outbreaks of mycotoxicoses in the state of Parana in Brazil. Fumonisin 1 and 2 were detected in 20 and 18 of the 21 feed samples, respectively.

Finally, tibial dyschondroplasia, recognized as a developmental abnormality in broilers subjected to dietary mineral imbalance, can also be caused by *Fusarium*-produced toxins (Haynes *et al.*, 1985; Haynes and Walser, 1986). Krogh *et al.* (1989) reported a 56% incidence of tibial dyschondroplasia on broiler farms in Denmark using feed contaminated with *Fusarium* spp. and containing fusarochromanone.

Ochratoxins

Ochratoxins were isolated as a result of feeding mouldy maize cultures to ducklings, rats and mice. Ochratoxin A (OA) is the major toxic principle from *Aspergillus ochraceus*. It has been detected in maize, barley, oats, peanuts, white beans and mixed feeds. Dwivedi and Burns (1986) reviewed the occurrence of OA and its effect in poultry and reported that OA-induced nephropathy has been recorded in natural field outbreaks in poultry from many countries. Ochratoxins do not appear to be as widely spread in the hot regions of the world as aflatoxins. Ozpinar *et al.* (1988) examined 74 samples by thin-layer chromatography from 15 feed factories in the Marmara area in Turkey and found 32 samples containing aflatoxin, but none contained ochratoxins. Ochratoxicosis has been reported in broilers, layers and turkeys. Field outbreaks covering seven different states in the USA were recorded, with levels of ochratoxin in feed ranging from 0.3 to 16.0 p.p.m. Doerr *et al.* (1974) reported that, on a comparative basis, ochratoxin is more toxic to chicks than either aflatoxin or T-2 toxins. The minimal dietary growth-inhibitory dose for young broiler chicks is 2 p.p.m. for ochratoxin, whereas 2.4 and 4 p.p.m. are required for growth inhibition by aflatoxin and T-2 toxin, respectively.

In broilers, gross pathological findings include severe dehydration and emaciation, proventricular haemorrhages and visceral gout, with white urate deposits throughout the body cavity and internal organs. The target organ is the kidney, since OA is primarily a nephrotoxin affecting kidney function through disruption of the proximal tubules (Wyatt, 1988a). Kidney damage mediated through nephrotoxic agents can result in hypertension.

Several studies have shown deleterious effects on broiler chicks from feeding graded levels of OA up to 8 p.p.m. (Huff and Hamilton, 1975; Huff and Doerr, 1981). Huff and Hamilton (1975) observed poor weight gains, decreased pigmentation and very poor efficiency of feed utilization. These workers found reduced levels of plasma carotenoids in chicks fed diets containing 4 or 8 mg kg^{-1} OA. This depression of pigmentation is of economic importance as underpigmented chickens have decreased market value in many countries. Another factor of economic importance in broiler production is bruising. Huff *et al.* (1983) compared the effects of OA with these of aflatoxin and found that both are implicated in increased susceptibility to bruising in broilers. Furthermore, they observed that the effects of OA are longer-lasting than those of aflatoxin. Bone strength in broilers has been studied in ochratoxicosis, and Huff *et al.* (1974) demonstrated that ochratoxin causes a decrease in bone strength and a rubbery condition of the bones related to increased tibial diameters and possibly poor mineralization of bone tissue.

Mature laying hens fed OA show lowered egg production, growth depression and high morbidity. Prior and Sisodia (1978) reported that levels as low as 0.5 p.p.m. of OA cause a decrease in egg production and feed consumption. Fertility and hatchability of eggs, however, were unaffected by OA.

Scholthyssek *et al.* (1987) fed levels of OA ranging from 0 to 5.2 p.p.m. to laying hens and observed a decrease in body-weight, feed consumption and egg production. They also observed some changes in functional traits of the eggs produced by those hens.

Ochratoxin has also been shown to affect the immune system in poultry. Reduced lymphoid organ size (bursa, spleen, thymus) has been observed after feeding the toxin. Chang *et al.* (1979) found a reduced number of lymphocytes in chickens fed 0.5 p.p.m. ochratoxin. Campbell *et al.* (1983) found a reduced number of lymphocytes when they fed 2 p.p.m. ochratoxin.

Dwivedi and Burns (1984) found a reduced number of immunoglobulin-containing cells in the lymphoid organs and reduced concentrations of immunoglobulins in sera from chickens fed 2 p.p.m. ochratoxin. Campbell *et al.* (1983) could not find any effect on antibody titres in chickens fed the same level of ochratoxin.

Residues of OA were found in liver, kidney and muscle, but not in eggs, fat or skin. Within 24 hours after ending ochratoxin feeding, no residues were found in muscle. Residues in liver and kidney persisted after 48 hours post-ochratoxin feeding (Huff *et al.*, 1974). Bohn (1993) suggested that ochratoxin content in feed should be kept as low as possible because it accumulates in poultry tissues and its carcinogenic effects cannot be excluded.

Oosporein

Oosporein is one of the mycotoxins that is considered nephrotoxic. In other words, affected birds will show marked swelling of the kidneys. It is also known to cause severe growth depression and high mortality. During oosporein toxicosis, urate accumulation in the joints and covering the visceral organs is very common. Proventriculitis with loss of proventricular rigidity has also been reported, along with slight haemorrhage at the junction of proventriculus and gizzard. Wyatt (1988a) reported on a case that was confirmed in Venezuela, as well as several cases of natural outbreaks in the eastern USA.

Ross *et al.* (1989) used chemical ionization mass spectrometry for the purposes of confirming thin-layer chromatography tests for the presence of oosporein in poultry feeds. Oosporein was isolated in pure form from a culture of *Chaetonium trilaterale*. This method was successfully applied to extracts of feed containing 5 p.p.m. oosporein. Rottinghaus *et al.* (1989) reported on a rapid screening procedure, involving extraction, column clean-up and detection by thin-layer chromatography, for oosporein in feed.

T-2 Toxin

This is a trichothecene produced by *Fusarium tricinctum*. In initial studies dealing with trichothecenes in poultry, Wyatt *et al.* (1972) found that necrotic lesions were produced in young broilers fed a diet containing purified T-2 toxin. These studies showed that the degree of oral necrosis was dose-related and that oral necrosis was the primary effect of T-2 toxicosis. Oral inflammation occurred with dietary levels of T-2 toxin that did not reduce growth or feed efficiency (Wyatt *et al.*, 1973a). The presence of oral lesions in poultry is regarded as the primary means of field diagnosis of T-2 toxicosis and other trichothecene toxicities. The oral lesions appear first on the hard palate and along the margin of the tongue and later there are large caseous discharges around the corner of the mouth and along the margin of the beak. They are raised lesions, yellowish-white in colour and caseous in texture (Wyatt *et al.*, 1972).

T-2 toxin also affects the nervous system of poultry (Wyatt *et al.*, 1973b). Abnormal positioning of wings, hysteroid seizures and impaired righting reflex have all been reported by these workers. Another symptom of T-2 toxicosis is abnormal feathering. At or above growth-inhibitory doses (4 p.p.m.), the feathering process is disturbed and the feathers present are short and protruding at odd angles. The mechanism responsible for this abnormal feathering is not known, but malnutrition may be a cause since feed intake in those birds is severely reduced.

Laying hens are also affected by T-2 toxin. Besides the oral response, there is reduced feed intake and body-weight loss. Egg production drops in 7–14 days after start of T-2 toxin feeding. Shell thickness and strength are reduced (Chi *et al.*, 1977). Speers *et al.* (1971) also demonstrated that, when laying hens are fed balanced rations with 2.5 and 5% maize invaded by *F. tricinctum* (8 and 16 p.p.m.), feed consumption, egg production and weight gain are reduced.

Chi *et al.* (1978) demonstrated the transmission of T-2 toxin to the egg, 24 hours after dosing. Hatchability and egg production were reduced by diets containing 4, 6 and 8 p.p.m. T-2 toxin. Allen *et al.* (1982) studied the effects of *Fusarium* cultures, T-2 toxin and zearalenone on reproduction of turkey females. Their results show that, although fertility was not affected by *Fusarium* mycotoxins, hatchability was decreased by each of the three *Fusarium* cultures fed. Pure zearalenone or T-2 toxin, however, did not affect hatchability, which suggests that other mycotoxins in these cultures are potent embryotoxic agents. In the same study, they undertook to assess the effects of *Fusarium* cultures on the immune responses of female turkeys to a killed ND virus vaccine. There was no indication of suppression of the immune response from feeding *Fusarium* mycotoxins. Richard *et al.* (1978) reported that T-2 toxin causes a reduction in lymphoid organ size (bursa, thymus, spleen). The effect on the immune system has been noted by Boonchuvit *et al.* (1975) to be an increase in sensitivity to *Salmonella* infections.

Vomitoxin (Deoxynivalenol – DON)

Vomitoxin is a potent mycotoxin to swine that causes vomiting in affected animals. It is a member of the trichothecenes, the same group of toxins to which the T-2 toxin belongs, and is thus produced by moulds belonging to the *Fusarium* genus.

Wyatt (1988a) summarized the most typical effects observed in the commercial poultry industry that are associated with tricothecene contamination of feedstuffs as follows:

1. Necrotic cream-coloured lesions in the oral cavity of affected birds.
2. Severe decreases in feed intake during the period of time the toxin is being consumed.
3. Increases in feed intake during recovery from the toxicosis.
4. Very small spleen size in affected birds.
5. Mild enteritis.

Harvey *et al.* (1991) investigated the effects of deoxynivalenol (DON)-contaminated wheat diets on haematological measurements, cell-mediated immune responses and humoral immune responses of Leghorn chickens. The authors concluded that subtle changes in haematological and immunological parameters could affect productivity or increase susceptibility to infection and thus caution should be exercised when utilizing DON-contaminated feedstuffs in poultry diets.

Bergs *et al.* (1993) reported that DON increases the incidence of developmental anomalies in chicks hatched from hens receiving the toxin in feed at levels ranging from 120 to 4900 μg kg^{-1}. The most frequent major malformations were cloacal atresia and cardiac anomalies, while minor malformations included unwithdrawn yolk sac and delayed ossification.

Behlow (1986) reported that DON was costing the poultry industry millions of dollars in the USA alone, due to depression of feed efficiency and growth rate, reproductive disorders and low product output.

Muirhead (1989) showed that, out of 1018 feed samples tested in the USA 406 (40%) were positive for vomitoxin.

Zearalenone

Zearalenone is a mycotoxin that is produced by many species of *Fusarium*, but the most common and well-known producers are *Fusarium graminearum* and *Fusarium culmorum*. They produce the toxin in the field during the growing stage and during storage of cereals at high moisture. Zearalenone is known to possess a strong oestrogenic activity in swine. Besides cereals, other feedstuffs, such as tapioca, manioc and soyabeans, can be contaminated (Gareis *et al.*, 1989).

Zearalenone has very low acute toxicity in chickens. Feeding high zearalenone levels ($\geq$300 p.p.m.) in the feed to female broiler chicks results in increased comb, ovary and bursa weights (Speers *et al.*, 1971; Hilbrich, 1986). The combs in male chicks were decreased when high levels of the toxin were fed.

The effects of zearalenone on laying hens are minimal, even when it is fed at high concentration (Allen *et al.*, 1981). Egg production is rarely reduced, with the exception of a few field cases where a reduction in egg production has been reported when both zearalenone and DON have been found in the feed. Bock *et al.* (1986), however, reported on a case of high mortality in broiler breeders due to salpingitis, with a possible role for zearalenone. The histopathological examination showed chronic salpingitis and peritonitis. Examination of feed extracts by radio-receptor assay indicated a high degree of oestrogenic activity. Examination by thin-layer and high-pressure liquid chromatography indicated that zearalenone was present in concentrations of up to 5 μg g^{-1} of feed.

Detection and Control of Mycotoxins

Although mycotoxins have been shown to produce numerous physiological and production changes in various species of poultry, none of these changes are specific enough to be used in a differential diagnosis. Furthermore, the continued effects of mycotoxins under field conditions in warm regions makes the clinical diagnosis very difficult. For example, Huff et al. (1986) demonstrated an additive effect of the simultaneous administration of aflatoxin and vomitoxin to broiler chicks. Ultrastructural lesions in the kidneys of Leghorn chicks were exacerbated when both citrinin and ochratoxin were fed and the severity of lesions was modified by the duration of administration (Brown et al., 1986). Therefore final diagnosis of toxicity should depend on measurement of the mould and the amount of toxin present in the feed.

In the case of aflatoxin, rapid screening methods of detection are now available. These include examination for the presence of the *Aspergillus flavus* mould in damaged peanut kernels and examination of cottonseed and maize for the bright greenish-yellow fluorescence frequently observable when material contaminated with aflatoxin is exposed to ultraviolet illumination. There are now also some rapid minicolumn chemical tests that can detect specified levels of aflatoxin in various materials. Tutour et al. (1987) gave limits of detection for aflatoxin B_1, using the minicolumn technique, of 5 μg kg^{-1} in wheat, maize and soyabean meal, 7 μg kg^{-1} in flax cake, 10 μg kg^{-1} in wheat bran and 15 μg kg^{-1} in mixed feeds.

Test kits for detection of a range of mycotoxins are available for thin-layer chromatography and can be used in diagnostic laboratories and for quality control of ingredients in hot countries (Gimeno, 1979).

Wyatt (1988b) studied the black light test for bright greenish-yellow fluorescence and concluded that there are major limitations to this test and that the ELISA technique is more accurate.

Immunoassay offers considerable promise in detecting, identifying and quantifying mycotoxins in feed ingredients. Commercial kits employing specific monoclonal antibodies are now available which simplify the processes of extraction and detection.

Biological assay using susceptible species such as ducklings can be used in areas where analytical equipment is unavailable. These methods, however, are often expensive and time-consuming. Furthermore, to confirm a diagnosis, a sufficient quantity of a suspect feed needs to be available for a controlled-feeding trial.

In areas where a quick evaluation needs to be made, visual examination of maize can be used and the absence of abnormal kernels indicates higher-quality grain since the highest levels of mycotoxins occur in abnormal kernels.

It has been observed that mould development in stored ground grain significantly reduces the fat content of these grains (Bartov et al., 1982). The fat content of diets containing these grains also decreased after a short period of storage. Therefore, the determination of dietary fat level, which is a relatively simple procedure, could be used as an additional estimate to evaluate fungal activity in poultry diets.

The best control of mycotoxin formation is to prevent the development of fungi in feedstuffs and complete feeds. Good conditions during harvest, transportation and storage of the feedstuffs are important in preventing the growth of moulds. Use of mould inhibitors is sometimes warranted when good management and handling conditions are not easy to attain. Propionic acid has been extensively used for that purpose and the level needed depends on the storage conditions of the feedstuffs and their moisture content. Smith *et al.* (1983) reported that, as the moisture content of maize increased, higher levels of propionic acid were needed to prevent mould growth and toxin production.

Propionic acid is now commonly used as an antifungal agent for maize and does not reduce the value of this grain as an animal feed. Both ammonia and propionic acid significantly reduce mould growth and subsequent formation of aflatoxin and ochratoxin and they both should have practical application for preventing the formation of mycotoxins in stored maize. Ammoniation treatment has been reported by several workers to be effective in minimizing the effects of aflatoxins (Ramadevi *et al.*, 1990; Phillips *et al.*, 1991). Maryamma *et al.* (1991) tested five different substances for the inactivation of aflatoxin, both *in vivo* and *in vitro*, and found that kaolin and bleaching powder were most effective and did not cause any deleterious effects in the White Pekin ducks that they used in those tests.

Different extraction procedures have been tried for the removal of aflatoxin. It has been shown that the most successful application is the removal of aflatoxin from oils during normal commercial processing. A variety of polar solvents have been found to be effective in extracting aflatoxins. Examples of these are 95% ethanol, 90% aqueous acetone, 80% isopropanol, hexane-ethanol, hexane-methanol, etc. Although these solvents are effective in removing all the aflatoxin from the meal, they add to the processing costs.

One approach to the detoxification of aflatoxin is the use of non-nutritive sorptive materials in the diet to reduce aflatoxin absorption from the gastrointestinal (GI) tract. Dalvi and Ademoyero (1984) and Dalvi and McGowan (1984) reported improvement in feed consumption and weight gain when activated charcoal was added to poultry diets containing aflatoxin. Hydrated sodium calcium aluminium silicate at a concentration of 0.5% of the diet significantly diminished many of the adverse effects caused by aflatoxin in chickens (Kubena *et al.*, 1990a, b) and turkeys (Kubena *et al.*, 1991). Kubena *et al.* (1993a), in a study on the efficacy of hydrated sodium calcium aluminium silicate to reduce aflatoxin, found that this compound gave total protection against the effects caused by aflatoxin. Scheideler (1993) compared four types of aluminosilicates for their ability to bind aflatoxin B_1 and prevent the effects of aflatoxicosis. Three of the four aluminosilicates tested by this worker alleviated the growth depression caused by aflatoxin B_1. Kubena et al. (1993b) confirmed the fact that hydrated sodium calcium aluminosilicates are protective against the effects of aflatoxin in young broilers and further emphasized that silicate-type solvents are not all equal in their ability to protect against aflatoxicosis. These authors concluded that sorbent compound use may be another tool for the preventive management of mycotoxin-contaminated feedstuffs in poultry. The mechanism of action appears to involve aflatoxin sequestration and chemisorption of hydrated sodium calcium aluminosilicates in the GI tract of poultry, resulting in a major reduction in bioavailability and toxicity (Phillips et al., 1989).

Another group of compounds that have been studied for their efficacy on aflatoxin toxicity are the zeolitic ores. Harvey et al. (1993) evaluated six of these compounds and found that zeolite mordenite ore reduced the toxicity of aflatoxin by 41% as indicated by weight gains, liver weight and serum biochemical measurements. Effectiveness of zeolites in inhibiting absorption of aflatoxin offers promise for hot climates to alleviate the effects of mycotoxins that are really damaging in those regions. In certain areas of the world it may be necessary to use a combination of antifungal compounds in the feed as well as a physical adsorbent. Ramkrishna et al. (1992) concluded that a combination of organic acids with physical adsorbent is most effective in combating toxicosis in India. Saxena (1992) feels that one of the major causes of aflatoxicosis in India is the use of contaminated groundnut (peanut) meals and, by monitoring this ingredient or the use of alternative protein sources in poultry diets, the incidence can be drastically reduced.

Several other methods of control have been under investigation. Heating, roasting and chemical inactivation have been studied and some of these methods appear to be promising for the future. Nearly all of these have dealt with aflatoxin and very few studies are being conducted on other mycotoxins. Gonzalez et al. (1991) evaluated the possible reduction of *Aspergillus flavus* contamination of grains by ozone and the effect of ozone on the nutritional value of maize. Treatment with ozone (60 mg l^{-1}) for 4 min killed conidial suspensions of *Aspergillus flavus*. These workers did not detect any significant difference in live weight of chickens consuming ozonized maize compared with control birds, nor did they find any pathological changes in those birds. Another possibility for alleviation of unfavourable effects of mould-contaminated feeds was the use of antibiotics in broiler diets. Ivandija (1989) tested several antibiotics in feeds that were contaminated with aflatoxins and zearalenone. Of the antibiotics tested, oxytetracycline at 50 p.p.m. gave the most favourable results.

Plant geneticists are working on the development of commercially acceptable varieties of grains and oil-seeds that would resist toxin-producing moulds or possibly inhibit the production of the toxin. Such a task may be more difficult than breeding for regular plant disease resistance.

Since in mycotoxicoses there is a decrease in energy utilization by birds, due to an impairment of lipid absorption or metabolism and in part to a decrease in energy content of the mouldy feed, Bartov (1983) studied the effects of adding both fat and propionic acid to diets containing mouldy maize. He concluded that the nutritional value of diets containing mouldy grain can be completely restored if their fat content is increased in proportion to the amount lost in the mouldy grains and propionic acid is added. Recently, Hazzele *et al.* (1993) showed that some of the detrimental effects of OA in the diet can be counteracted by dietary supplementation of 300 mg ascorbic acid kg^{-1} diet.

Chickens rapidly recover from chronic poor performance when changed to diets free from moulds or mycotoxins. Growing chickens recover in about 1 week and production returns to normal within 3–4 weeks in adult birds.

Conclusions

This chapter has dealt with the problems associated with the contamination of feeds with various mycotoxins. The mycotoxins presented were aflatoxins, citrinin, fumonisins, ochratoxins, oosporein, T-2 toxin, vomitoxin and zearalenone. Since aflatoxins have been researched most and since they appear to be the most prevalent in hot climates, a detailed description of their occurrence, effects on broilers, layers and breeders, residues in eggs and poultry meat, influence on resistance and immunity and relationship to certain vitamins and trace minerals has been presented. A general description of methods of detection and control of these mycotoxins has also been included

The most significant conclusion of this chapter is that all those concerned in the production, processing and marketing of feedstuffs, as well as those concerned in the formulation, production, processing and marketing of complete poultry feeds, should make every effort so that the poultry producer will be provided with feeds that not only are complete from a nutrient requirement standpoint, but also have no or minimal levels of mycotoxins in them. Furthermore, poultry producers in those regions should purchase ingredients from reputable suppliers against presentation of analytical certificates from reliable laboratories.

References

Allen, N.K., Mirocha, C.J., Aakhus, A.S., Bitgood, J.J., Wedner, G. and Bates, F. (1981) Effect of dietary zearalenone on reproduction of chickens. *Poultry Science* 60, 1165–1174.

Allen, N.K., Jerne, R.L., Mirocha, C.J. and Lee, Y.W. (1982) The effect of a

Fusarium roseum culture and diacetoxyscirpenol on reproduction of White Leghorn females. *Poultry Science* 61, 2172-2175.

Ames, D.D., Wyatt, R.D., Marks, H.L. and Washburer, K.W. (1976) Effect of citrinin, a mycotoxin produced by *Penicillium citrinum*, on laying hens and young broiler chicks. *Poultry Science* 55, 1294-1301.

Anonymous (1988) Mycotoxin testing: not 'whether' grain is contaminated but 'how much'. *Feed Management* 39, 11.

Bababunmi, E.A. and Bassir, O. (1982) A delay in blood clotting of chickens and ducks induced by aflatoxin treatment. *Poultry Science* 61, 166-168.

Babila, A. and Acktay, B. (1991) Determination of total fungal numbers and toxins in feeds and raw feed materials in outbreaks of mycotoxicosis in poultry. *Pendik Hayran Hastalklar Merkez* 22, 63-85.

Bartov, I. (1983) Effects of propionic acid and copper sulfate on the nutritional value of diets containing moldy corn for broilers. *Poultry Science* 62, 2195-2200.

Bartov, I., Paster, N. and Lisker, N. (1982) The nutritional value of moldy grains for broiler chicks. *Poultry Science* 61, 2247-2254.

Behlow, R.F. (1986) Deoxynivalenol said to be costing millions. *Poultry Digest* 45, 512-513.

Bergs, B., Herstad, O. and Naystad, I. (1993) Effects of feeding deoxynivalenol contaminated oats on reproduction performance in White Leghorn hens. *British Poultry Science* 34, 147-159.

Bird, F.H. (1978) The effect of aflatoxin B_1 on the utilization of cholecalciferol by chicks. *Poultry Science* 57, 1293-1296.

Blaney, B.J. and Williams, K.C. (1991) Effective use in livestock feeds of mouldy and weather damaged grain containing mycotoxins - case histories and economic assessments pertaining to pig and poultry industries of Queensland. *Australian Journal of Agriculture Research* 42, 993-1012.

Bock, R.R., Shore, L.B., Samberg, Y. and Perl, S. (1986) Death in broiler breeders due to salpingitis: possible role of zearalenone. *Avian Pathology* 15, 495-502.

Bohn, J. (1993) On the significance of mycotoxins deoxnivalenol, zearalenone and ochratoxin A for livestock. *Archives of Animal Nutrition* 42, 95-111.

Boonchuvit, B. and Hamilton, P.B. (1975) Interaction of aflatoxin and paratyphoid infection in broiler chickens. *Poultry Science* 54, 1567-1573.

Boonchuvit, B., Hamilton, P.B. and Burmiester, H.R. (1975) Interaction of T-2 toxin with salmonella infections in chickens. *Poultry Science* 54, 1693-1696.

Brown, T.P., Manning, R.O., Fletcher, O.J. and Wyatt, R.D. (1986) The individual and combined effects of citrinin and ochratoxin A on renal ultrastructure in layer chicks. *Avian Diseases* 30, 191-196.

Brown, T.P., Rottinghaus, G.E. and Williams, M.E. (1992) Fumonisin mycotoxicosis in broilers: performance and pathology. *Avian Diseases* 36, 450-454.

Campbell, M.L., May, J.D., Huff, W.E. and Doerr, J.A. (1983) Evaluation of immunity of young broilers during simultaneous aflatoxicosis and ochratoxicosis. *Poultry Science* 62, 2138-2144.

Chang, C.F. and Hamilton, P.B. (1982) Increased severity and new symptoms of IBD during aflatoxicosis in broiler chicks. *Poultry Science* 61, 1061-1068.

Chang, C.F., Huff, W.E. and Hamilton, P.B. (1979) A leucocytopenia induced in chickens by dietary ochratoxin A. *Poultry Science* 58, 555-558.

Chi, M.S., Mirocha, C.J., Kurtz, H.F., Weaver, G., Bates, F. and Shimoda, W. (1977) Effects of T-2 toxin on reproductive performance and health of laying hens. *Poultry Science* 56, 628–637.

Chi, M.S., Robinson, T.S., Mirocha, C.J. and Reddy, K.R. (1978) Acute toxicity of 12, 13-epoxytrichothecenes in one-day-old broiler chicks. *Applied Environmental Microbiology* 35, 636–640.

Daghir, N.J., Hajj, R. and Akrabawi, S.S. (1966) Studies on peanut meal for broilers. *Proceedings 13th World's Poultry Congress*, pp. 238–246.

Dalvi, R.R. (1986) An overview of aflatoxicosis of poultry: its characteristics, prevention and reduction. *Veterinary Research Communication* 10, 429–443.

Dalvi, R.R. and Ademoyero, A.A. (1984) Toxic effects of aflatoxin B1 in chickens given feed contaminated with *Aspergillus flavus* and reduction of the toxicity by activated charcoal and some chemical agents. *Avian Diseases* 28, 61–69.

Dalvi, R.R. and McGowan, C. (1984) Experimental induction of chronic aflatoxicosis in chickens by purified aflatoxin B1 and its reversal by activated charcoal, phenobarbitol and reduced glutathione. *Poultry Science* 63, 485–491.

Doerr, J.A. and Hamilton, P.B. (1981) Aflatoxicosis and intrinsic coagulating function in broiler chickens. *Poultry Science* 60, 1406–1411.

Doerr, J.A., Huff, W.E., Tung, A.T., Wyatt, R.D. and Hamilton, P.B. (1974) A survey of T-2 toxin, ochratoxin and aflatoxin for their effects on the coagulation of blood in young broilers. *Poultry Science* 53, 1728–1734.

Doerr, J.A., Huff, W.E., Wabeck, C.J., Choloupka, G.W., May, J.D. and Murkley, J.W. (1983) Effects of low level chronic aflatoxicosis in broiler chickens. *Poultry Science* 62, 1971–1977.

Dwivedi, P. and Burns, R.B. (1984) Effect of ochratoxin A on immunoglobulins in broiler chicks. *Research in Veterinary Science* 36, 117–121.

Dwivedi, P. and Burns, R.B. (1986) The natural occurrence of ochratoxin A and its effects in poultry. A review. I. Epidemiology and toxicity. *World's Poultry Science Journal* 42, 32–47.

Edds, G.T. (1973) Acute aflatoxicosis: a review. *Journal of the American Veterinary Medical Association* 162, 304–309.

Elzanaty, K., El-Maraghy, S.S.M., Abdel-Motelib, T. and Salem, B. (1989) Effect of aflatoxins on Newcastle disease vaccination in broilers. *Assiut Veterinary Medical Journal* 22, 184–189.

Ewaidah, E.H. (1988) Survey of poultry feeds for aflatoxins from the Riyadh region. *Arab Gulf Journal of Scientific Research* 6, 1–7.

Gareis, M., Bauer, J., Enders, C. and Gedek, B. (1989) Contamination of cereals and feed with *Fusarium* mycotoxins in European countries. In: Chalkowski, J. (ed.) *Fusarium Mycotoxins, Taxonomy and Pathogenecity*. Elsevier, Amsterdam, pp. 441–472.

Garlich, J.D., Tung, H.T. and Hamilton, P.B. (1973) The effects of short term feeding of aflatoxin on egg production and some plasma constituents of the laying hen. *Poultry Science* 52, 2206–2211.

Gelderblom, W.C.A., Marasas, W.F.O., Vieggar, R., Thiel, P.G. and Cawooe, M.E. (1992) Fumonisins: isolation, chemical characterization and biological effects. *Mycopathologia* 117, 11–16.

Giambrone, J.J., Ewert, D.L., Wyatt, R.D. and Eidson, C.S. (1978) Effect of aflatoxin on the humoral and cell-mediated immune system of the chicken. *American Journal of Veterinary Research* 39, 305-308.

Gimeno, A. (1979) Thin layer chromatographic determination of aflatoxins, ochratoxins, sterigmatocystin, zearalenone, citrinia, T-2 toxin, diacetoxyscirpenol, penicillic acid, patulin and penitren, *Journal of the Association of Official Analytical Chemists* 62, 579-586.

Ginting, N. (1984) Aflatoxin in broiler feed from Daerah Khusus Ibukata, Jakarta region. *Penyakit Hewan* 16, 212-214.

Ginting, N. and Barleau, B.I. (1987) Correlation between aflatoxin and zinc in maize and chicken feed. *Penyakit Hewan* 19, 94-96.

Goldblatt, L.A. (1969) *Aflatoxins.* Academic Press, New York, 472 pp.

Gonzalez, M., Malerio, J. and Muroz, C. (1991) Ozone action on *Aspergillus flavus* cultures and on feeds for broiler chickens. *Revista de Salud Animal* 13, 193-198.

Gumbman, M.R., Williams, G.N., Booth, A.N., Vohra, P., Ernst, R.A. and Bettard, M. (1970) Aflatoxin susceptibility in various breeds of poultry. *Proceedings of the Society of Experimental Biology and Medicine* 34, 683-686.

Hamilton, P.B. (1990) Problems with mycotoxins persist, but can be lived with. *Feedstuffs* 62, 22-23.

Hamilton, P.B. and Garlich, J.D. (1971) Aflatoxin as a possible cause of fatty liver syndrome in laying hens. *Poultry Science* 50, 800-804.

Harvey, R.B., Kubena, L.F., Huff, W.E., Elissalde, M.H. and Phillips, T.D. (1991) Hematologic and immunologic toxicity of deoxynivalenol-contaminated diets to growing chickens. *Bulletin of Environmental Contamination and Toxicology* 46, 410-416.

Harvey, R.B., Kubena, L.F., Elissalde, M.H. and Phillips, T.D. (1993) Eficacy of zeolitic ore compounds on the toxicity of aflatoxin to growing broiler chickens. *Avian Diseases* 37, 67-73.

Haynes, J.S. and Walser, M.M. (1986) Ultrastructure of *Fusarium*-induced tibial dyschondroplasia in chickens: a sequential study. *Veterinary Pathology* 23, 499-502.

Haynes, J.S., Walser, M.M. and Lowler, E.M. (1985) Morphogenesis of *Fusarium* spp.-induced tibial dyschondroplasia in chickens. *Veterinary Pathology* 22, 629-633.

Hazzele, F.M., Guenter, W., Marquart, R.R. and Frohlich, A.A. (1993) Beneficial effects of dietary ascorbic acid supplement on hens subjected to ochratoxin A toxicosis under normal and high temperature. *Canadian Journal of Animal Science* 73, 149-157.

Hegazy, S.M., Azzam, A. and Gabal, M.A. (1991) Interaction of naturally occurring aflatoxins in poultry feed and immunization against fowl cholera. *Poultry Science* 70, 2425-2428.

Hilbrich, P. (1986) Abnormal comb growth in female chicks due to gonadotropic substances in the feed. *Deutch Tierarztliche Wochenschrift* 93, 39-40.

Huff, W.E. (1980) Evaluation of tibial dyschondroplasia during aflatoxicosis and feed restriction in young broiler chickens. *Poultry Science* 59, 991-995.

Huff, W.E. and Doerr, J.A. (1981) Synergism between aflatoxin and ochratoxin A in broiler chickens. *Poultry Science* 60, 550-555.

Huff, W.E. and Hamilton, P.B. (1975) The interaction of ochratoxin A with some environmental extremes. *Poultry Science* 54, 1659–1662.

Huff, W.E., Wyatt, R.D. and Tucker, T.L. (1974) Ochratoxicosis in the broiler chicken. *Poultry Science* 53, 1585–1591.

Huff, W.E., Doerr, J.A., Wabeck, C.J., Choloupka, G.W., May, J.D. and Merkley, J.W. (1983) Individual and combined effects of aflatoxin and ochratoxin A on bruising in broiler chickens. *Poultry Science* 62, 1764–1771.

Huff, W.E., Kubena, L.F., Harvey, R.B., Hagler, W.M., Swanson, S.P., Phillips, T.D. and Greger, C.R. (1986) Individual and combined effects of aflatoxin and deoxynivalenol (DON – vomitoxin) in broiler chickens. *Poultry Science* 65, 1291–1295.

Ivandija, L. (1989) Possibilities for allevation of unfavourable effects of mould contaminated feeds in broiler chicks. *Krmiva* 31, 39–45.

Jacobson, W.C. and Wiseman, H.G. (1974) The transmission of aflatoxin B_1 into eggs. *Poultry Science* 53, 1743–1745.

Jelinek, C.F., Pohland, A.E. and Wood, G.E. (1989) Worldwide occurrence of mycotoxins in foods and feeds: an update. *Journal of the Association of Analytical Chemists* 72, 223–230.

Johri, T.S., Agarwal, R. and Sadagopan, V.R. (1986) Surveillance of aflatoxin B_1 content of poultry feedstuffs in and around Bareilly district of Uttar Pradesh. *Indian Journal of Poultry Science* 21, 227–230.

Johri, T.S., Agarwal, R. and Sadagopan, V.R. (1987) Aflatoxin B_1 contamination of poultry feedstuffs of north eastern region of Uttar Pradesh. *Indian Veterinary Journal* 64, 433–435.

Jones, F.T. and Hamilton, P.B. (1986) Factors influencing fungal activity in low moisture poultry feeds. *Poultry Science* 65, 1522–1525.

Jones, F.T., Hogler, W.M. and Hamilton, P.B. (1984) Correlation of aflatoxin contamination with zinc content of chicken feed. *Applied and Environmental Microbiology* 47, 478–480.

Kan, C.A., Rump, R. and Kosutzky, J. (1989) Low level exposure of broilers and laying hens to aflatoxin B1 from naturally contaminated corn. *Archiv fur Geflugelkunde* 53, 204–206.

Krogh, P., Christensen, D.H., Hald, B., Harlou, B., Larsen, C., Pedersen, E. and Thane, U. (1989) Natural occurrence of the mycotoxin fusarochromanone, a metabolite of *Fusarium equiseti* in cereal feed associated with tibial dyschondioplasia. *Applied and Environmental Microbiology* 55, 3184–3188.

Kubena, L.F., Harvey, R.B., Huff, W.E., Corrier, D.E., Phillips, T.D. and Rottinghaus, G.E. (1990a) Efficacy of a hydrated sodium calcium aluminosilicate to reduce the toxicity of aflatoxin and T-2 toxin. *Poultry Science* 69, 1078–1086.

Kubena, L.F., Harvey, R.B., Phillips, T.D., Corrier, D.E. and Huff, W.E. (1990b) Diminution of aflatoxicosis in growing chickens by dietary addition of a hydrated sodium calcium aluminosilicate. *Poultry Science* 69, 727–735.

Kubena, L.E., Huff, W.E., Harvey, R.B., Yersin, A.G., Elissaide, M.H., Witzel, D.A., Giroir, L.E., Phillips, T.D. and Petersen, H.D. (1991) Effects of a hydrated sodium calcium aluminosilicate on growing turkey poults during aflatoxicosis. *Poultry Science* 70, 1823–1830.

Kubena, L.F., Harvey, R.B., Phillips, T.D. and Clement, B.A. (1993a) Effect of

hydrated sodium calcium aluminosilicates on aflatoxicosis in broiler chicks. *Poultry Science* 72, 651–657.
Kubena, L.E., Harvey, R.B., Huff, W.E., Elissalde, M.H., Yersin, A.G., Phillips, T.D. and Rottinghaus, G.E. (1993b) Efficacy of a hydrated sodium calcium aluminosilicate to reduce the toxicity of aflatoxin and diacetoxy-cirpenol. *Poultry Science* 72, 51–59.
Ledoux, D.R., Brown, T.P., Weibking, T.S. and Rottinghaus, G.E. (1992) Fumonisin toxicity in broiler chicks. *Journal of Veterinary Diagnostic Investigation* 4, 330–333.
Mahipal, S.K. and Kaushik, R.K. (1983) A note on the prevalence of aflatoxicosis in poultry in Haryana. *Haryana Veterinarian* 22, 51–52.
Maryamma, K.I., Rajan, A., Gangadharan, B. and Manmohan, C.B. (1991) In-vivo and in-vitro studies on aflatoxin B1 neutralization. *Indian Journal of Animal Science* 61, 58–60.
Mohiuddin, S.M. (1992) Effects of aflatoxin on immune response in viral diseases. *Proceedings 19th World's Poultry Congress*, Vol. 2 (Suppl.), pp. 50–53.
Muirhead, S. (1989) Studies show cost of mycotoxin to poultry firms. *Feedstuffs* 61, 10.
National Research Council (NRC) (1984) *Nutrient Requirements of Domestic Animals*, No. 7. *Nutrient Requirements of Poultry*, 8th edn. NAS–NRC, Washington D.C.
Osborne, D.J., Wyatt, R.D. and Hamilton, P.B. (1975) Fat digestion during aflatoxicosis in broiler chickens. *Poultry Science* 54, 1802 (Abstract).
Oyejide, A., Tewe, O.O. and Okosum, S.E. (1987) Prevalence of aflatoxin B_1 in commercial poultry rations in Nigeria. *Beitrange zur Tropischen Landwirtchaft und Veterinarmedicin* 25, 337–341.
O'zpnar, H., O'zpnar, A. and Seael, H.S. (1988) Determination of aflatoxin and ochratoxin in poultry feed components and raw feedstuffs in the Marmara area. *Veteriner Facultesi Dergisi (Istanbul)* 14, 11–18.
Pettersson, H. (1991) Mycotoxins in poultry feed and detoxification. *Proceedings of the 8th European Symposium on Poultry Nutrition*, pp. 27–41.
Phillips, T.D., Clement, B.A., Kubena, L.F. and Harvey, R.B. (1989) Prevention of aflatoxicosis in animals and aflatoxin residues in food of animal origin with hydrated sodium calcium aluminosilicates. Healthy animals, safe foods, healthy man. *Proceedings World Association of Veterinary Food Hygienists Xth International Symposium*, pp. 103–108.
Phillips, T.D., Sarr, B.A., Clement, B.A., Kubena, L.F. and Harvey, R.B. (1991) Prevention of aflatoxicosis in farm animals via selective chemosorption of aflatoxin. In: Bray, G.A. and Ryan, D.H.J. (eds) *Mycotoxins, Cancer and Health*. Louisiana State University, Baton Rouge, pp. 223–237.
Pier, A.C. (1986) Immunomodulation in aflatoxicosis. In: Richard, J.L. and Thurston, J.R. (eds), *Diagnosis of Mycotoxicosis*. Martinus Nijhoff, Dordrecht, the Netherlands, pp. 143–146.
Prior, M.G. and Sisodia, C.S. (1978) Ochratoxicosis in white Leghorn hens. *Poultry Science* 57, 619–623.
Purwoko, H.M., Hold, B. and Wolstrup, J. (1991) Aflatoxin content and number of fungi in poultry feedstuffs from Indonesia. *Letters in Applied Microbiology* 12, 212–215.

Raina, J.S. and Singh, B. (1991) Prevalence and pathology of mycotoxicosis in poultry in Punjab. *Indian Journal of Animal Sciences* 61, 671–676.

Ramadevi, V., Rao, P.R. and Moorthy, V.S. (1990) Pathological effects of ammoniated and sundried aflatoxin contaminated feed in broilers. *Journal of Veterinary and Animal Sciences* 21, 108–112.

Ramkrishna, G.T., Devegouda, G., Umesh, D. and Gagendragod, M.R. (1992) Evaluation of mold inhibitors in broiler diets and their influence on the performance of broilers. *Indian Journal of Poultry Science* 27, 91–94.

Rao, D.G., Naidu, N.R.G. and Rao, R.R. (1985) Observations on the concomitant incidence of aflatoxicosis and aspergillosis in khaki Campbell ducklings. *Indian Veterinary Journal* 62, 461–464.

Richard, J.L., Cysewski, S.J., Pier, A.C. and Booth, G.D. (1978) Comparison of effects of dietary T-2 toxin on growth, immunogenic organs, antibody formation, and pathologic changes in turkeys and chickens. *American Journal of Veterinary Research* 39, 1674–1679.

Rizvi, A., Shakoori, A.R. and Rizvi, S.M. (1990) Aflatoxin contamination of commercial poultry feeds in Punjab. *Pakistan Journal of Zoology* 22, 387–398.

Roberts, W.T. and Mora, E.C. (1978) Toxicity of *Penicillium citrinum* AUA 532 contaminated corn and citrinin in broiler chicks. *Poultry Science* 57, 122–1226.

Rosiles, M.R. (1987) Mycotoxicoses in farm animals. In: Zuber, M.S., Lillehaj, E.B. and Reufro, B.L. (eds) *Proceedings of Workshop on Aflatoxin in Maize.* CIMMYT, El-Bata, Mexico, pp. 66–70.

Ross, P.F., Osheim, D.L. and Rottinghaus, G.E. (1989) Mass-spectral confirmation of oosporein in poultry rations. *Journal of Veterinary Diagnostic Investigation* 1, 271–272.

Rottinghaus, G.E., Sklebor, H.T., Seuter, L.H. and Brown, T.P. (1989) A rapid screening procedure for the detection of the mycotoxin oosporein. *Journal of Veterinary Diagnostic Investigation* 1, 174.

Samvanshi, R., Mohanty, G.C., Verma, K.C. and Kataria, J.M. (1992) Spontaneous occurrence of aflatoxicosis, infectious bursal disease and their interaction in chicken. *Indian Veterinary Medical Journal* 16, 11–17.

Saxena, H.C. (1992) Two major problems of poultry farms in India: colibacillosis and mycotoxicosis. *Misset World Poultry* 8, 47.

Scheideler, S.E. (1993) Effects of various types of aluminosilicates and aflatoxin B1 on aflatoxin toxicity, chick performance, and mineral status. *Poultry Science* 72, 282–288.

Scholtyssek, S., Niemiec, J. and Baver, J. (1987) Ochratoxin A in layer's feed. 1. Report: influence on laying performance and egg quality. *Archiv fur Geflugelkunde* 51, 234–240.

Selvasubramanian, S., Chandrasekaran, D., Viswanathan, K., Balachandran, C., Pumiamurthy, N. and Veerapandian, C. (1987) Surveillance of aflatoxin B_1 in various livestock and poultry feeds. *Indian Veterinary Journal* 64, 1033–1034.

Shaaban, A.A., Ibrahim, T.A., Mokhtar, Z. and Shehata, A. (1988) Aflatoxin residues in rations of chicken laying flocks in Assiut province. *Assiut Veterinary Medical Journal* 19, 195–199.

Shaaban, F.E., Abou-Hadeed, A.H., El-Shazly, M.O. and Gammel, A.A. (1991)

Detection and estimation of aflatoxin in stored grains and the pathological effect of its feeding in rats and chickens. *Egyptian Journal of Comparative Pathology and Clinical Pathology* 4, 339-357.

Smith, J.W. and Hamilton, P.B. (1970) Aflatoxicosis in the broiler chicken. *Poultry Science* 49, 207-215.

Smith, P.A., Nelson, T.S., Kirby, L.K., Johnson, Z.B. and Beasly, J.N. (1983) Influence of temperature, moisture and propionic acid on mold growth and toxin production in corn. *Poultry Science* 62, 419-423.

Speers, G.M., Meronuck, R.A., Ames, D.M. and Mirocha, C.J. (1971) Effect of feeding *Fusarium roseum* f. sp. *graminearum* contaminated corn and the mycotoxin F-2 on the growing chick and laying hen. *Poultry Science* 50, 627-633.

Sydenham, E.W., Marasas, W.F.O., Shephard, G.S., Thiel, P.G. and Hirooka, E.Y. (1992) Fumonisin concentrations in Brazilian feeds associated with field outbreaks of confirmed and suggested animal mycotoxicoses. *Journal of Agricultural and Food Chemistry* 40, 994-997.

Thaxton, J.P., Tung, H.T. and Hamilton, P.B. (1974) Immunosuppression in chickens by aflatoxin. *Poultry Science* 53, 721-725.

Trucksess, M.W., Staloff, L. and Young, K. (1983) Aflatoxicol and aflatoxin B1 and M1 in eggs and tissues of laying hens consuming aflatoxin contaminated feed. *Poultry Science* 62, 2176-2182.

Tung, H.T., Smith, J.W. and Hamilton, P.B. (1971) Aflatoxicosis and bruising in the chicken. *Poultry Science* 50, 795-800.

Tung, H.T., Donaldson, W.E. and Hamilton, P.B. (1973) Decreased plasma carotenoid during aflatoxicosis. *Poultry Science* 52, 80-83.

Tung, H.T., Cook, F.W., Wyatt, R.D. and Hamilton, P.B. (1975) The anaemia caused by aflatoxin. *Poultry Science* 54, 1962-1969.

Tutour, B., El-yazgi, R., and Tantawi-Elaraki, A. (1987) Rapid detection by minicolumn of aflatoxin B1 in poultry feeds. *Actes de l'institut Agronomique de Vétérinaire Hassan II* 7, 91-100.

Vasenko, S.V. and Karteeva, E.A. (1987) The influence of feeds containing fungi on the immune system in poultry. *Stornik Nonchnyhk Trudov - Moskovskaya Veterinarnaya Akademiya* 126, 26-29.

Weibking, T.S., Ledoux, D.R., Bermudez, A.J., Turk, J.R., Rottinghaus, G.E., Wang, E. and Merrill, A.H. (1993) Effects of feeding *Fusarium moniliforme* culture material, containing known levels of fumonisin B1, on the young broiler chick. *Poultry Science* 72, 456-466.

Westlake, K. and Dutton, M.F. (1985) The incidence of mycotoxins in litter, feed and livers of chickens in Natal. *South African Journal of Animal Science* 15, 175-177.

Widiastuti, R., Maryam, R., Blaney, B.J., Staltz, S. and Stoltz, D.R. (1988) Corn as a source of mycotoxins in Indonesian poultry feeds and the effectiveness of visual examination methods for detecting contamination. *Mycopathologia* 102, 45-49.

Wu, W., Cook, M.E. and Smalley, E.B. (1991) Decreased immune response and increased incidence of tibial dyschondroplasia caused by fusaria grown on sterile corn. *Poultry Science* 70, 293-301.

Wyatt, R.D. (1979) Biological effects of mycotoxins on poultry. In: NAS

Interaction of Mycotoxins in Animal Production. National Academy of Sciences, Washington D.C., pp. 87–95.

Wyatt, R.D. (1988a) Mycotoxins in feedstuffs, a hazard to flock health. *Vineland Update*, No. 23.

Wyatt, R.D. (1988b) How reliable is the black light test in the detection of aflatoxin contamination? *Feedstuffs* 60, 12–14.

Wyatt, R.D., Weeks, B.A., Hamilton, P.B. and Burmeister, H.R. (1972) Severe oral lesions in chickens caused by ingestion of dietary fusariotoxin T-2. *Applied Microbiology* 24, 251–257.

Wyatt, R.D., Hamilton, P.B. and Burmeister, H.R. (1973a) The effects of T-2 toxin in broiler chickens. *Poultry Science* 52, 1853–1859.

Wyatt, R.D., Harris, J.R., Hamilton, P.B. and Burmeister, H.R. (1973b) Possible outbreaks of fusariotoxicosis in avians. *Avian Diseases* 16, 1123–1130.

Wyatt, R.D., Hamilton, P.B. and Burmaister, H.R. (1975) Altered feathering of chicks caused by T-2 toxins. *Poultry Science* 54, 1042–1045.

8

Broiler Feeding and Management in Hot Climates

N.J. Daghir
Faculty of Agricultural Sciences, UAE University, Al-Ain, UAE.

Introduction	185
Nutritional Manipulations during Heat Stress	186
Energy and protein	187
Minerals and vitamins	189
Seasonal Effects on Broiler Performance	192
Temperature and Body Composition	194
Broiler Management in Hot Climates	196
Feeding programmes	196
Feed withdrawal	198
Drug administration	198
Vaccination	199
Beak trimming	200
Broiler house management	201
Water Consumption	205
Acclimatization to Heat Stress	207
Lighting Programmes	209
Conclusions	210
References	211

Introduction

As shown in Chapter 1, the broiler industry has undergone very rapid developments during the past decade and it is expected that this development will continue and be more pronounced in the warm regions of the world than in the temperate regions. This is because most temperate regions have already reached self-sufficiency, while most countries in the warm regions have not. Furthermore, the implications of good health practices in relation to poultry meat consumption are now beginning to reach the developing warm regions of the world and therefore it is expected that this will greatly enhance the growth of the broiler industry in those areas.

This chapter deals mainly with feeding and management aspects of broilers that are unique to hot climates or the summer months in the temperate regions. Even in that respect, it is in no way exhaustive of everything that has been researched on the subject. Recent developments on the feeding of broilers in hot regions have been covered, as well as the effects of temperature and season on broiler performance, and the effects of temperature on body composition and carcass quality. Certain management practices, such as feed withdrawal, drug administration, vaccination, beak trimming, litter management and others, have been discussed. The importance of water quality and quantity as well as lighting programmes to reduce detrimental effects of heat stress have been reviewed. Finally, a section on the possible benefits of gradual acclimatization of broilers to heat stress has been included.

Nutritional Manipulations during Heat Stress

Environmental temperature is considered to be the most important variable affecting feed intake and thus body-weight gain of broilers. Figure 8.1, adapted from data presented by North and Bell (1990), shows the effects of house temperature on growth and feed consumption of straight-run broilers. The increase in house temperature from 32 to 38°C causes a drop in feed intake of 21.3 g per bird per day and a reduction in body-weight at 8 weeks of age of 290 g. Several workers have tried to quantitate the extent to which reduced feed intake limits broiler performance at high environmental temperatures. Fuller and Dale (1979) studied two diurnal temperature cycles, a hot one consisting of 24–35°C and a cool one of 13–24°C. Birds were fed the experimental diets *ad libitum* in both environments and an additional group of birds in the cool environment were limited to the amount of feed consumed by the birds in the hot environment. Their results showed that growth was depressed by 25% in birds maintained in the hot environment. When birds maintained in the cool environment were fed the same amount of feed as that consumed by birds in the hot environment, their performance was reduced by only 16% compared with those fed on *ad libitum* bases. The results of this study indicate that reduced feed intake is not the only factor causing reduced broiler performance in hot climates. More recently, Fattori *et al.* (1990) demonstrated this in a comprehensive study in which they simulated broiler grower conditions by incorporating environmental variation into the research design. Thus, they were able to partition growers into homogeneous production environments, which allowed a more accurate

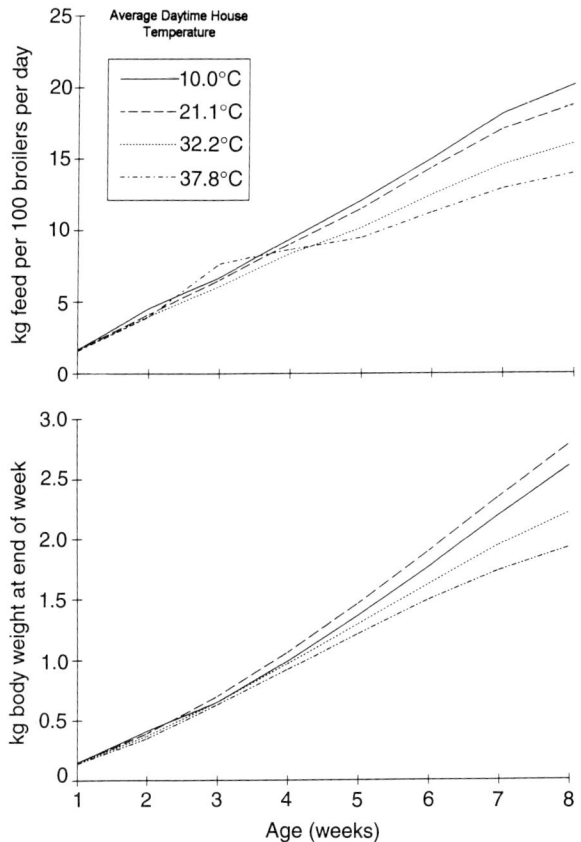

Fig. 8.1. Effect of house temperature on growth (lower graph) and feed consumption (upper graph) of straight-run broilers. Adapted from North and Bell, 1990.

evaluation of different feed treatments. This resulted in feeding recommendations appropriate to specific groups of broiler growers. Environmental temperature was the most important variable found by these workers to affect feed intake and thus weight gain.

Energy and protein

Farrel and Swain (1977a, b) conducted a series of studies on several aspects of energy metabolism in broilers. They varied environmental temperature from 2 to 35°C and determined the influence of environmental temperature on fasting heat production, metabolizable energy intake and efficiency of metabolizable energy utilization. They found that heat production in both fasting and fed birds decreased as environmental temperature increased from 2 to 30°C and plateaued at about 30–35°C. Although the efficiency of metabolizable energy (ME) retention was not affected by temperature, the amount of ME retained was maximum at 16–22°C. It declined at both higher and lower temperatures.

Hoffman *et al.* (1991) conducted several studies on the energy metabolism of growing broilers kept in groups in relation to environmental temperature. Energy metabolism was measured by those workers by indirect calorimetry. It was found that energy utilization was dependent on dietary protein content and decreased from 75 to 69% with an increase in protein content from 20 to 40%. Energy requirement (ME) for maintenance was independent of dietary protein and increased from 433 kJ kg^{-1} at 35°C to 693 kJ kg^{-1} daily at 15°C. With equal live weight, feed intake increased with decreasing temperature (Hoffman, 1991). This worker evaluated temperatures of 15 to 35°C and found feed intake to be similar between 20 and 25°C up to 57 days of age.

It has been known for some time that the reduction in energy intake of broilers is responsible for the reduced growth rate associated with heat stress (Pope, 1960). Several researchers have shown that increasing the energy content of the diet can partially overcome this growth depression. It is common practice now in formulating broiler feeds for hot regions to boost the energy level of these diets by the addition of fat. This practice not only increases the energy intake, but also reduces the specific dynamic effect of the diet, which helps birds to cope better with heat stress (Fuller and Rendon, 1977; Fuller, 1978). Dale and Fuller (1979) demonstrated a beneficial effect of isocaloric substitution of fat for carbohydrate in broilers reared from 4 to 7 weeks at constant temperatures of either 20 or 30°C. These results were confirmed by the same workers (Dale and Fuller, 1980) in broilers subjected to cyclic temperature fluctuation (22 to 33°C). McCormick *et al.* (1980) noted that survival time of chicks fed a carbohydrate-free diet was greater than those fed a diet containing glucose.

Besides energy, consideration must be given to the amino acid balance of the diet during heat stress. If the energy content of the diet is increased, then it must be kept in mind that there must be a proportional increase in all other nutrients. Minimizing excesses of amino acids usually improves feed intake. Waldroup *et al.* (1976) improved the performance of heat-stressed broilers by minimizing amino acid excesses in their diets. During hot periods, diets containing lower protein levels and supplemented with the limiting amino acids, methionine and lysine, gave better results than high-protein diets. Ait-Tahar and Picard (1987), in a study on protein requirements of broilers at high temperature, showed that raising the protein level of the diet to 26% does not improve performance at 33°C versus 20°C. Furthermore, if the protein level of the diet is raised at high temperature without raising the level of the essential amino acid lysine, there is a significant reduction in body-weight gain and in feed conversion.

These workers concluded that increasing the essential amino acid levels of low-protein diets in hot climates does not compensate for the reduced growth. This does not seem to be in agreement with Waldroup (1982), who recommended such a practice to counter the effect of high temperature. Yaghi and Daghir (1985) showed that, if National Research Council (NRC) recommended levels of methionine and lysine are met, the protein levels of starter, grower and finisher rations can be reduced to 20, 17 and 15%, respectively, without a significant reduction in performance. Baghel and Pradhan (1989a) studied energy and protein requirements during the hot and humid season in India at maximum temperatures of 42–45°C and minimum temperatures of 14–19°C with relative humidities of 63–68%. They reported that the best feed conversion was obtained when starter diets had 3200 kcal kg^{-1} and 24.9% protein, grower diets 3000 kcal kg^{-1} and 24.1% protein and finisher diets 2800 kcal kg^{-1} and 21.1% protein. The same workers in another study (Baghel and Pradhan, 1989b) found optimum daily gain and feed conversion to be at 2800 kcal kg^{-1} and 25, 24 and 21% protein with lysine levels of 1.2, 1.0 and 0.85% and methionine and cystine at 0.93, 0.72 and 0.6%. Koh *et al.* (1989) studied the effects of environmental temperature on protein and energy requirements of broilers in Taiwan. Feed intake and weight gain in the hot season (30°C) were 15–20% lower than in the cool season (16–18.6°C). Higher dietary energy improved feed efficiency and lowered energy and protein requirements per unit weight gain. Abdominal fat content was higher in broilers reared in the cool season and those fed on the high-protein diet had decreased abdominal fat.

Research at the Pig and Poultry Research and Training Institute in Singapore demonstrated that the most favourable growth and profit were obtained from broilers fed rations with 3100 kcal kg^{-1} and 22% protein in the starter and 3000 kcal kg^{-1} and 19.5% protein in the finisher (Anonymous, 1982). Lysine and methionine levels recommended were 1.34 and 0.44% in the starter and 1.08 and 0.38% in the finisher, respectively.

Minerals and vitamins

Several acid–base imbalances occur in heat-stressed broilers. The occurrence of alkalosis in heat-stressed birds has been known for a long time. Bottje and Harrison (1985) corrected the alkalotic state in birds by carbonating the drinking-water and observed improved weight gain and survival compared with those birds given tap-water. Teeter *et al.* (1985) improved weight gain by adding ammonium chloride to the diet of broilers. They were also able to accomplish the same thing by supplementing the drinking-water with ammonium chloride (Teeter and Smith, 1986). These authors also observed beneficial effects on mortality by ammonium chloride supplementation. Smith and Teeter (1987) studied potassium retention in heat-stressed broilers and observed that potassium output was significantly increased in birds at 35°C versus those at 24°C. Drinking-water supplementation with potassium chloride significantly increased survival from 15 to 73%. Gorman (1992) evaluated the performance of finishing broilers (3-6 weeks of age) at high temperature (32°C), using micromineral supplementation at either normal or at half-normal levels, with or without sodium bicarbonate supplementation (17 g sodium bicarbonate kg^{-1}). The results of this study indicated that commercial finisher rations may contain microminerals in excess of the requirement of broilers grown at high temperature. Furthermore, dietary supplementation with sodium bicarbonate at high temperature stimulated water and feed intakes and improved weight gain.

Smith (1994) investigated the possible interactions between electrolytes and photoperiod in broilers grown at elevated temperatures. They found that male birds that received sodium chloride gained 10.5% more weight than those receiving no water additive. The use of electrolytes had no effects on carcass characteristics.

McCormick *et al.* (1980) reported that much of the heat-induced mortality could be alleviated if the proper Ca : P ratio is maintained in the diet. Garlich and McCormick (1981) showed a direct relationship with plasma calcium. Survival time in fasted chicks was greater when their previous diet contained 0.3% calcium and 0.55% phosphorus rather than 1% calcium and 0.50% phosphorus. This may be important to consider when feed withdrawal is practised to reduce mortality during high-temperature spells. More recently, Orban and Roland (1990) reported on the response of four broiler strains to dietary phosphorus when brooded at two temperatures (29.4 versus 35°C). There was a temperature–phosphorus interaction on bone strength, bone ash and mortality. Cier *et al.* (1992b) evaluated the effects of different levels of available phosphorus on broiler performance in a hot region in Israel. Available phosphorus levels above 0.45% in the starter and 0.40% in the grower did not improve growth rate or feed utilization. Levels of 0.55% in the starter and 0.51% in the grower caused very wet litter.

Ait-Boulahsen *et al.* (1992), in studying the relationship between blood ionized calcium and body temperature of broiler chickens during acute heat stress, observed that exposure to high temperature elicited in these birds a significant elevation in plasma sodium to calcium ion ratio. The magnitude of this elevation appeared to determine the degree of hyperthermia. A solution of potassium biphosphate produced the highest plasma sodium to calcium ion ratio. This was followed in decreasing order by ammonium chloride solution, sodium chloride solution, unsupplemented water (control) and, finally, potassium chloride solution. There was a highly significant positive correlation between plasma sodium to calcium ion ratio and body temperature. These authors concluded that changes in body temperature response to heat stress parallel those of the plasma sodium to calcium ion ratio. In other words, a high sodium to calcium ion ratio impairs heat tolerance and vice versa.

Based on the above, mineral therapy and manipulation appear to be effective means of reducing detrimental effects of heat stress in broilers.

Since heat stress always depresses appetite and therefore reduces nutrient intake, the use of a vitamin and electrolyte pack in the drinking-water for 3–5 days, or until a heat wave passes, has been commonly practised. Morrison *et al.* (1988) reported the results of a survey conducted on 132 broiler flocks that either used or did not use a vitamin and electrolyte water additive during heat stress. They found that additives in water resulted in significantly lower mortality of broilers during heat stress. Although the feeding of extra amounts of vitamins during heat stress has been suggested for a long time (Faber, 1964), investigations into the effects of heat stress on specific vitamin requirements have not been very conclusive (see Chapter 5). Vitamin C supplementation during heat stress has been shown to have beneficial effects. Blood ascorbic acid decreases with an increase in environmental temperature (Thornton, 1961). Ascorbic acid limits the increase of body temperature during heat stress (Lyle and Moreng, 1968). Njokn (1984) reported that broilers reared under tropical conditions benefited from dietary supplementation of vitamin C during heat stress. Nakamura *et al.* (1992) exposed 24–26-day-old chickens to temperatures of 20 and 40°C and gave them ascorbic acid supplements of 150 or 1500 mg kg^{-1} for 7 days. Ascorbic acid supplements at 40°C reduced the severity of heat stress as measured on body-weight gain, feed intake and feed efficiency. Rajmane and Ranade (1992) tested the effect of a mixture of 1% ammonium chloride, 1000 p.p.m. ascorbic acid and 0.25 mg 100 g^{-1} α-methyl-*p*-tyrosine on performance of broilers during the summer season in India. Temperatures ranged from 38 to 42°C and relative humidity was about 75%. Their results showed a significant decrease in mortality rate and improved performance. They also observed reduced plasma corticosteroids and increased plasma potassium and ascorbic acid. They proposed that ascorbic acid supplements might have altered plasma corticosterone levels and helped to maintain concentration of potassium. α-Methyl-*p*-tyrosine probably acts as a catecholamine synthesis blocker and helps to maintain higher values of ascorbic acid. Cier *et al.* (1992a) investigated the use of ascorbic acid at 150, 300 and 600 p.p.m. in the feed for broilers under summer conditions in Israel. They concluded that ascorbic acid supplements significantly improved growth of both males and females, but the improvement was greater in males. No significant improvement was obtained in feed conversion. Several nutritionists, therefore, recommend the administration in the drinking-water of 1 g ascorbic acid l^{-1} or 4 g gal.$^{-1}$ all through the heat wave period.

Since many countries in hot regions import their vitamin and trace mineral mixes, and since there are often delays in acquisition and transport of ingredients, the problem of vitamin stability is of primary concern. Various data show that appreciable loss of vitamin activity in feed can occur during storage. Temperature, moisture and oxidation by polyunsaturated fatty acids, peroxides and trace minerals are the most critical factors that affect vitamin stability in poultry feeds. Coelho (1991) presented data on vitamin stability in both complete feeds and vitamin–trace mineral premixes. Table 8.1 presents average industry vitamin stability in vitamin–trace mineral premixes as well as in complete feeds. It is observed that the use of a premix without choline considerably improves its vitamin stability. This is why it has been industry practice for many years to add the choline separately to the feed. The data in Table 8.1 show that vitamin stability in feeds is somewhat similar to vitamin stability in trace mineral premixes. In either case, there is substantial loss with time and, considering the conditions that feed ingredients and complete feeds are subjected to in hot climates, the percentage loss of vitamin activity would be much greater than that shown in Table 8.1. Therefore, vitamin activity in feeds can be preserved by the incorporation of antioxidants, selecting gelatin-encapsulated vitamins, appropriate storage conditions, adding choline separate from the vitamin and trace mineral premix, delaying the addition of fats until just before the use of the feed and, finally, using feeds as soon as possible after mixing.

Finally, it is important to emphasize that nutritional manipulations can reduce the detrimental effects of high environmental temperature on broilers, but cannot fully correct them.

Seasonal Effects on Broiler Performance

Nir (1992) reported that in a relatively mild hot climate, like that of certain areas of the Middle East where the days are hot and the nights are cool, a strong relationship was found between the growing season and broiler body-weight. Two years' data from an experimental farm in the Jerusalem area showed a similar curvilinear pattern. The best performance was obtained during the winter and the worst during the summer. Those of spring and autumn were intermediate. Nir (1992) suggested that there are factors other than environmental temperature involved in growth regulation throughout the year. Within the ambient temperature range recorded in the poultry house, a 1°C increase in mean maximal ambient temperature from 4 to 7 weeks of age was accompanied by a decrease of 23 g in body-weight. Corresponding values for the average and minimal ambient temperatures were 32 g and 43 g, respectively. Nir's observations agree with other workers (Hurwitz *et al.*, 1980; Cahaner and Leenstra,

Table 8.1. Average industry vitamin stability in vitamin-trace mineral premixes and in complete feeds (adapted from Coelho, 1991).

	Loss per month (%)		
	Premixes		
Vitamin	With choline	Without choline	Complete feeds
A 650 (beadlet)	8.0	1.0	9.5
D3 325 (beadlet)	6.0	0.6	7.5
E acetate	2.4	0.2	2.0
E alcohol	57.0	35.0	40.0
MSBC[a]	38.0	2.2	17.0
MPB[b]	34.0	1.6	15.0
Thiamine HC1	17.0	0.5	11.0
Thiamine mono	9.6	0.4	5.0
Riboflavin	8.2	0.3	3.0
Pyridoxine	8.8	0.4	4.0
B_{12}	2.2	0.2	1.4
Ca pantothenate	8.4	0.3	2.4
Folic acid	12.2	0.4	5.0
Biotin	8.6	0.3	4.4
Niacin	8.4	0.3	4.6
Ascorbic acid	40.0	3.6	30.0
Choline	2.0	–	1.0

[a] MSBC = menadione sodium bisulphite complex.
[b] MPB = menadione dimethylpyrimidol bisulphate.

1992), who reported that maximal growth of broilers and turkeys is achieved at ambient temperatures much below the thermoneutral zone.

The response of broilers at high temperatures differs with different relative humidities. High temperature accompanied with high humidity is more detrimental to broiler growth than high temperature with low humidity. Nir (1992) reported that the response of broilers to excessive heat differed between Jerusalem and Bangkok. While 1°C increase in maximal ambient temperature was accompanied by a 23 g decrease in body-weight in the Jerusalem area, in Bangkok the corresponding decrease was 77 g. The higher relative humidity in the latter area must have partially contributed to this difference. It is well established that constant high temperature is more deleterious to poultry than cyclic or alternating temperatures. Osman et al. (1989) studied the influence of a constant high temperature (30–32°C) compared with an alternating temperature of 30–32°C by day and 25°C by night on broiler growth up to 12 weeks of age. High temperature decreased growth significantly starting at 4 weeks of age and the effect increased with age. Constant high temperature decreased growth more than an alternating temperature. Deaton *et al.* (1984) conducted five trials with commercial broilers to determine whether lowering the temperature during the cool portion of the day in the summer affected their performance. Treatments used were 24 h linear temperature cycles ranging from 35°C to 26.7°C or 21.1°C. Results obtained showed that lowering the low portion of the temperature cycle from 26.7°C to 21.1°C significantly increased broiler body-weight at 48 days of age.

Feed conversion in broilers is subject to marked fluctuations because of seasonal as well as ambient temperature changes. Poultry producers in the state of Florida found that 0.09 kg more feed was required to produce a unit of gain in broilers in the period of June to August than in November to April (McDowell, 1972). The influence of high temperature on efficiency of feed utilization has been studied by several workers and these studies all indicate that high temperatures bring about a reduction in efficiency in the utilization of feed energy for productive purposes. Animals eat less and they return less per unit of intake. McDowell (1972) summarized the reason for reduced feed efficiency at high temperature by saying 'in warm climates, generally, chemical costs for a unit of product are higher than in cooler climates because a portion is siphoned off for the processes required to dissipate body heat'.

Temperature and Body Composition

Kleiber and Dougherty (1934) were among the earliest workers to report on the effects of environmental temperature on body composition in the chicken. They observed that maximum fat synthesis occurred at an environmental temperature of 32°C, while no effect was noted on protein synthesis. This has been confirmed by Chwalibog and Eggum (1989). Olson *et al.* (1972) reported that carcass dry matter, fat and energy were increased with increasing temperature or dietary energy level, but protein decreased. The ME required per kcal of carcass was increased with lower temperature, reflecting the increased requirement for body temperature maintenance. These workers observed that a 1° drop in temperature reduced ME efficiency by about 1%.

El-Husseiny and Creger (1980) studied the effects of raising broilers at 32°C versus 22°C on their carcass energy gain. Their results indicated that energy per gram of dry tissue increased with decreasing ambient temperature and the energy-gained value for birds reared at 22°C was higher than at 32°C. This was confirmed by Sonaiya (1989), who showed further that broilers reared at 21°C retained more energy in the carcass as fat than did broilers reared at 30°C. Sonaiya (1988) studied the effects of temperature, dietary ME, age and sex on fatty acid composition of broiler abdominal fat. Chickens reared at high temperature (30°C) had a significantly lower proportion of polyunsaturated fatty acids in their abdominal fat between 34 and 54 days than had birds at low temperature (21°C). The depot fat contents of oleic, linoleic and linolenic acids were all reduced by high temperature at 54 days. For chickens slaughtered at 54 days, saturated fatty acids content was much higher in females than in males at high temperature, while at low temperature polyunsaturated fatty acids were much lower in males than in females. Sonaiya (1988) recommended the early finishing of broilers from the viewpoint of fatty acid composition because the polyunsaturated fatty acid to saturated fatty acid ratio declines significantly with age, regardless of temperature.

Environmental temperature has been shown to influence carcass amino acid content of broilers. Tawfik *et al.* (1992) reported that broilers kept at 18°C from 4 to 12 weeks of age had higher glycine and proline concentrations in breast muscle than those kept at 32°C. Bertechini *et al.* (1991) did not observe any changes in carcass characteristics of broilers raised at 17.1, 22.2 or 27.9°C and given a maize–soyabean meal base diet. Sonaiya *et al.* (1990) reported that broiler meat flavour was improved by age (34 vs. 54 days) and high environmental temperature (21°C vs. cycling 21 to 30°C).

Today's broilers are selected and managed with the aim of increasing meat yield and decreasing fat deposition. Several studies have shown that environmental temperature has an effect on carcass composition and meat yield. At high temperature, meat yield and especially yield of breast meat are reduced (Howlider and Rose, 1989; Tawfik *et al.*, 1989). Leenstra and Cahaner (1992) found temperature to have a negative effect on breast meat yield. Males were more affected by high temperature than females.

Geraert *et al.* (1992) studied the effect of high ambient temperature and dietary protein concentration on growth performance, body composition and energy metabolism of genetically lean (LL) and fat (FL) male chickens between 3 and 9 weeks of age. Heat exposure reduced feed intake in both lines (−28%). Lean broilers gained significantly more weight between 5 and 7 weeks of age when exposed to 32°C than fat chickens (+20%). At 9 weeks of age, body dry matter, mineral or lipid contents were not affected by high temperature, while body protein content was significantly decreased. The main effect of high temperature is a reduction in ME intake to decrease metabolic heat production, leading to a lower energy retention. Gross protein efficiency was depressed in hot conditions in the LL compared with the FL. Smith (1993) conducted a study in which carcass parts from broilers reared under different growing temperature regimens were examined for crude protein, fat, calcium, phosphorus, potassium and sodium. Protein content of thighs and drumsticks from birds grown at elevated temperatures were higher than those from birds grown in a constant temperature environment. The fat content of these parts was higher from birds in the cooler environment. The sodium content of the breast was higher at elevated temperatures. The breast portion of carcasses from heat-stressed birds had the greatest amount of potassium.

Broiler Management in Hot Climates

Feeding programmes

Recommended nutrient levels of broiler starter, grower and finisher rations are shown in Table 8.2. They have been modified to better suit conditions in hot climates. Protein levels recommended are 1–2% lower than what is normally used in the temperate regions because of what has been presented earlier in this chapter as well as in Chapter 5. Energy levels have been also adjusted to protein levels, but kept higher than those currently used in many hot regions. The potassium level has been increased to 0.6% versus the 0.4% normally recommended in the temperate regions. Levels of certain amino acids have been raised higher than those normally used at that protein level.

Table 8.2 specifies that a starter is fed for the first 3 weeks of life to straight-run chicks and a grower from 3 to 6 weeks of age. The finisher is used from 6 weeks to market. Under certain conditions in hot climates, it may be beneficial to extend the feeding of the grower to 7 weeks, if growth rate is slow. Decuypere *et al.* (1992) studied different protein levels for broilers in Zaïre on isocaloric diets and found that, in the hot and humid conditions of that country, slight increases in dietary protein late in the broiler cycle are beneficial for growth and feed efficiency. Another advantage of increases in dietary protein observed by those workers is that abdominal fat content is reduced with these higher protein levels.

Providing adequate levels of each nutrient and a good balance of these nutrients is a prerequisite to successful feeding. Using feed with good-quality ingredients is a second prerequisite and using only fresh feed is the third prerequisite. It is very important to order fresh feed, preferably once a week, in a hot, humid climate, where nutrient deterioration is more rapid and fats tend to go rancid quickly.

Broilers are usually fed either crumbles or pellets. The broiler starter is provided as crumbles, while grower and finisher rations are provided as pellets. Crumbles and pellets tend to reduce feed wastage and improve feed efficiency.

Broilers are normally fed *ad libitum*. However, in recent years, intermittent feeding programmes have been used on modern, well-managed broiler farms. If intermittent feeding is used in hot climates, then broilers should be fed for a period of 1.5 h, after which the lights go off, allowing the broilers to rest and digest their meal. After 3 hours of darkness, the

Table 8.2. Recommended broiler ration specifications.

	Starter	Grower	Finisher
Protein (%)	22	20	18
Metabolizable energy			
(kcal kg^{-1})	3000	3050	3100
(kcal lb^{-1})	1365	1385	1410
Calcium (%)	1.0	0.90	0.80
Available phosphorus (%)	0.45	0.42	0.40
Sodium (%)	0.18	0.18	0.18
Chloride (%)	0.18	0.18	0.18
Potassium (%)	0.60	0.60	0.60
Lysine (%)	1.20	1.15	0.95
Methionine (%)	0.50	0.45	0.40
Methionine + cystine (%)	0.85	0.80	0.70
Tryptophan (%)	0.23	0.20	0.18
Arginine (%)	1.45	1.35	1.15
Histidine (%)	0.35	0.35	0.28
Phenylalanine (%)	0.75	0.75	0.65
Phenylalanine + tyrosine (%)	1.35	1.35	1.15
Threonine (%)	0.80	0.75	0.70
Leucine (%)	1.35	1.35	1.15
Valine (%)	0.85	0.80	0.70
Glycine + serine (%)	1.50	1.25	1.00
Males	0-21 days	3-6 weeks	6 weeks to market
Females	0-14 days	2-5 weeks	5 weeks to market
Straight run	0-21 days	3-6 weeks	6 weeks to market

lights are turned on again for another period of feeding. When intermittent feeding is used, 20-30% more feeder and drinker space is provided so that all birds can eat and drink during the 1.5 h when the lights are on. This method should have a possible application in hot climates since the hours of darkness provide minimum activity on the part of the birds and therefore reduced heat production.

Fast growth rate in the modern broiler has not been without its problems. It has contributed to increased mortality due to heart attacks, increased leg problems and increased incidence of ascites. By restricting feed and/or nutrient intake early in life, it has been possible to reduce some of these problems. Body fat, an undesirable component in the modern broiler, can also be reduced by feed and/or nutrient restriction. Several methods have been used to reduce very rapid growth rate in the early life of the broiler. One method is by altering the lighting programme, another is by actual feed restriction and the third is by reducing energy and/or protein content of the diet by adding to it extra fibre.

Sex-separate feeding of broilers has become quite popular in many countries. In areas where uniformity in the final product is important and where uniformity has been hard to attain, this practice is highly recommended. Furthermore, this practice can save in feed costs since rations for the females can be reduced by 1–2% in protein. This can be done by shifting from the starter to the grower and from the grower to the finisher at earlier ages (see Table 8.2). Males and females have the same nutrient requirements during the first 2 weeks of life. After this period, females tend to respond differently to certain nutrients in the feed. This is why reducing the plane of nutrition for females may save in feed costs. Sex-separate feeding also allows for marketing the males earlier than the females and thus provides for greater yearly output from the same farm.

Feed withdrawal

McCormick *et al.* (1979) reported that fasting of 5-week-old broilers for 24, 48 or 72 h resulted in progressively increased survival time when exposed to heat stress.

Feed withdrawal during the hottest hours of the day has become a common management practice in many broiler-producing areas. Short-term feed withdrawal can lower the bird's body temperature and increase its ability to survive acute heat stress. Smith and Teeter (1987, 1988) studied the influence of fasting duration on broiler gain and survival. They showed that fasting intervals beginning 3–6 h prior to heat stress initiation and totalling up to 12 hours daily during significant heat stress (up to 37°C) reduce mortality significantly. One suggested practice when a heat wave is expected is to remove feed at 8.0 a.m. and return it at 8.0 p.m. Fasting will probably reduce weight gains. Therefore the producer will have to weigh the importance of a more rapid growth rate versus a greater mortality risk.

Drug administration

Drug activity under heat stress may not be the same as in a temperate environment. Furthermore, drug administration through the drinking-water needs much more careful calibration than under normal temperature conditions.

Producers tend to avoid the use of nicarbazine as an anticoccidial drug during heat stress periods. Reports of nicarbazine-induced mortality during heat stress vary greatly. Keshavarz and McDougal (1981) reported that the incorporation of nicarbazine at the regular dose rate of 125 p.p.m. can result in excessive broiler mortality following heat stress. This relationship between nicarbazine and high environmental temperature had been reported earlier by McDougal and McQuinston (1980). These workers showed that environmental temperatures exceeding 38°C for 11 days during a 50-day broiler cycle, caused a mortality of 91% in broilers fed nicarbazine compared with 27% in non-medicated controls. This effect of nicarbazine on broiler mortality at high temperature had also been previously reported in South Africa (Buys and Rasmussen, 1978). At temperatures of 40°C, broilers aged 3–5 weeks and receiving nicarbazine showed from 60 to 75% mortality, compared with only 10–25% in the corresponding age groups which were medicated with 125 p.p.m. amprolium. Smith and Teeter (1988) reported on the adverse effects of nicarbazine, which caused a 16% mortality in broilers. In a second study, birds were exposed to a hot climate prior to the start of the experiment. Nicarbazine-treated birds and controls had the same mortality (20%). The same workers showed that the high mortality noted with nicarbazine was reduced when potassium chloride was added to the drinking-water. Macy *et al.* (1990) studied the effect of Lasalocid on broilers raised in a hot cyclic (26.7–37.8°C) versus a moderate (21°C) environment. In the hot cyclic environment, broilers fed Lasalocid gained more weight and had better feed conversion than those fed the basal diet.

Inclusion of amprolium in feed or water at very high levels may cause thiamine deficiency since this drug functions as a specific antagonist of thiamine (Polin *et al.*, 1962). It is important therefore to check that adequate levels of this vitamin are present in the feed whenever amprolium is used.

Vaccination

Vaccination is the main method for the control of infectious diseases in poultry and, when properly used and administered, can reduce the overall weight of infection on a farm and result in great benefits from improved performance. Vaccination programmes used in several countries in the hot regions of the world have been described by Aini (1990). This section will present some suggestions and recommendations designed to aid in reducing vaccine failures in hot climates.

Water vaccination is one of the most widely used methods of vaccine administration in broiler production. Vaccines that are given in the drinking-water are live vaccines and are sensitive to high temperatures. Mutalib (1990) recommended the following for handling vaccines:

1. Use ice or a thermos to transport vaccines from source to farm.

2. Avoid exposure of vaccine to heat and direct sunlight (do not leave vaccine bottles in car).
3. Store vaccines refrigerated until use.
4. Once reconstituted, the vaccine should be used within the hour.
5. It is advisable to vaccinate early in the morning to avoid the heat of the day.
6. Do not use outdated vaccines.
7. Dilute vaccines only in clean non-chlorinated water.

In hot climates, it is important to determine the amount of water consumed by the flock and to use that figure for diluting the vaccine rather than a book figure. Another precaution to take in hot climates is not to deprive the birds of water for a prolonged period of time, usually not more than 1 h prior to vaccination. Every effort should be made to reduce stress in birds before and after vaccination. Therefore avoid vaccination on days when there are exceptionally high temperatures and try to provide as much comfort for the birds as possible. All other practices used in temperate regions for vaccination should also be applied in hot regions.

Beak trimming

Beak trimming is a common practice with the young commercial laying pullet, but is not as commonly used for broilers. However, in hot climates, where feather picking and cannibalism may be frequent, producers resort to beak trimmming as a means of reducing bird losses. Furthermore, when the broiler house is open-sided in hot climates, the light intensity cannot be controlled. This high light intensity necessitates beak trimming to prevent feather picking and cannibalism.

Results obtained by different researchers have varied, but in general no effect on final body-weight has been observed when broiler chicks were beak-trimmed at 1 day of age (Harter-Dennis and Pescatore, 1986; Trout et al., 1988). Even an improvement in feed efficiency was observed by Harter-Dennis and Pescatore (1986) when birds were beak-trimmed at 1 day of age. Andrews (1977), however, reported that, in broilers that had been beak-trimmed at 1 or 10 days of age, both procedures resulted in reduced 8-week body weights when compared with controls that had not been beak-trimmed. Christmas (1993) compared the performance of spring- and summer-reared broilers as affected by beak trimming at 7 days of age. He observed that beak trimming resulted in significantly lower body-weights and feed intake in summer-reared broilers, but not in spring-reared broilers. Broiler producers in hot climates should therefore weigh the relative importance of a higher growth rate versus higher mortality due to cannibalism. Although beak trimming at 6–7 days of age may produce better results than at 1 day of age, there is the added labour involved with catching the birds at 6–7 days. If broilers must have their beaks trimmed, then it is most appropriate to trim them at the hatchery.

Broiler house management

The principles of housing and housing conditions for optimum performance in hot climates have been covered in detail in Chapter 4. In this section, therefore, selected aspects of broiler house management that have not been covered earlier will be presented. It cannot be overemphasized that, in high-temperature conditions, all practices that aid in reducing heat production by the bird and those that aid in the elimination of heat already produced in that house need to be implemented. There is, of course, heat produced by lighting and by machinery present in the broiler house, but these are minor sources of heat in comparison with that coming from the birds. From the standpoint of heat production, the easiest practice to implement is to reduce the density of birds in the broiler house. Navahari and Jayaprasad (1992) evaluated space requirements of broilers on three floor types (deep litter, wire floor and slatted floor) during three seasons (summer, monsoon and winter). They concluded that, for a broiler weighing 1.1 kg at market, a litter floor space of 400 cm^2 or a wire-slatted floor area of 280 cm^2 was sufficient for that region in India. North and Bell (1990) recommended 500 cm^2, for a broiler weighing 1.3 kg at market, raised on a litter floor during the summer. As for the elimination of heat already produced, the most economical method is to increase ventilation rates in the house.

The optimum temperature and humidity within a poultry house depend on the type of stock and age of the flock. In broiler houses, the climate has to be properly adjusted to the age of the flock. Special attention should be paid not only to air temperature and humidity but also to purity and draught-free flow. Afsari (1983) suggested maximum rates, in percentage by volume, for the three important chemical factors that should not be exceeded. These are carbon dioxide (3.5), ammonia (0.05) and hydrogen sulphide (0.01).

Litter management

The control of litter quality is important to the successful management of broiler flocks. High-moisture litter due to improper operation of drinkers or to poor ventilation will lead to accumulation of ammonia, which increases incidence of respiratory problems and ocular lesions, as well as favouring intestinal infections.

It is common practice to reduce litter depth in hot weather for more bird comfort. This is a good practice, provided litter depth is not reduced to less than 5 cm (2 in.). Litter should be kept in good crumbly condition.

Both quantity and quality of litter affect breast blister incidence and thus broiler condemnation rate.

Excess of sodium in the diet for broilers increases water intake and moisture content of the droppings (Damron and Johnson, 1985). However, too low a level can have a negative effect on performance. Therefore, it is important to meet the exact sodium requirement in broiler diets. Vahl and Stappers (1992) reported that it was possible to decrease the dietary sodium content without negative effects on performance during the latter part of the broiler cycle. They recommended a sodium level of $0.9 \, g \, kg^{-1}$ diet after 5 weeks of age.

Broiler house cooling devices

Cooling systems for both naturally ventilated and closed, fan-ventilated houses have been described in Chapter 3. This section covers a few selected systems that have been tested and used for broiler houses.

In hot dry climates, broiler producers have for many years used desert-type air coolers after the 2–3-week brooding period. These coolers have been found to be the cheapest and most efficient means of lowering air temperature and increasing humidity in hot, dry climates (see Fig. 8.2). They are still used for small operations and for naturally ventilated houses, while large integrated operations in hot climates use closed, environmentally controlled housing with different sophisticated evaporative cooling systems.

Several studies have been conducted to show the effects of using these desert air coolers on broiler performance and economic efficiency. Al-Zujajy *et al.* (1978) studied the effect of air coolers on broilers housed at different densities in two naturally ventilated houses in Iraq. Differences in weight gain between birds in cooled and uncooled houses reached 385 g per bird at market at a density of 16 birds per square metre. Feed conversion efficiency, meat yield and carcass conformation were much better in the cooled house than in the uncooled house and financial returns per square metre were 40% higher in the cooled house.

Another method of reducing heat stress in naturally ventilated broiler houses is by roof insulation. Several workers have shown that roof insulation can considerably reduce the death losses from heat prostration in broilers. Reece *et al.* (1976) indicated that the radiation from a hot roof alone has no effect on body-weight gain or feed conversion of broilers. However, when the radiation was combined with a 3°C increase in ambient temperature, which was the increase attributed to absence of insulation in a minimally ventilated broiler house, male body-weight gain and feed conversion were adversely affected.

Reilly *et al.* (1991) conducted a study to determine whether water-cooled floor perches could be used by broilers exposed to a constant hot

Fig. 8.2. Naturally ventilated house equipped with desert-type air cooler.

ambient environment and whether utilization of these perches would improve performance beyond those provided uncooled floor perches. Their results indicate that water-cooled perches are beneficial in improving broiler performance during periods of high environmental temperatures.

Wet versus dry mashes

Behavioural responses and growth of broilers were investigated by Abasiekong (1989) in three seasons using two wet mashes with feed : water ratios of 2 : 1 and 3 : 1. Wet mashes given in the hot dry season caused significantly lower rates of panting. In the hot dry season, 60% of chickens preferred wet mashes, whereas 80% preferred dry meal in the cool dry season.

Whitewashing house roof

Reflective materials applied to the roof and their effectiveness in reducing house temperature have been described in Chapter 4. Heat load can be reduced by whitewashing the broiler house roof and house temperatures can be reduced by several degrees in uninsulated houses and to a lesser extent in roof-insulated buildings. The temperature differences can range

Table 8.3 Formula for whitewashing poultry house roofs (from Casey, 1983).

Formula	Area covered (sq. ft)	Years effective
20 lbs hydrated lime + 5 gals water	600	1
20 lbs hydrated lime + 6 gals water + 20 lbs white cement + 0.5 cup blueing	700	3
20 lbs hydrated lime + 5 gals water + 1 qt polyvinyl acetate	700	4

from 3 to 9 °C (5 to 15 °F) on hot days. Since the effectiveness of whitewash is due to its ability to reflect radiant heat off a surface, this reflective property is greatly reduced by dust accumulation. The dust may need to be removed periodically by sprinkling the roof. Table 8.3 gives the formula proposed by Casey (1983) to be used for whitewashing poultry house roofs.

Disinfectants used in hot climates

Biosecurity is a very important consideration in hot climates. The all-in, all-out system is highly recommended and preferred, and, for thorough decontamination, special disinfectants may be needed for use between batches. Effective poultry house disinfection is not always achieved in hot climates. High ambient temperatures usually lead to rapid evaporation of the disinfectant solution, hence not allowing sufficient contact time for the disinfectant to be effective. Disinfectants used in developing countries are often traditional formulations that are not always designed to be effective against the specific disease problem at hand. In some areas, the poultry industry still uses outdated housing with dirt or sand floors. These can harbour infection several centimetres below the floor surface which ordinarily disinfectants cannot reach.

Therefore, in order to overcome some of these problems, a disinfectant with the following properties is recommended for use:

1. A disinfectant that has the ability to mix with both diesel oil and water. A stable emulsion can be formed of the oil, water and disinfectant. This emulsion will dry slowly and allow disinfectant time to be effective.
2. The disinfectant should be broad-spectrum – capable of killing bacteria, viruses, fungi and insects.
3. The disinfectant must have detergent properties to enable it to penetrate and emulsify the organic matter which harbours infection in the sand or dirt floors.

Dewinged broilers

Feathers of birds are an effective barrier to heat loss from the skin. The wings cover a wide body surface area and their presence reduces the efficiency of heat loss during elevated ambient temperatures. Al-Hassani and Abrahim (1992) conducted two experiments to evaluate the effect of dewinging on performance of broilers reared under cyclic temperatures of 25 to 35°C. Although body temperature and mortality were lower in the dewinged groups than in controls, the body-weight of controls was higher. There were no differences in feed intake, feed conversion, dressing percentage and boneless meat percentage between the two groups. The authors concluded that the dewinged birds were more heat-tolerant than controls.

Water Consumption

Water is an important coolant. Drinking-water plays an important role in cooling broilers. The cooler the water, the better the birds can tolerate high environmental temperatures. The warmer the water, the more they need to drink. To help them meet this increased need for water, growers provide broilers with approximately 25% more drinker space than the standard temperate climate recommendation. Where possible, wide and deep drinkers, permitting not only the beak, but all the face to be immersed, are used. Vo and Boone (1977) observed that birds consuming water from drinkers large enough to allow head dunking survived longer under heat stress than those not provided with such drinkers. Morrison *et al.* (1988) in a survey on broiler farms reported that drinker type had significant effect on percentage increase in mortality during the summer months. These workers found that, among four types of drinkers, the trough drinker that allowed head dunking had the lowest mortality rate.

Ambient temperature is probably the most important factor affecting water intake in broilers. Figure 8.3, adapted from data presented by North and Bell (1990), shows the effect of house temperature on water consumption of straight-run broilers. Broilers at a house temperature of 38°C consume four times as much water as those at 21°C. Harris *et al.* (1975) demonstrated the adverse effect of providing drinking water at 35–40°C compared with 17–23°C. Water at low temperature functions as a heat sink in the intestinal tract. Insulation of header tanks and supply piping is necessary if the temperature of the water at the point of consumption exceeds 25°C. If possible, bury water pipes at least 60 cm (2 ft) underground and shade the area where the pipes run. The water : food intake ratio increases from approximately 2 : 1 at moderate temperatures to about 5 : 1 at 35°C (Balnave, 1989). This can have drastic effects on the intake of nutrients, drugs or any other substance present in

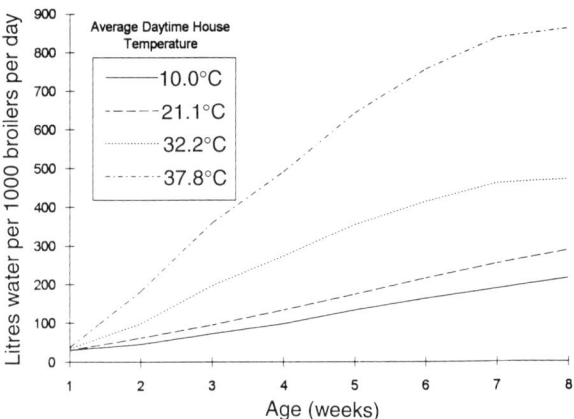

Fig. 8.3. Effect of house temperature on water consumption of straight-run broilers.

the drinking-water. The most important of these are the minerals, and many producers believe that the mineral content of drinking-water can influence broiler performance, especially when well water is used. Barton (1989) examined the effect of well water in a survey on broiler performance in Arkansas. The only mineral ion to show a significant effect on performance was nitrate, with lower nitrate concentrations in well water being associated with better performance. Increased concentrations of calcium, on the other hand, were associated with increased growth and improved feed conversion, but with increased mortality and downgrading. The overall results of the study were not very conclusive and more controlled tests on this subject need to be conducted.

Although ambient temperature is the major factor affecting water intake, other factors have also been shown to be important. Austic (1985) summarized these factors and stated that electrolyte content of both diet and water is a major determinant of water intake. Since the chicken has a limited capacity to concentrate urine, excessive dietary levels of soluble mineral elements can increase water consumption. In many hot regions of the world, the water used on poultry farms is brackish water with high levels of total dissolved solids. These contribute to the solute load which must be excreted by the bird. The intake of excessive levels of minerals therefore increases water requirements of broilers and contributes to wet litter problems. High-protein diets also cause increased water intake because of the need for excretion of by-products of protein catabolism. If chlorinated water is being used on the farm, it is recommended to discontinue chlorination on extremely hot days (Murphy, 1988) and to clean drinkers at least once daily using a disinfectant rinse to help reduce slime build-up. Not all studies on the use of brackish water available for poultry in hot regions have been negative. Ilian *et al.* (1982) showed that brackish water containing up to 3000 p.p.m. of total dissolved solids did not adversely affect the overall performance of broilers.

The common practice of 'walking the birds' during heat stress has always been shown to increase water consumption by about 10%. Walk birds early, before temperatures reach critical levels. One of the best ways of increasing water consumption is by lowering its temperature. Header tanks and water piping should be insulated to ensure that water temperature does not exceed 25°C. One suspended drinker should be provided per 100 broilers. Water lines should be flushed every few hours during the high heat period of the day to keep water cool. Add extra waterers to the pens and, if emergency conditions occur, consider adding ice or mechanically cooled water. Do not disturb birds when air temperatures are high. Burger (1988) emphasized that, during heat stress, anything that can be done to keep the broilers quiet will reduce mortality. This has also been emphasized by Van Kampen (1976), who cautioned against disturbing birds during heat stress.

May and Lott (1992) studied feed and water consumption patterns of broilers at a constant 24°C or at a cyclic 24, 35 and 24°C. Their data show that the increased water intake and decreased feed intake observed due to high cyclic temperature arise from changes that occur during certain times of the day and no changes at other times. The increase in water intake precedes the reduction in feed intake. Maintenance of both carbon dioxide and blood pH is critical to the heat-stressed broiler and the beneficial effects of adding ammonium chloride and potassium chloride to the drinking-water are well documented (see Chapter 5). Increased consumption of cool drinking-water is crucial to survival of heat-stressed broilers.

Acclimatization to Heat Stress

It has been known for some time that chickens can adapt to climatic changes. Van Kampen (1981) classified responses to climatic changes into three phases: neuronal, hormonal and morphological. The neuronal phase is the first and is seen immediately, while the morphological response is observed much later. Changes in metabolic rate occur fairly quickly since they involve neuronal and hormonal aspects. Morphological changes, such as enlarged combs and wattles, less body fat and less feather cover, which have been known to occur in chickens reared under high temperature, require a much longer time to appear.

Based on these earlier studies, attempts have been made at reducing heat stress mortality in broilers by acclimatization. Raising house temperature prior to the onset of a heat wave has been shown to reduce mortality.

This is partly attributed to a reduction in feed intake in response to the stress. Reece *et al.* (1972) demonstrated that broilers can acclimatize in 3 days and resist extremely high mortality from heat prostration. The treatment used by these workers consisted of birds taken from 21°C and exposed for 3 days to a 24-hour cycle of 24, 35 and 24°C before a temperature stress of 40.6°C was imposed. This treatment reduced mortality from 33 to 0%. Ernst *et al.* (1984), however, showed that exposing chicks at 1 day of age for 2 h to 43°C reduced performance up to 16 days of age. Arjona *et al.* (1990) observed that exposure for 24 h at 35–38°C at 5 days of age reduced mortality when these birds were heat-stressed for 8 h at 44 days of age. Teeter and Smith (1986) reported that at least 50% of the hypothermic effect of acclimatization immediately prior to heat stress may be attributed to a reduction in feed intake in response to the stress.

Another metabolic adaptation of broilers exposed to high temperature has been reported by Hayashi *et al.* (1992). These workers studied the effects of increasing ambient temperature on protein turnover and oxygen consumption in skeletal muscle. They found that rates of protein synthesis and breakdown and oxygen consumption were higher in birds raised at 20°C than in those raised at 30°C. They suggested that skeletal muscle may function as a regulator of body temperature by changing the rate of skeletal muscle protein turnover in response to ambient temperature changes.

Several adaptations involving the cardiovascular and respiratory systems take place when birds are exposed to high temperatures. These changes have previously been described in Chapter 3. Furthermore, birds undergo certain metabolic adaptations (Chwalibog, 1990). Heat acclimatization involves certain specific adaptations that regulate dehydration and hypovolaemia. These adaptations tend to maximize body water reserves needed for evaporative cooling and blood volume maintenance (Van Kampen, 1981). Belay and Teeter (1993) showed that water consumption increases to match increased respiratory water loss during heat stress.

Arjona *et al.* (1988) showed that exposing broilers to elevated environmental temperature at 5 days of age improved their tolerance to heat stress without reducing subsequent performance. Lott (1991) exposed broilers to high temperature for 3 consecutive days at 21 days of age and was able to decrease mortality resulting from high ambient temperature later in life. This worker observed that acutely heat-acclimatized birds consume more water than non-acclimatized controls during heat stress. Realizing that heat acclimatization may include adaptations in kidney function, Wideman *et al.* (1994) conducted a study on renal function in broilers at normal and high temperature and found that heat-acclimatized birds have significantly lower glomerular filtration rates, filtered loads of sodium and tubular sodium reabsorption rates than the respective control groups. These changes in kidney function, according to these workers, would minimize urinary fluids and solute loss when heat-acclimatized broilers consume large quantities of water to support evaporative cooling. Although this practice of acclimatization is still in the experimental stage, it has definite potential for the broiler industry in hot climates.

Lighting Programmes

The majority of broilers the world over are grown on a light programme of 23 h of light each day. At some time during the night, the lights are turned off for 1 h. This is done to allow the birds to become accustomed to the lights suddenly being turned off. Thus, in the event of a power failure, the flock will accept the loss of light without panic, and crowding and possible suffocation is avoided. This practice is very important because power failures are more common in the developing hot regions than in the developed temperate areas of the world.

The use of intermittent light has been shown to increase feed consumption during the cooler part of the day. Very few studies have been conducted on broiler lighting programmes in hot tropical climates. A study on the integration of lighting programmes and operation of feeders has been reported from Singapore (Ngian, 1982), using convection-ventilated housing in which improved performance was obtained by illumination and feeding during the night. This programme produced the highest live weight at 56 days, but feed conversion and mortality were lower than other combinations examined. Diab *et al.* (1981) at the Kuwait Institute of Scientific Research showed that sequential 7-hour periods of darkness and illumination provided better growth and feed conversion in broilers when compared with continuous feeding. Renden *et al.* (1992) compared performance of light-restricted broilers (16 h light, 8 h darkness (16L : 8D)) with those reared under an extended lighting programme (23 L: 1D) during the summer. Although relative growth was greater from 28 to 48 days of age for light-restricted broilers, final body-weights were similar. Smith (1994) studied the effects of electrolytes and lighting regimens (23L : 1D vs. 16L : 8D) on broilers grown at elevated temperatures. This worker found that photoschedule had no effect on body-weight gain, feed consumption or carcass characteristics.

Hulan and Proudfoot (1987) studied the effects of light source, ambient temperature (20–34°C) and dietary energy source on the general performance and incidence of leg abnormalities of roaster chickens. They found that light source had no significant effect on mortality, body-weight or feed conversion. Incidence of angular deformity (AD) and total leg abnormalities (TLA) was lower under fluorescent as opposed to incandescent light. Cooler ambient rearing temperature increased linearly the incidence of mortality, curly toes (CT), AD, enlarged hocks (EH) and TLA. Mean financial returns were not significantly better for roasters reared under fluorescent versus incandescent light and were better for birds reared under warm versus cool ambient temperature.

Contamination from the digestive tract is a persistent problem in broiler processing. May *et al.* (1990) studied the effects of light and environmental temperature on quantity of crop, gizzard and small intestine contents during feed withdrawal. Results indicated that crop clearance is improved by lighting before and after cooping. No differences were observed in broilers maintained at 18 or 27°C.

Very few studies have been conducted on the effects of light quality on broiler performance in hot climates. In the USA, incandescent lamps have been the standard for many years and are still used in many broiler houses, especially those with side-wall curtains, which can use daylight as a supplement to artificial lighting (Weaver, 1992). Boshouwers and Nicaise (1992), in a study on the effects of light quality on broiler performance, reported that light source significantly affected physical activity, energy expenditure and body-weight. Physical activity was lowest under 100 Hz fluorescent light and highest under incandescent. Since in hot climates we may want to reduce the activity of birds, fluorescent light may have an advantage over incandescent lights.

Conclusions

Several studies have been conducted during the past two decades on various techniques to improve broiler performance in hot climates. Nutritional manipulation, such as the addition of fat and the reduction of excess protein and amino acids, has been widely adopted. Birds should also be fed during the cool hours of the day. Feed withdrawal prior to heat stress initiation has become a viable management tool. Fasting intervals beginning 3–6 h prior to maximum heat stress and totalling up to 12 h daily during significant heat stress reduces broiler mortality. Fasting will probably reduce weight gains. Therefore, the producer will have to weigh the importance of a more rapid growth rate versus a greater mortality risk.

Maintenance of both carbon dioxide and blood pH is critical to the heat-stressed broiler and the beneficial effects of adding ammonium chloride and potassium chloride to the drinking-water are well documented. Acclimatization to heat stress by exposing birds to high temperature before the onset of the heat wave has potential, but more research and testing are needed before large-scale field application becomes a reality. The addition of extra vitamins and electrolytes to the drinking-water has been helpful in most tests and under most situations. The use of ascorbic acid in the feed or in the drinking-water has become a common practice in many hot regions of the world.

Microingredient premixes should be purchased from reputable sources and vitamins should be in gelatine-encapsulated and stabilized forms. Antioxidants should be incorporated into vitamin premixes and into rations to prevent production of toxic free radicals as a result of oxidation. This is particularly important in rations with added fats.

Besides ambient temperature, season *per se* has an effect on broiler performance and the best performance is usually attained in winter while the worst is in summer. High temperatures not only reduce growth, but also cause a reduction in efficiency in the utilization of feed energy for productive purposes.

Environmental temperature has been shown to influence body composition in broilers. Broilers reared at normal temperatures (21°C) retain more energy in their carcasses as fat than those reared at high temperature (30°C). High ambient temperatures tend to reduce the proportion of polyunsaturated fatty acids in abdominal fat. Polyunsaturated to saturated fatty acid ratio in broiler carcass declines with age regardless of temperature. Reports on the effect of temperature on carcass proteins are not conclusive.

Biosecurity is a very important consideration in hot climates. The all-in, all-out system is highly recommended and for thorough decontamination special disinfectants may be needed for use between batches. Properties of such disinfectants and adequate methods for their application have been proposed.

Intermittent lighting is potentially a sound practice for hot climates, particularly when used to increase feed consumption during the cooler parts of the day. The provision of cool drinking-water plays an important role in reducing heat stress. Increased consumption of cool drinking-water is crucial to the survival of heat-stressed broilers. Electrolyte content of both diet and water increases water consumption. High-protein diets also cause increased water intake because of the need for excretion of by-products of protein catabolism. Be sure to discontinue chlorination of water on extremely hot days if this practice is used. The above practices can improve growth and feed efficiency of broilers in hot climates. They can also help in reducing mortality due to heat stress.

References

Abasiekong, S.F. (1989) Behavioural and growth responses of broiler chickens to dietary water content and climatic variables. *British Poultry Science* 29, 563–570.

Afsari, N. (1983) Air conditioning of houses in hot climates. *Poultry International* 22, 64–73.

Aini, I. (1990) Control of poultry diseases in Asia by vaccination. *World's Poultry Science Journal* 46, 125–132.

Ait-Boulahsen, A., Garlich, J.D. and Edens, F.W. (1992) Relationship between blood ionized calcium and body temperature of chickens during acute heat stress. *Proceedings 19th World's Poultry Congress*, Vol. 2, pp. 87–92.

Ait-Tahar, N. and Picard, M. (1987) Influence of ambient temperature on protein requirements of broilers. *Research Note*, INRA Laboratoire de Recherches Avicoles, Nouzilly, France, pp. 1–12.

Al-Hassani, D.H. and Abrahim, D.K. (1992) Performance of normal and dewinged broiler chickens under high ambient temperatures. *Proceedings 19th World's Poultry Congress*, Vol. 2, pp. 709–711.

Al-Zujajy, R.J., El-Hammady, H. and Abdulla, M.A. (1978) The use of air-coolers in broiler houses under subtropical conditions in Iraq. *British Poultry Science* 19, 731–735.

Andrews, L.D. (1977) Performance of broilers with different methods of debeaking. *Poultry Science* 56, 1689–1690.

Anonymous (1982) Energy and protein requirements for broilers in the tropics. *Poultry International*, August, pp. 62–63.

Arjona, A.A., Denbow, D.M. and Weaver, W.D. (1988) Effect of heat stress early in life on mortality of broilers exposed to high environmental temperatures just prior to marketing. *Poultry Science* 67, 226–231.

Arjona, A.A., Denbow, D.M. and Weaver, W.D. (1990) Neonatally induced thermotolerance: physiological responses. *Comparative Biochemistry and Physiology* 95A, 393–399.

Austic, R.E. (1985) Feeding poultry in hot and cold climates. In Yousef, M.K. (ed.) *Stress Physiology in Livestock*, Vol. 3, Poultry. CRC Press, Boca Raton, Florida, pp. 123–136.

Baghel, R.P.S. and Pradhan, K. (1989a) Energy, protein and limiting amino acid requirement of broilers in their different phases of growth during hot-humid season. *Indian Journal of Animal Science* 59, 1467–1473.

Baghel, R.P.S. and Pradhan, K. (1989b) Studies on energy and protein requirement of broilers during hot-humid season at fixed level of limiting amino acids. *Indian Journal of Poultry Science* 24, 127–132.

Balnave, D. (1989) Mineral in drinking water and poultry production. *Monsanto Nutrition Update* 7(3), 1–8.

Barton, T.L. (1989) Effects of water quality on broiler performance. *Zootecnica International*, March, 44–46.

Belay, T. and Teeter, R.G. (1993) Broiler water balance and thermobalance during thermoneutral and high ambient temperature exposure. *Poultry Science* 72, 116–124.

Bertechini, A.G., Rostagno, H.S., Fonseca, J.B. and Oliveira, A.I.G. (1991) Effects of environmental temperature and physical form of diet on performance and carcass quality of broiler fowls. *Revista da Sociedade Brasileira de Zootecnia* 20, 257–256.

Boshouwers, F.M.G. and Nicaise, E. (1992) Light quality affects physical activity, energy expenditure and growth of broilers. *Proceedings 19th World's Poultry Congress*, Vol. 3, pp. 178–181.

Botte, W.G. and Harrison, P.C. (1985) Effect of carbonated water on growth performance of cockerels subjected to constant and cyclic heat stress temperatures. *Poultry Science* 64, 1285–1292.

Burger, R.E. (1988) Bird death at high temperatures: is there anything we can do? *California Poultry Newsletter*, April, 6–7.

Buys, S.B. and Rasmussen, R.M. (1978) Heat stress mortality in nicarbazine-fed chickens. *Journal of South African Veterinary Association* 49, 127–131.

Cahaner, A. and Leenstra, F.R. (1992) Effects of high temperature on growth and efficiency of male and female broilers from lines selected for high weight gain, favorable feed conversion and high or low fat content. *Poultry Science* 71, 1237–1250.

Casey, J.M. (1983) White wash formula for the poultry house roof. *Poultry Tips*, P.S. 1. Cooperative Extension Service, University of Georgia, Athens, Georgia, USA.

Christmas, R.B. (1993) The performance of spring and summer-reared broilers as affected by percision beak trimming at seven days of age. *Poultry Science* 72, 2358–2360.

Chwalibog, A. (1990) Heat production, performance and body composition in chickens exposed to short time high temperature. *Archives fur Geflugelkunde* 54, 167–172.

Chwalibog, A. and Eggum, B.O. (1989) Effect of temperature on performance, heat production, evaporative heat loss, and body composition in chickens. *Archiv fuer Geflugelkunde* 53, 179–184.

Cier, D., Rimsky, Y., Rand, N., Polishuk, O., Gur, N., Benshoshan, A., Frisch, Y. and BenMoshe, A. (1992a) The effects of supplementing ascorbic acid on broiler performance under summer conditions. *Proceedings 19th World's Poultry Congress*, Vol. 1, pp. 586–589.

Cier, D., Rimsky, Y., Rand, N., Polishuk, O., Frisch, Y., Gur, N., Benshoshan, A. and BenMoshe, A. (1992b) The effects of different dietary levels of available phosphorus on broiler performance. *Proceedings 19th World's Poultry Congress*, Vol. 2, pp. 264–265.

Coelho, M.B. (1991) Effects of processing and storage on vitamin stability. *Feed International*, December, 39–45.

Dale, N.M. and Fuller, H.L. (1979) Effects of diet composition on feed intake and growth of chicks under heat stress. I. Dietary fat levels. *Poultry Science* 58, 1529–1534.

Dale, N.M. and Fuller, H.L. (1980) Effect of diet composition on feed intake and growth of chicks under heat stress. II. Constant versus cycling temperatures. *Poultry Science* 59, 1434–1440.

Damron, B.L. and Johnson, W.L. (1985) Relation of dietary sodium chloride to chick performance and water intake. *Nutrition Reports International* 31, 805–810.

Deaton, J.W., Reece, F.N. and Lott, B.D. (1984) Effect of differing temperature cycles on broiler performance. *Poultry Science* 63, 612–615.

Decuypere, E., Buyse, J., VanIsterdael, J., Michels, H. and Hermans, A. (1992) Growth, feed conversion and carcass quality in broiler chickens in hot and humid tropical conditions. *Proceedings 19th World's Poultry Congress*, Vol. 2, pp. 97–100.

Diab, M.F., Husseini, M.D. and Salman, A.J. (1981) Effect of thermal stress, lighting and feeding regimens on performance of broilers. *Poultry Science* 60, 1464 (Abstract).

El-Husseiny, O. and Creger, C.R. (1980) The effect of ambient temperature on carcass energy gain in chickens. *Poultry Science* 59, 2307-2311.

Ernst, R.A., Weathers, W.W. and Smith, J. (1984) Effects of heat stress on day-old broiler chicks. *Poultry Science* 63, 1719-1721.

Faber, H.V. (1964) Stress and general adaptation syndrome in poultry. *World's Poultry Science Journal* 20, 175-182.

Farrell, D.J. and Swain, S. (1977a) Effects of temperature treatments on the heat production of starving chickens. *British Poultry Science* 18, 725-734.

Farrell, D.J. and Swain, S. (1977b) Effects of temperature treatments on the energy and nitrogen metabolism of fed chickens. *British Poultry Science* 18, 735-743.

Fattori, T.R., Mather, F.B. and Hilderbrand, P.E. (1990) Methodology for partitioning poultry producers into recommendation domains. *Agricultural systems*, 32, 197-205.

Fuller, H.L. (1978) The extra value of fat and reduced heat increment. *Proceedings of the Florida Nutrition Conference*, pp. 21-35.

Fuller, H.L. and Dale, N.M. (1979) Effect of diet on heat stress in broilers. *Proceedings Georgia Nutrition Conference*, University of Georgia, Athens, pp. 56-60.

Fuller, H.L. and Rendon, M. (1977) Energetic efficiency of different dietary fats for growth of young chicks. *Poultry Science* 56, 549-557.

Garlich, J.D. and McCormick, C.C. (1981) Interrelationship between environmental temperature and nutritional status of chicks. *Proceedings of the Federation of American Societies for Experimental Biology* 40, 73-76.

Geraert, P.A., Guillaumin, S. and Leclercq, B. (1992) Effect of high ambient temperature on growth, body composition and energy metabolism of genetically lean and fat male chickens. *Proceedings 19th World Poultry Congress*, Vol. 2, pp. 109-110.

Gorman, I. (1992) Dietary mineral supplementation of broilers at high temperature. *Proceedings 19th World's Poultry Congress*, Vol. 3, p. 651.

Harris, G.C., Nelson, G.S., Seay, R.L. and Dodgen, W.H. (1975) Effects of drinking water temperature on broiler performance. *Poultry Science* 54, 775-779.

Harter-Dennis, J.M. and Pescatore, A.J. (1986) Effect of beak trimming regimen on broiler performance. *Poultry Science* 65, 1510-1515.

Hayashi, K., Kaneda, S., Otsuka, A. and Tomita, Y. (1992) Effects of ambient temperature and thyroxine on protein turnover and oxygen consumption in chicken skeletal muscle. *Proceedings 19th World's Poultry Congress*, Vol. 2, pp. 93-96.

Hoffman, L. (1991) Energy metabolism of growing broiler chickens kept in groups in relation to environmental temperature. 1. Feed intake, heat production, and energy utilization. *Archives of Animal Nutrition* 41, 245-255.

Hoffman, L., Schiemann, R. and Klein, M. (1991) Energy metabolism of growing broiler chickens kept in groups in relation to environmental temperature. *Archives of Animal Nutrition* 41, 167-181.

Howlider, M.A.R. and Rose, S.P. (1989) Rearing temperature and the meat yield of broilers. *British Poultry Science* 30, 61-67.

Hulan, H.W. and Proudfoot, F.G. (1987) Effects of light source, ambient temperature and dietary energy source on the general performance and incidence of leg abnormalities of roaster chickens. *Poultry Science* 66, 645–651.

Hurwitz, S., Weiselberg, N., Eisner, U., Bartov, I., Riesenfield, G., Shareit, M., Nir, A. and Bornstein, S. (1980) The energy requirements and performance of growing chickens and turkeys as affected by environmental temperature. *Poultry Science* 59, 2290–2299.

Ilian, M.A., Diab, M.F., Husseini, M.D. and Salman, A.J. (1982) Effects of brackish water utilization by broilers and growing pullets on performance. *Poultry Science* 60, 2374–2379.

Keshavarz, K. and McDougal, L.R. (1981) Influence of anticoccidial drugs on losses of broiler chickens from heat stress and coccidiosis. *Poultry Science* 60, 2423–2426.

Kleiber, M. and Dougherty, J.E. (1934) The influence of environmental temperature on the utilization of food energy in baby chicks. *Journal of General Physiology* 17, 701–726.

Koh, M.T., Wei, H.W. and Shen, T.F. (1989) The effects of environmental temperature on protein and energy requirements of broilers. *Journal of Taiwan Livestock Research* 22, 23–41.

Leenstra, F. and Cahaner, A. (1992) Effects of low and high temperature on slaughter yield of broilers from lines selected for high weight gain, favorable feed conversion and high or low fat content. *Poultry Science* 71, 1994–2006.

Lott, B.D. (1991) The effect of feed intake on body temperature and water consumption of male broilers during heat exposure. *Poultry Science* 70, 756–759.

Lyle, G.R. and Moreng, R.E. (1968) Elevated environmental temperature and duration of post-exposure ascorbic acid administration. *Poultry Science* 47, 410–417.

McCormick, C.C., Garlich, J.D. and Edens, F.W. (1979) Fasting and diet affect the tolerance of young chickens exposed to acute heat stress. *Journal of Nutrition* 109, 1089–1097.

McCormick, C.C., Garlich, J.D. and Edens, F.W. (1980) Phosphorus nutrition and fasting: interrelated factors which affect survival of young chickens exposed to high ambient temperatures. *Journal of Nutrition* 110, 784–790.

McDougal, L.R. and McQuinston, T.E. (1980) Mortality in heat stress in broiler chickens influenced by anticoccidial drugs. *Poultry Science* 59, 2421–2425.

McDowell, R.E. (1972) *Improvement of Livestock Production in Warm Climates*. W.H. Freeman, San Fransisco, 110 pp.

Macy, L.B., Harris, G.C., Delee, J.A., Waldroup, P.W., Izat, A.L., Gwyther, M.J. and Eoff, H.J. (1990) Effects of feeding Lasalocid on performance of broilers in moderate and hot temperature regimens. *Poultry Science* 69, 1265–1270.

May, J.D. and Lott, B.D. (1992) Feed and water consumption patterns of broilers at high environmental temperature. *Poultry Science* 71, 331–336.

May, J.D., Lott, B.D. and Deaton, J.W. (1990) The effect of light and environmental temperature on broiler digestive tract contents after feed withdrawal. *Poultry Science* 69, 1681–1684.

Morrison, W.D., Braithwaite, L.A. and Leeson, S. (1988) Report of a survey of poultry heat stress losses during the summer of 1988. Unpublished report from the Department of Animal and Poultry Science, University of Guelph, Guelph, Ontario, Canada.

Murphy, D.W. (1988) Non-nutritional nutrition effects. *Proceedings, Monsanto Technical Symposium*, Fresno, California, pp. 49–55.

Mutalib, A. (1990) How to reduce water vaccination failures. *Poultry Digest*, March, 14–16.

Nakamura, Y., Aoyagi, Y. and Nakaya, T. (1992) Effect of ascorbic acid on growth and ascorbic acid levels of chicks exposed to high ambient temperature. *Japanese Poultry Science* 29, 41–46.

Navahari, D. and Jayaprasad, I.A. (1992) Influence of season, floor type and space on broiler performance in humid tropics. *Proceedings 19th World's Poultry Congress*, Vol. 2, p. 138.

Ngian, M.F. (1982) Night feeding of broilers optimizes feed efficiency. *Poultry International*, November, 48.

Nir, I. (1992) Optimization of poultry diets in hot climates. *Proceedings 19th World's Poultry Congress*, Vol. 2, pp. 71–76.

Njokn, P.C. (1984) The effect of ascorbic acid supplementation on broiler performance in a tropical environment. *Poultry Science* 63 (Suppl. 156).

North, M.O. and Bell, D.D. (1990) *Commercial Chicken Production Manual*, 4th edn, 457 pp.

Olson, D.M., Sunde, M.L. and Bird, H.R. (1972) The effect of temperature on ME determination and utilization by the growing chick. *Poultry Science* 51, 1915–1922.

Orban, J.I. and Roland, D.A. (1990) Response of four broiler strains to dietary phosphorus above and below the requirement when brooded at two temperatures. *Poultry Science* 69, 449–455.

Osman, A.M.K., Tawfik, E.S., Klein, F.W. and Hebeler, W. (1989) Effect of environmental temperature on growth, carcass traits, and meat quality of broilers of both sexes and different ages. *Archiv fur Geflugelkunde* 53, 163–175.

Polin, D., Wynosky, E.R. and Porter, C.C. (1962) Amprolium: influence of egg yolk thiamine concentration on chick embryo mortality. *Proceedings Society of Experimental Biology and Medicine*, 110, 844–848.

Pope, D.L. (1960) Nutrition and environmental studies with broilers. *Proceedings University of Maryland Nutrition Conference for Feed Manufacturers*, pp. 48–51.

Rajmane, B.V. and Ranade, A.S. (1992) Remedial measures to control high mortality during, summer season in tropical countries. *Proceedings 19th Worlds Poultry Congress*, Vol. 1, pp. 343–345.

Reece, F.N., Deaton, J.W. and Harwood, F.W. (1976) Effect of roof insulation on the performance of broiler chickens reared under high temperature conditions. *Poultry Science* 55, 395–398.

Reece, F.N., Deaton, J.W. and Kubena, L.F. (1972) Effects of high temperature and humidity on heat prostration of broiler chickens. *Poultry Science* 51, 2021–2025.

Reilly, W.M., Koelkebeck, K.W. and Harrison, P.C. (1991) Performance

evaluation of heat stressed commercial broilers provided water cooled floor perches. *Poultry Science* 70, 1699-1703.
Renden, J.A., Bilgili, S.F. and Kincaid, S.A. (1992) Live performance and carcass yield of broiler strain crosses provided either 16 or 23 hours of light per day. *Poultry Science* 71, 1427-1435.
Smith, M.O. (1993) Nutrient content of carcass parts from broilers reared under cycling high temperatures. *Poultry Science* 72, 2166-2171.
Smith, M.O. (1994) Effects of electrolytes and lighting regimen on growth of heat-distressed broilers. *Poultry Science* 73, 350-353.
Smith, M.O. and Teeter, R.G. (1987) Potassium balance of the 5 to 8 week-old broiler exposed to constant heat or cycling high temperature stress and the effects of supplemental potassium chloride on body weight gain and feed efficiency. *Poultry Science* 66, 487-492.
Smith, M.O. and Teeter, R.G. (1988) Practical application of potassium chloride and fasting during naturally occurring summer heat stress. *Poultry Science* 67 (Suppl. 1), 36.
Sonaiya, E.B. (1988) Fatty acid composition of broiler abdominal fat as influenced by temperature, diet, age and sex. *British Poultry Science* 29, 589-595.
Sonaiya, E.B. (1989) Effects of environmental temperature, dietary energy, sex, and age on nitrogen and energy retention on the edible carcass of broilers. *British Poultry Science* 30, 735-745.
Sonaiya, E.B., Ristic, M. and Klein, F.W. (1990) Effect of environmental temperature, dietary energy, age and sex on broiler carcass portions and palatability. *British Poultry Science* 31, 121-128.
Tawfik, E.S., Osman, A.M.A., Ristic, M., Hebeler, W. and Klein, F.W. (1989) Einflus der Stalltemperatur auf Mastleistung, Schlachtkorperwert und Fleischbeschaffeaheit von Broiler Unterschiedlichen alters und geschlechts. 2. Mitteilung Schlachtkorperwert. *Archiv fur Geflugelkunde* 53, 235-244.
Tawfik, E.S., Osman, A.M.A., Hebeler, W., Ristic, M. and Freudenreich, P. (1992) Effect of environmental temperature, sex and fattening period on amino acid composition of breast meat of broilers. *Archiv fur Geflugelkunde* 56, 201-205.
Teeter, R.G. and Smith, M.O. (1986) High chronic ambient temperature stress effects on broiler acid-base balance and their response to supplemental NH_4Cl, KCl and K_2CO_3. *Poultry Science* 65, 1777-1781.
Teeter, R.G., Smith, M.O., Owens, F.N., and Arp, S.C. (1985) Chronic heat stress and respiratory alkalosis: occurrence and treatment in broiler chicks. *Poultry Science* 64, 1060-1064.
Thornton, P.A. (1961) Increased environmental temperature influences on ascorbic acid activity in the domestic fowl. *Proceedings of the Federation of American Societies for Experimental Biology* 20, 210A.
Trout, J.M., Bierlmaier, S.J. and Mashaly, M.M. (1988) Effect of beak trimming on performance of broiler chicks. *Poultry Science* 67 (Suppl. 1), 166.
Vahl, H.A. and Stappers, H.P. (1992) Effect of lower sodium levels in broiler diets. *Proceedings 19th World's Poultry Congress*, Vol. 1, pp. 598-602.
Van Kampen, M. (1976) Activity and energy expenditure in laying hens: the

energy cost of eating and posture. *Journal of Agricultural Science, Cambridge*, 87, 85–88.

Van Kampen, M. (1981) Water balance of colostomized hens at different ambient tempratures. *British Poultry Science* 22, 17–23.

Vo, K.V. and Boone, M.A. (1977) Effect of water availability on hen survival time under high temperature stress. *Poultry Science* 56, 375–377.

Waldroup, P.W. (1982) Influence of environmental temperature on protein and amino acid needs of poultry. *Federation Proceedings* 41, 2821–2823.

Waldroup, P.W., Mitchell, R.J., Payne, J.R. and Hazen, K.R. (1976) Performance of chicks fed diets formulated to minimize excess levels of essential amino acids. *Poultry Science* 55, 243–253.

Weaver, W.D. (1992) Broiler housing in the USA. *Proceedings 19th World's Poultry Congress*, Vol. 3, pp. 161–163.

Wideman, R.F., Ford, B.C., May, J.D. and Lott, B.D. (1994) Acute heat acclimatization and kidney function in broilers. *Poultry Science* 73, 75–88.

Yaghi, A. and Daghir, N.J. (1985) Protein requirement for broiler starter, grower and finisher rations. *Poultry Science* 64 (Suppl.), 201.

9

Replacement Pullet and Layer Feeding and Management in Hot Climates

N.J. Daghir
Faculty of Agricultural Sciences, UAE University, Al-Ain, UAE.

Introduction	219
Replacement Pullets	220
Body-weights of replacement pullets	220
Feeding the replacement pullet	221
Water consumption	225
Acclimatization	225
Replacement pullet management	226
Layers	228
Feeding the laying hen	228
Water quality and quantity for layers	237
Acclimatization	239
Effects of temperature on egg quality	240
Layer management practices	241
Conclusions and Recommendations	245
References	247

Introduction

This chapter covers selected aspects of feeding and management of the replacement pullet as well as the laying hen in hot climates. It presents a combination of nutrition and management strategies because the author believes that this is the most adequate approach to overcome problems of heat stress in laying stock. The chapter starts with a section on pullet body-weights at housing, since this is probably the most challenging pullet production practice in hot climates. Feeding the young pullet and the laying hen, acclimatization of both growing pullets and layers, lighting programmes, water quality and quantity and several other management tips for hot climates have been covered.

A section has been included in this chapter on the detrimental effects of high temperature on egg weight, shell quality and interior egg quality, because it is fairly well documented that these effects on egg quality are somewhat independent of the effect of reduced feed intake at high temperature. This has previously been discussed in Chapter 5 where it was pointed out that the detrimental effects of heat stress on performance can be divided into those that are due to high temperature *per se* and those due to reduced feed intake.

Replacement Pullets

Body-weights of replacement pullets

The success of a table egg production enterprise depends, to a very large extent, on the quality of pullets at housing. A quality pullet can be defined as one of optimum body-weight and condition required for optimum performance in the laying house. Jensen (1977) reported that a major problem in rearing pullets in the southern USA was obtaining acceptable body-weights at sexual maturity for pullets reared in the hot months of the year. A study by Bell (1987a) on over 100 commercial flocks in the USA showed that April, May and June hatches had the lowest production per hen housed. He indicated that this was probably due to lighter pullets grown in hot weather. Payne (1966) reared pullets from 6 to 21 weeks of age at mean temperatures of either 20 or 33°C. Birds reared at 33°C were 118 g lighter at 21 weeks of age and their eggs were consistently smaller throughout the laying period. Stockland and Blaylock (1974), in a similar study, observed a difference of 130 g in body-weight between birds reared at 29.4°C and those reared at 18.3°C. Vo *et al.* (1978) reared Leghorn pullets at constant temperatures of 21, 29 and 35°C. They showed that birds reared at 35°C weighed 20–30% less than those reared at 21°C. Vo *et al.* (1980) also demonstrated that sexual maturity was significantly delayed for pullets reared at 35°C as compared with those reared at 21 or 29°C. Escalante *et al.* (1988) studied the effect of body-weight at 18 weeks of age on performance of white Leghorn pullets in Cuba. Egg production from 21 to 66 weeks was higher for the heavier-weight birds. Body-weight at 18 weeks, however, had no effect on age at sexual maturity or age at 50% production.

Considering the above studies, it is therefore recommended that pullets should be weighed frequently, starting as early as 4 weeks of age, and that weight and uniformity be watched closely during hot weather. Flock uniformity is very important in obtaining optimum performance and the greatest profitability. In situations where uniformity is a problem, growers should sort out all the small birds and pen them separately at about

5 weeks. One practice used when it is known that pullets are going to be grown during warm weather is to start about 10% fewer pullets for a given space as compared with normal temperature conditions. The result will mean increased floor space per pullet along with more water and feeder space. It is important to take body-weight measurements every 2 weeks, from 6 weeks of age and on to housing, in order to determine if the pullets are growing satisfactorily during this critical period.

In hot weather, it is desirable to get pullets as heavy as possible before the onset of egg production. It is therefore recommended that pullets at housing be above the breeders' recommended target weight. Since adequate feed consumption is of primary concern in hot climates, heavier hens will consume more feed, which will result in higher peaks and better persistency in production.

Feeding the replacement pullet

It has been shown for many years that house temperature is one of the most important factors affecting feed consumption. There is a change in feed consumption as house temperatures increase or decrease, but the relationship is not constant at various house temperatures. Table 9.1 shows that the percentage change in feed consumption is much larger during hot weather than during cold weather.

The influence of temperature on the nutrient requirements of replacement pullets has not been widely investigated. Stockland and Blaylock (1974) in their study on rearing pullets at 29.4°C and 18.3°C concluded that protein requirement as percentage of diet was increased in a hot environment. McNaughton *et al.* (1977), on the other hand, reported that neither dietary protein nor energy influenced body-weights at 20 weeks of age under high temperature conditions. These workers, however, used a cyclic temperature of 24 to 35°C between 12 and 20 weeks of age for the

Table 9.1. Temperature and feed consumption for growing pullets (from North and Bell, 1990).

Average daytime house temperature		% Change in feed consumption for each 1°F (0.6°C) change in temperature
°F	°C	
90-100	32.2-37.8	3.14
80-90	26.7-32.2	1.99
70-80	21.1-26.7	1.32
60-70	15.6-21.1	0.87
50-60	10.0-15.6	0.55
40-50	4.4-10.0	0.3

duration of the study. They also reported that increasing lysine and methionine plus cystine levels, above National Research Council (NRC) recommendations, did not influence body-weights under high temperature conditions. Henken *et al.* (1983) confirmed this by showing in controlled metabolic studies that protein deposition in the growing pullet was not influenced by temperature and that protein anabolism was relatively independent of environmental temperature.

In general, however, pullet growth can be improved at high temperatures by increasing nutrient density. Leeson and Summers (1981) found that pullets reared at 26°C after brooding did respond to increased nutrient density up to 8 weeks of age in terms of improved weight gains. The same workers (Leeson and Summers, 1989) tested pullets reared at a constant 22°C or at cyclic temperatures of 22 to 32°C and given diets ranging from 2650 to 3150 kcal kg^{-1} with 15–19% crude protein. Increasing the metabolizable energy (ME) of the diet increased body-weight and the effect was most pronounced in the hot environment. Increasing dietary protein increased body-weight initially, but, at 140 days of age, body-weights were not affected by dietary protein. They concluded that, given adequate protein, pullet growth is most responsive to energy intake. Rose and Michie (1986) concluded that high-protein or nutrient-dense rearing feeds increased body-weights of pullets reared under high temperatures and decreased the time taken by the pullets to reach sexual maturity.

Researchers at the University of Florida have studied for many years feeding programmes for raising pullets in hot climates. Douglas and Harms (1982) showed that, when low-protein diets are fed to pullets to be housed in hot weather, birds did not consume sufficient layer feed to gain optimum body-weight and maintain maximum egg weight. Douglas *et al.* (1985) conducted three experiments using a step-down protein programme for commercial pullets. Maximum body-weight was obtained by these workers at 20 weeks when a 20% protein diet was fed from 0 to 8 weeks and 17% from 8 to 12 weeks, followed by reducing the protein level by one percentage point at biweekly intervals. Douglas and Harms (1990) published a similar study on dietary amino acid levels for commercial replacement pullets. On the control diet, winter-reared pullets averaged 1443 g and summer-reared pullets 1322 g at 20 weeks of age, both groups receiving 21% protein starter and 18% protein from 8 to 20 weeks. These authors recommended the use of a step-down, amino acid programme with linear reductions at biweekly intervals of the following ranges: 0.54–0.42% total sulphur amino acids (TSAA) (0.28–0.21% methionine), 0.64–0.48% lysine, 0.18–0.14% tryptophan and 0.96–0.78% arginine. Such a programme would allow the protein source to be gradually lowered and the energy source to be raised, which is needed in most hot weather situations. El-Zubeir and Mohammed (1993) fed commercial egg-type pullets from 10 to 19 weeks of age diets containing 13, 15 or 17% protein and ME of

10, 11 or 12 MJ kg^{-1}, followed by a layer diet containing 16% protein and 11.7 MJ kg^{-1}. At summer temperatures in Sudan ranging between 26 and 45°C, the diet containing 15% protein and 10 MJ kg^{-1} was best for optimal growth, minimum number of days required to reach point of lay and age at 25 and 50% egg production.

Under hot weather conditions, optimum pullet growth cannot be achieved with low-energy diets and, in many cases, high-energy diets containing fat are essential. Although growing rations of 2750–2900 kcal kg^{-1} are adequate under most conditions, during hot weather birds will not eat enough feed and body-weights will be low. Metabolizable energy levels of 3000–3100 kcal kg^{-1} of diet may be necessary to get enough energy into the bird and increased protein is needed to maintain the same calorie to protein ratio, particularly early in the rearing period. The use of a 'broiler type' starter (0–3 weeks) can help in getting early growth. If birds are doing well, then one can change to a normal starter at 21 days. Table 9.2 presents suggested ration specifications for the starter, grower and prelay, while Table 9.3 presents recommended levels for vitamins and trace minerals per tonne of complete feed. Normally, a starter is fed up to 6 weeks of age,

Table 9.2. Ration specifications for chick starter, grower and prelay.

Nutrients	Starter	Grower I	Grower II	Prelay
Protein (%)	19–20	16–17	15–16	16–17
Metabolizable energy				
(kcal kg^{-1})	2850–2950	2750–2850	2700–2800	2750–2850
(kcal lb^{-1})	1295–1340	1250–1295	1225–1275	1250–1295
Megajoules (MJ kg^{-1})	11.9–12.3	11.5–11.9	11.3–11.7	11.5–11.9
Calcium (%)	0.90	1.00	1.00	2.00
Available phosphorus (%)	0.45	0.40	0.35	0.40
Sodium (%)	0.18	0.18	0.18	0.18
Chloride (%)	0.17	0.17	0.17	0.17
Potassium (%)	0.60	0.60	0.60	0.60
Lysine (%)	1.00	0.75	0.70	0.75
Methionine (%)	0.42	0.36	0.34	0.36
Methionine + cysteine (%)	0.72	0.62	0.58	0.65
Tryptophan (%)	0.20	0.17	0.16	0.18
Arginine (%)	1.05	0.90	0.84	0.94
Histidine (%)	0.40	0.34	0.32	0.36
Phenylalanine (%)	0.70	0.65	0.56	0.65
Threonine (%)	0.70	0.60	0.56	0.63
Leucine (%)	1.44	1.22	1.15	1.30
Isoleucine (%)	0.84	0.72	0.67	0.76
Valine (%)	0.86	0.73	0.69	0.77

Table 9.3. Recommended vitamin-trace mineral levels per tonne of complete feed.

		Starter	Grower	Layer
Vitamin A	(IU)	9,500,000	8,500,000	10,000,000
Vitamin D_3	(ICU)	2,000,000	1,000,000	2,200,000
Vitamin E	(IU)	15,000	15,000	15,000
Vitamin K_3	(g)	2	1.5	2
Thiamine	(g)	2.2	1.5	2.2
Riboflavin	(g)	5.0	5.0	6.5
Panthotenic acid	(g)	12.0	10.0	15.0
Niacin	(g)	40.0	30.0	40.0
Pyridoxine	(g)	4.5	3.5	4.5
Biotin	(g)	0.2	0.15	0.2
Folic acid	(g)	0.8	0.5	1
Vitamin B_{12}	(g)	0.012	0.010	0.014
Choline	(g)	1300	1000	1200
Iron	(g)	96	96	96
Copper	(g)	10	10	10
Iodine	(g)	0.4	0.4	0.4
Manganese	(g)	66	66	66
Zinc	(g)	70	70	70
Selenium	(g)	0.15	0.15	0.15

Note: Antioxidants should be added at levels recommended by the manufacturer. Antioxidants are especially important in hot climates and where fats are added to the ration.

and the change from starter to the first grower takes place during the 42–49-day period. If body-weights are not up to those specified by the breeder, then feed change should be postponed and the starter is continued until standard weight for age is reached. Usually, a starter is not fed beyond the tenth week of age. If house temperatures exceed 30°C, an increase in these suggested specifications may be needed. The extent of this increase would depend on the relative decrease in feed intake.

Several methods have been suggested for increasing feed consumption in growing pullets. Feeding crumbled feeds has been shown to help. These feeds are eaten faster and digested more easily than mash. However, the quality of the crumble should be checked for too much dust. Feed consumption can be encouraged by increasing the frequency of feeding and by stirring the feed between feedings. Spraying the feed with water can help encourage eating, but care needs to be taken to avoid mould growth.

Water consumption

Water consumption in growing pullets varies with age, breed, ambient temperature, humidity, density of the feed and several nutritional factors. Ambient temperature is by far the most important factor affecting water intake. Figure 9.1 shows weekly water consumption of growing Leghorn pullets at four house temperatures. It is seen that Leghorn pullets drink about twice as much water per day at 38°C as at 21°C. House temperatures below 21°C, however, do not significantly reduce water intake.

Acclimatization

There are indications that rearing pullets at a relatively high temperature (26-29°C) will help in getting them acclimatized to subsequent high-temperature exposure in the laying house. Although there has been a lot of work on acclimatization of broilers (May et al., 1987) and laying hens (Sykes and Fataftah, 1980; Bohren et al., 1982), very little has been done in this area on replacement pullets. Sykes and Fataftah (1986a) studied the effect of acclimatization of chicks to a hot, dry climate, by evaluating changes in rectal temperature during regular daily exposure to an ambient temperature of 42°C and 26% relative humidity (RH). They reported that young chicks (5-47 days of age) were able to acclimatize and showed a progressive reduction in the rate of increase in rectal temperature over the period of exposure and the ability to survive conditions that initially would

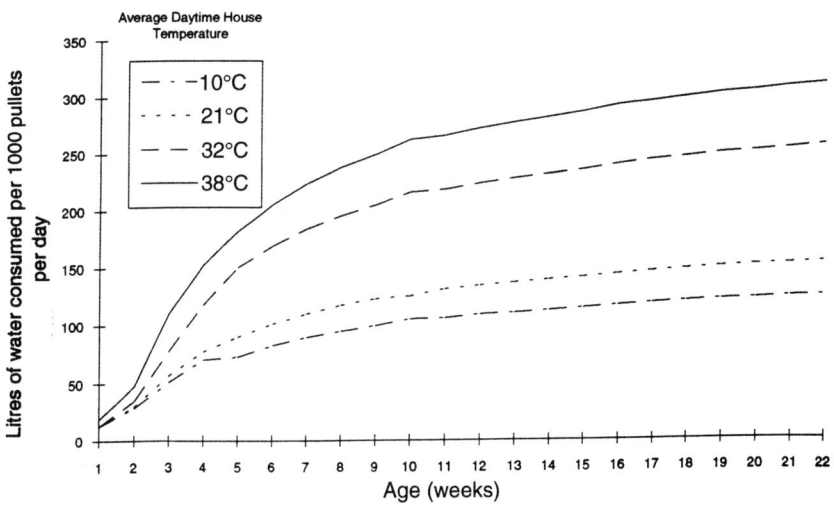

Fig. 9.1. Water consumption of Leghorn growing pullets.

have been fatal. The heat tolerance of the youngest chickens was high and these acclimatized rapidly to an ambient temperature of 42°C. This initial heat tolerance was reduced at 19 days of age and more so at 47 days. Both of these age groups were able to acclimatize to this very hot climate within 5 and 8 days respectively. The question to be raised here is how long will these birds in practice retain this ability to withstand heat stress and what effect will this acclimatization have on body-weight at housing and subsequent laying performance. This area needs further investigation.

Replacement pullet management

Managing the growing pullet is one of the most important activities of a table egg production enterprise. If pullets are grown well, then the prerequisite for success of the enterprise is satisfied. A great deal has been written on the management of growing Leghorn pullets. This section, however, will cover only a few selected aspects that have some bearing on hot climates and that can help in reducing the detrimental effects of heat stress. High environmental temperatures have been shown to create a severe stress on young growing pullets. Pullets will eat less and drink more water than they do at normal temperatures. It is important, therefore, that, in hot climates, adequate feed and water space be provided and additional waterers be used during the very hot periods of the year. This increased water consumption will aggravate litter problems, and ventilation and air movement through the house become more critical. In severe wet litter cases it is recommended to add 2.5 kg of superphosphate to each 10 m^2 of floor space and mix the phosphate with the litter by stirring.

Growing pullets do not perform well at high temperatures. Their growth is reduced, they feather poorly, flocks are not uniform and feed conversion is poor. Therefore, every effort should be made to reduce house temperature and thus reduce heat stress on the bird.

Beak trimming

Beak trimming stress is a likely cause of underweight pullets and its detrimental effects are more severe on birds reared in hot climates than on those in temperate areas. Research has suggested that beak trimming at an early age will reduce this problem (Bell, 1987b). Precision beak trimming at 7–10 days of age is highly recommended in hot regions because it is less stressful than at an older age (6–9 weeks). Retrimming, if necessary, should be done no later than 12 weeks.

Light

The rearing of pullets between latitudes 30° N and 30° S requires special light consideration. These areas present a special problem for the producer. Producers need not only to take account of the amount of natural daylight and the amount of light needed for maximum production, but also to consider adding light during the coolest part of the night to stimulate feed consumption. Intermittent lighting of pullets from 2 to 20 weeks of age has resulted in improved weights (Ernst *et al.*, 1987). Biomittent lighting for pullets has also been recommended by Purina Mills Inc. (1987) in hot weather. They claim that pullets perform better during heat stress on this programme than on standard lighting. They are less underweight due to lessened activity and have better feed utilization. The programme consists of 24 h light during the first week. At 2 weeks of age, light is reduced to 8 h daily. From 3 to 18 weeks, lights are maintained at 8 h daily. During these 8 h, the lights are usually on 15 min h^{-1} (15 min light and 45 min darkness) (15 L : 45 D). The exception is the last hour of each day when the pattern is 15 L : 30 D : 15 L. One recommended programme for pullets in light-tight houses is shown in Table 9.4.

The light phase of short-day programmes should be started early in the day to encourage feed consumption. One hour light in the dark period for open-sided houses can be beneficial to feed consumption.

Handling pullets

Pullets should not be serviced (vaccinated, beak-trimmed, etc.) or moved on hot days. Always move pullets on cool days, whenever possible, or, better still, during the night. Moving is less stressful if it is done before pullets reach sexual maturity. Put 30% fewer birds per crate for moving and move birds at least 2 weeks before the first egg is expected. Stress can increase vitamin needs in birds. Nockels (1988) pointed out that vitamin levels, particularly vitamins A, C and E, required for stressed birds are greater than those needed for birds under normal environmental conditions. Furthermore, in hot and humid areas, vitamin stability in feeds is considerably reduced (McGinnis, 1989). It is a good practice, therefore, to

Table 9.4 Programme for pullets in light-tight housing.

Age (weeks)	Programme
1	24 h constant
2	8 h constant
3-18	8 h (15 L : 45 D)

give vitamins for 3 days in the drinking-water prior to moving and electrolytes for another 3 days after moving.

Housing

When the time comes to house pullets, many growers house by body-weight and maturity. This makes it possible to provide the lighter and more immature birds with a separate lighting and feeding programme to speed up their development and maturity.

The watering system in the laying house should be similar to that in the rearing house. This is very critical in hot climates, because, if pullets fail to locate the water immediately after housing, they can be severely stressed and damaged for a long time.

Very little work has been done on methods of housing pullets in hot climates. Owoade and Oduye (1992) tested the use of a slatted floor in an open-sided house for pullet rearing in Nigeria. They were able to significantly reduce rearing mortality and suggested that this was a good option of housing for the prevention of disease during rearing in a hot environment. The environmentally controlled house, however, offers a better means of creating a good environment for rearing pullets in hot climates for large integrated operations. The open-sided house suggested by the above workers may be feasible for small-scale operations in a hot and humid environment.

Layers

Feeding the laying hen

Just as with the growing pullet, house temperature is one of the most important factors affecting feed consumption in laying hens. Sykes (1976) reported that the average decrease in energy intake was 1.6% per degree centigrade as environmental temperature increased above 20°C. Reid (1979) indicated that ME intake declined 2.3% per degree centigrade as environmental temperature increased from 20 to 30°. Above 30°C, both feed intake and egg production were markedly reduced. Smith and Oliver (1972) found that feed intake declines an average of 5% per degree centigrade between 32 and 38°C. Therefore, the value of a 1.5% drop in food intake per degree rise in temperature reported by Austic (1985) is only valid up to 30°C. It is very important, therefore, to monitor daily feed consumption in hot weather to ensure an adequate intake of nutrients on a per bird basis. This is particularly important early in the production cycle. Petersen *et al.* (1988), in a study on the effect of heat stress on performance of hens with different body-weights, reported that a permanent laying stop

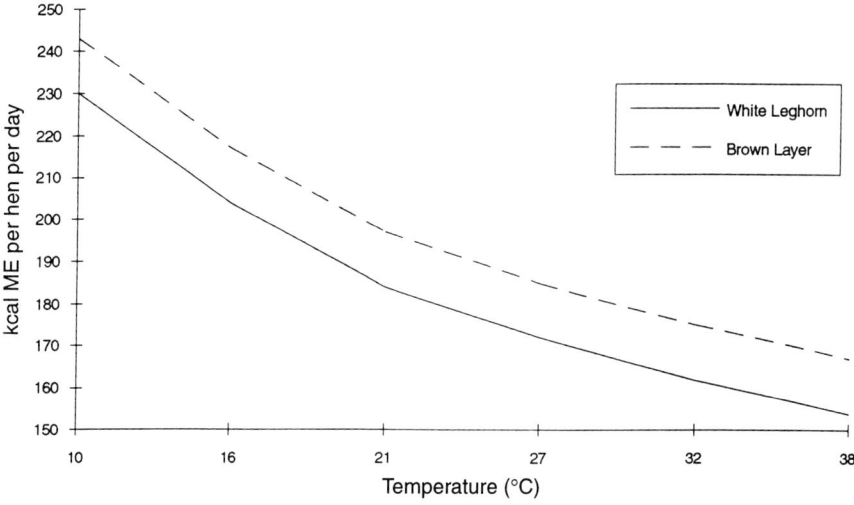

Fig. 9.2. Effect of ambient temperature on the ME requirements for maintenance of laying hens.

is observed in heavy birds and in hens with a low food intake during the first months of laying. Data presented by North and Bell (1990) show that feed consumption is reduced by half when house temperatures increase from 21.1 to 37.8°C. Most of this reduction in feed intake is due to reduced maintenance requirement. Figure 9.2 shows that the maintenance requirement of White Leghorns as well as brown layers is reduced by 30 kcal day^{-1} when ambient temperatures rise from 21 to 38°C.

Energy and protein

It has long been suggested that feed for laying hens should contain more protein in hot weather than in cold weather (Heywang, 1947). Reid and Weber (1973) studied different methionine levels for laying hens at 21 or 32°C and did not detect any limitations in egg production due to sulphur amino acids. They used regression equations to estimate the requirement, which was found to be 498 and 514 mg per hen per day, respectively, at 21 and 32°C. Valencia *et al.* (1980) varied protein levels from 12 to 20% at 21 or 32°C. They concluded that the protein requirement was not affected by temperature. Since the work of Heywang (1947), there has been no clear-cut evidence that the protein and/or amino acid requirements of laying hens are higher under high temperature conditions. Several nutritionists advocate, however, protein- and amino acid-rich programmes for layers in hot climates. The important practice to follow is to adjust density of amino acids in the feed to ensure the same daily intake of these nutrients as that normally consumed at 21°C. DeAndrade *et al.* (1977) fed laying hens under heat stress a diet containing 25% more of all nutrients except energy, which was increased by only 10%. They found that these dietary adjustments overcame most of the detrimental effects of high temperature on pecentage egg production and led to a limited improvement in egg weight. Eggshell quality was not improved by dietary adjustments.

Some workers have raised the hypothesis of a harmful effect of high-protein diets under high temperature conditions (Waldroup *et al.*, 1976). The explanation here is twofold. One is that excess amino acids in the bloodstream may depress food intake because of their effect on the hypothalamus. The second reason is the high heat increment of protein: thus a reduction in protein catabolism would result in a decrease in heat production and help the bird to maintain its energy balance under high temperature conditions. What is being advocated now is a low-protein diet balanced with commercial feed-grade methionine and lysine.

Scott and Balnave (1989) studied the influence of hot and cold temperatures and diet regimen (complete vs. self-select) on feed and nutrient intake and selection and egg mass output. Although pullets maintained under high temperatures consumed less food and nutrients and produced less egg mass, there were no differences between the protein : ME ratio selected by self-select-fed pullets under either temperature treatment. The authors point out that, for those concerned with feeding pullets at high temperatures, their work shows that pullets fed by self-selection are able to consume up to 20 g protein day^{-1} by the third week following sexual maturity, whereas pullets fed the complete diets were only consuming 12 g of protein day^{-1}. NRC (1984) estimates the daily protein requirement for egg-laying hens to be between 16 and 17 g hen^{-1} day^{-1}, while Scott and Balnave (1989) showed actual protein intakes by pullets at hot and cold temperatures to be respectively 19 and 30 g hen^{-1} day^{-1} by 3 weeks after laying the first egg. Austic (1985) proposed that protein and amino acid levels as percentage of diet be increased linearly as environmental temperatures increase from 20 to 30°C. However, beyond that temperature, no further increase is needed since rate of egg production begins to decline.

It has also been observed that energy consumption during the summer months drops significantly in comparison with winter or spring (Daghir, 1973). Energy intake during the summer can be 10–15% lower than during the winter according to work conducted by the author over a period of 3 years in Lebanon. Chawla *et al.*'s (1976) data show a difference in energy consumption of 10–25% between summer and winter in India (Punjab Agriculture University). Results of work at the University of California on 100 commercial Leghorn flocks are illustrated in Fig. 9.3, which shows the

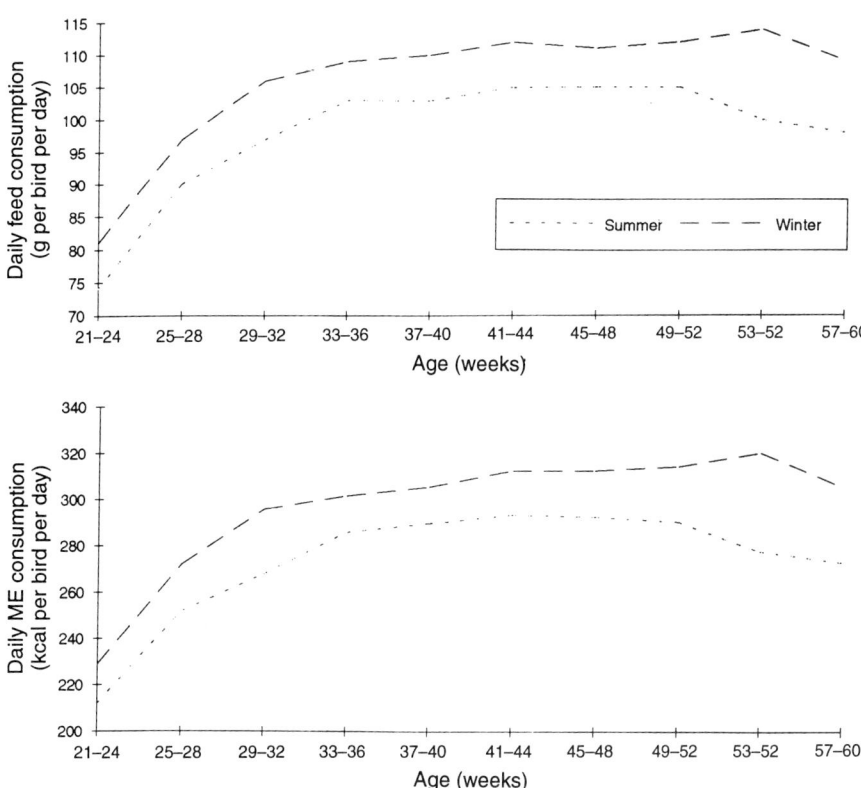

Fig. 9.3. Feed and ME consumption of Leghorn layers in relation to age and season.

effect of season and age on feed and ME consumption. Feed consumption varied over the whole year between 98 and 106 g per bird per day between summer and winter, while ME consumption varied from 274 to 297 kcal per bird per day. This is expected because it has been known for a long time that hens eat mainly to satisfy their energy requirements, and thus energy requirements for maintenance decrease as environmental temperatures rise. We have learned since that time, however, that this relationship holds true only within the zone of thermoneutrality. At very low temperatures, birds tend to overeat and at high temperatures they undereat. Above 30°C, feed intake decreases more rapidly and the hens' energy requirements begin to increase. This increase reflects the body's effort to get rid of the extra heat burden caused by high temperature. Figure 9.4 shows the relationship between temperature, energy content of the diet and feed consumed per 100 hens per day. At high temperature, the

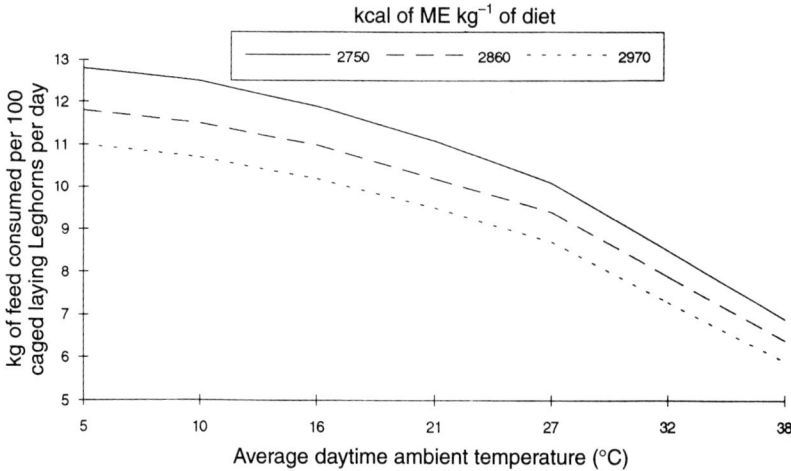

Fig. 9.4. Relationship between ambient temperature, energy content of the diet, and kg feed consumed per 100 caged laying Leghorns per day.

difference in feed consumption is greater for each degree change in temperature than at low ambient temperatures. Thus hens could become energy-deficient when subjected to high temperature. In practice, such a deficiency could be aggravated by increasing protein at the expense of energy in the diet. Protein has a relatively high heat increment and thus increases the heat burden of the laying hen. Realizing this energy shortage under high temperature conditions, nutritionists have tested high-energy diets during the summer and found that the addition of fat stimulates feed and ME consumption (Reid, 1979). The use of high-energy layer rations is now fairly common practice in hot areas, particularly during the early production period (20–30 weeks) when feed consumption is still low. We have shown that the addition of 5% fat not only improves feed intake at high temperature (30°C) but also improves egg weight and shell thickness (Daghir, 1987).

Ramlah and Sarinah (1992) in experiments with laying hens in Malaysia found that hens had higher performance in terms of egg production, egg mass and better feed efficiency when offered choices of feeds that included diets with supplemented fat and with a higher level of both protein and energy, when compared with choice-fed hens with lower levels of both protein and energy in their diets. They observed that hens under their conditions tended to select more of the feed with supplemental fat. Nir (1992) cautioned that, with the use of dietary fat in hot climates, special care needs to be taken to prevent its oxidation.

He suggested the use of saturated rather than polyunsaturated fats under those conditions.

Marsden and Morris (1987) summarized data of 30 published experiments in an attempt to examine the relationship between environmental temperature and ME intake, egg output and body-weight change in laying pullets. They concluded that the relationship between temperature and ME intake is curvilinear, with food intake declining more steeply as ambient temperature approaches body temperature. They calculated that the energy available for production is at a maximum at 23°C for brown birds and at 24°C for White Leghorns. Although gross energetic efficiency is at a maximum at 30°C, egg output is reduced at this temperature. The authors concluded that the optimum operating temperature for laying houses will depend upon the local cost of modifying ambient temperature and on the cost of supplying diets of appropriate protein content. Marsden *et al.* (1987), in two experiments with laying hens, found that it was not possible to maintain egg weight or egg output at 30°C by feeding a high-energy, high-protein diet. Peguri and Coon (1991) in a study on the effect of temperature and dietary energy on layer performance found that feed intake was 5.9 g lower when dietary ME was increased from 2645 to 2976 kcal kg^{-1} and was 21.7 g lower when temperatures were increased from 16.1 to 31.1°C. Egg production was not affected by either temperature or dietary energy density. Egg weight increased 0.78 g with increases in dietary energy from 2675 to 2976 kcal kg^{-1} and decreased 3.18 g when temperatures were increased from 16.1 to 31.1°C.

Researchers are looking at other means of increasing energy intake under high temperature conditions. Picard *et al.* (1987) suggested using a low-calcium diet along with a marine shell source fed separately and with free choice. At 33°C there was a clear-cut improvement in energy intake and in calcium intake and a significant increase in egg output and shell weight. Uzu (1989) also recommeded separate calcium feeding in the form of oyster shell, offered in the afternoon, along with a low calcium level in the diet. There is a tendency in hot regions of the world to give excessively high levels of calcium and this has a negative effect on feed intake not only because of a physiological effect on appetite, but also because of reduced palatability of the diet. This has been confirmed by Mohammed and Mohammed (1991) in a study on commercial layers in a hot tropical environment; they reported that feed intake was significantly reduced when dietary calcium increased from 2.5 to 3.9%. Devegowda (1992) reported that, in India, separate calcium feeding along with reducing the calcium level in the diet to about 2% improved feed intake and increased egg production and shell quality. Another suggestion for improving feed intake and thus energy intake in hot climates is pelleting of the diet. Since, in many hot regions, low-energy, high-fibre feedstuffs are used, Picard (1985) suggested that feeding pellets rather than mash should be considered for laying hens. Devegowda (1992) also recommended feeding pellets where low-energy and high-fibre diets are used.

Minerals and vitamins

Some aspects of laying hen mineral nutrition have been covered in Chapter 5. As for vitamin nutrition, apart from the work on ascorbic acid there has been very little research done on laying hens in hot climates. Heywang (1952), in a study on the level of vitamin A in the diet of laying and breeding chickens during hot weather, indicated that the absolute requirement of laying hens for vitamin A is not altered by high temperature. Thornton and Moreng (1959) were among the first to report the improvement in performance of laying hens receiving vitamin C in a hot environment. Although there is no evidence that vitamin C is an essential nutrient for poultry maintained under moderate environmental temperatures, numerous early reports have shown beneficial effects on laying hens receiving this vitamin in hot climates (Perek and Kendler, 1963; Kechik and Sykes, 1974). Other more recent studies have also shown that vitamin C helps laying hens maintain adequate performance under high temperature conditions. Cheng *et al.* (1990) report that laying hen mortality due to heat stress can be significantly reduced by using as little as 100 p.p.m. ascorbic acid. Manner *et al.* (1991) studied the influence of different sources of vitamin C on performance of layers at varying environmental temperatures. Three forms of vitamin C were used, namely crystalline ascorbic acid, protected ascorbic acid (Cuxavit C 50) or phosphate–ascorbic acid ester. Performance and eggshell quality of treated hens improved only at 34°C and not at 20°C. The best results were obtained with the protected ascorbic acid and the phosphate ester. The authors suggested that the wide variations in the recommendations for supplementing layer diets with ascorbic acid might be due to higher losses during storage. Morrison *et al.* (1988) reported the results of a survey conducted on 162 layer flocks that either received or did not receive a vitamin and electrolyte water additive during heat stress. Flocks receiving water additives experienced a smaller drop in egg production during heat stress and this drop in production was shorter in duration than that of those with no additives. Balnave and Zhang (1992b) showed that dietary supplementation with ascorbic acid is effective in preventing a decline in eggshell quality when laying hens are given saline drinking-water. Shell quality characteristics were maintained at control values with a dietary ascorbid acid supplement of 2 g kg^{-1} diet. Results of this study indicate that supplementation of the diet with ascorbid acid prevents the decline in eggshell quality caused by saline drinking-water.

The effects of dietary sodium zeolite (Ethacal) on poultry have been investigated extensively. Because of its high ion selectivity for calcium, a great deal of research has been done on the effects of this product on calcium and phosphorus utilization, eggshell quality and bone developement. Sodium zeolite has been shown to be beneficial in reducing the effects of heat stress on laying hens (Ingram and Kling, 1987). The exact way in which this product works is unknown but it may be acting as a buffer in the gut and reducing the alkalosis associated with panting.

Feeding Programmes

Since a major problem in rearing pullets in hot climates is obtaining acceptable body-weights at housing, the use of a prelay ration is recommended. During the 2–3 weeks prior to the first egg, the liver and reproductive system increase in size in preparation for egg production. At the same time, calcium reserves are being built up to meet the future demands for shell formation. Table 9.2 gives the specifications for a prelay ration, which is usually similar to a layer ration except for a 2.0–2.5% level of total calcium. Such a ration is usually fed until 5% production is reached and helps pullets to attain the desired body-weight at this early stage of production.

Phase feeding has become common practice for layer operations all over the world. A layer ration is usually fed when a flock reaches 5% production. The first layer feed is a high-nutrient-density ration to ensure that birds receive the required nutrients for sustained production and early egg size. Table 9.5 gives recommended specifications for a phase-feeding programme consisting of four formulae. Changes from a high-nutrient-density ration to a lower-density ration should be made on the basis of daily egg mass. Daily egg mass output is calculated by multiplying the actual hen-day rate of egg production by the average egg weight in grams. For example, a flock laying at 90% with an average egg weight of 55.6 g has a daily egg mass output of 50 g per bird. Normally, a shift from the first layer ration to the second layer ration is made when daily egg mass reaches a peak and starts declining. The change from the second to the third layer ration is not usually made before daily egg mass is down to 50 g and the change from the third to the fourth layer ration is made when daily egg mass goes below 47 g.

Energy intake of the laying hen is often more limiting than protein or amino acid intake, and this is especially true in warm climates and at onset of production when feed intake is low. The energy level as well as density of all other nutrients in the ration should be adjusted in accordance with actual intake of feed. Energy densities between 2700 kcal kg^{-1} (11.3 MJ kg^{-1}) and 2950 kcal kg^{-1} (12.3 MJ kg^{-1}) are suitable for the different phases of production.

The formulae suggested in Table 9.5 differ in the level of protein and other nutrients. Protein and amino acid requirements are greatest from the onset of production up to peak egg mass. This is the period when

Table 9.5. Ration specifications for the laying period.

Nutrients	Layer I	Layer II	Layer III	Layer IV
Crude protein (%)	18-19	17-18	16-17	15-16
ME (kcal kg^{-1})	2850-2950	2800-2900	2750-2850	2700-2800
(kcal lb^{-1})	1295-1340	1275-1320	1250-1295	1225-1275
Megajoules (MJ kg^{-1})	11.93-12.34	11.72-12.34	11.50-11.93	11.30-11.72
Calcium (%)	3.25	3.50	3.75	4.00
Available phosphorus (%)	0.45	0.43	0.40	0.37
Sodium (%)	0.18	0.18	0.17	0.16
Chloride (%)	0.17	0.17	0.17	0.17
Potassium (%)	0.60	0.60	0.60	0.60
Methionine (%)	0.40	0.38	0.36	0.34
Methionine + cysteine (%)	0.70	0.67	0.63	0.60
Lysine (%)	0.84	0.80	0.75	0.70
Tryptophan (%)	0.20	0.19	0.18	0.17
Threonine (%)	0.70	0.65	0.63	0.59
Leucine (%)	1.40	1.32	1.25	1.18
Isoleucine (%)	0.80	0.76	0.71	0.67
Valine (%)	0.82	0.78	0.73	0.69
Arginine (%)	1.00	0.95	0.89	0.84
Phenylalanine (%)	0.85	0.80	0.76	0.72
Histidine (%)	0.40	0.38	0.36	0.34

body-weight, egg weight and egg numbers are all increasing. The attainment of adequate egg size is one of the problems of the egg industry in hot climates. If satisfactory egg size is not attained with 19% protein in the ration, the levels of the most critical amino acids should be checked, particularly that of methionine. The best way of correcting a methionine limitation is by adding a feed-grade form of methionine. Small egg size can be due to low energy intake as well as low protein and amino acid intake. The use of fat in layer rations has been shown to be helpful not only because of its energy contribution, but also because it can increase the linoleic acid level, which should be over 1.2% in the ration. The importance of calcium has previously been discussed in this chapter as well as in Chapter 5. If separate feeding of calcium, as recommended earlier, is not feasible, then at least 50% of the calcium in the feed should be in granular form rather than all in powder form. Table 9.5 shows that available phosphorus levels vary from 0.45% to 0.37%. It is important not to overfeed phosphorus since it has been shown that excessive levels are detrimental to eggshell quality, particularly in hot climates.

Water quality and quantity for layers

Underground water supplies, which are high in total dissolved solids, are an important source of drinking-water for poultry in many countries of the hot regions. Information on the effects that these types of water have on the performance of laying hens is limited. Balnave (1993) reviewed the literature on the influence of saline drinking-water on eggshell quality and formation. He indicated that saline drinking-water supplied to mature laying hens with contents similar to those found in underground well water has an adverse effect on eggshell quality. Furthermore, these effects on shell quality can occur without adverse effect on egg production, feed intake or egg weight. Sensitivity of hens to saline drinking-water has been shown to increase with age of the hen and with increases in egg weight (Yoselewitz and Balnave, 1989a). Yoselewitz and Balnave (1990) also showed strain differences in sensitivity to saline drinking-water and considerable variation in response of hens within a strain. The incidence of poor shell quality also increases with higher concentrations of sodium chloride in the water (Balnave and Yoselewitz, 1987).

Balnave et al. (1989) suggested that one of the major causes of poor eggshell quality in laying hens receiving sodium chloride in the drinking-water may be a reduced supply of bicarbonate ions to the lumen of the shell gland. Furthermore, Yoselewitz and Balnave (1989b) showed that specific activity of carbonic anhydrase was significantly lower in hens receiving saline drinking-water than in hens receiving regular water. These studies brought Balnave (1993) to conclude that the primary metabolic lesion associated with the poor eggshell quality resulting from intake of saline drinking-water is related to the supply of bicarbonate rather than calcium to the lumen of the shell gland for shell formation.

Balnave (1993) recommends two treatments for this problem, besides using desalination of the drinking-water. One is the use of ascorbic acid supplements in the diet or in the drinking-water and the other is the use of zinc–methionine supplements in the diet. These recommendations are based on studies conducted by Balnave et al. (1991) and Balnave and Zhang (1992a) on ascorbic acid and Moreng et al. (1992) with zinc-methionine. Balnave (1993) cautions, however, that these treatments are preventive rather then remedial in nature and should be applied from the first time sexually mature hens are exposed to saline drinking-water. With the present economically feasible systems of desalination of water, the installation of desalination units on poultry farms with this problem may be the best solution.

The early work on the effects of water temperature on laying hen performance has been reviewed by Adams (1973), who reported that providing drinking-water at 35–40°C has a detrimental effect on performance. Cooling the drinking-water has been shown to improve performance of layers in many tests. North and Bell (1990) presented data from work at the University of Guelph which show that water temperature at 35° compared with water at 3°C reduced egg production by 12% and daily feed intake by 12 g per hen. It is always helpful to raise pullets on the same type of watering system that is going to be used in the laying house. In a study conducted by Odom *et al.* (1985) it was shown that, during periods of high environmental temperature (35°C), birds given carbonated water to drink had a significant relief from the reduction in eggshell quality as a result of a delay in time for the decline to occur.

Laying hens drink twice as much water per day when the temperature is 32°C (90°F) compared with 24°C (75°F). Figure 9.5 shows the effect of house temperature on water consumption of laying hens in cages. It also shows the variation in water consumption throughout the laying period, which is the result of changes in body-weight as well as rate of egg production. Maximum water consumption at all temperatures is shown to occur at 6–7 weeks of production, which coincides with peak production. For cage operations, one cup per cage of up to five commercial layers is recommended for hot climates. Egg production drops when hens are not able to drink enough water. The amount of production loss is proportional to the amount of water not consumed. Savage (1983) proposed certain practices for a closed watering system as well as for a flow-through type waterer to ensure that hens receive enough water during hot weather.

1. For a closed watering system (cup or nipple), in-line water filters should be checked and cleaned and may have to be replaced often. Some wells pull more sediment than others and require that the filters be changed much more often. It is good to have functional water pressure gauges on both sides of the water filter. A 3 to 8 lb differential between incoming and outgoing water pressure should be maintained. Line pressure gauges should be checked often to see that water pressure is maintained. During the hottest part of the day, at least one cup or nipple per line should be checked for pressure at the far end of the laying house. The drinker should be triggered and held in open position for a few seconds to observe the rate of water flow. A cup should fill in 2–3 seconds and a nipple should flow in a steady stream.
2. For a flow-through-type waterer, water troughs should be checked for flow rate at both ends of the house. The amount of waste water at the far end of the water trough should be checked closely during the hot part of the day to ensure that all birds have access to water when they need it most. To ensure that hens have good access to the water, check the spacing above the water trough. Sometimes, sagging feeders or cage fronts restrict the space available for the hens to reach the water. At least monthly, walk slowly down each aisle and observe the hens drinking. You can often find several cages where the hens must struggle to get their heads through a

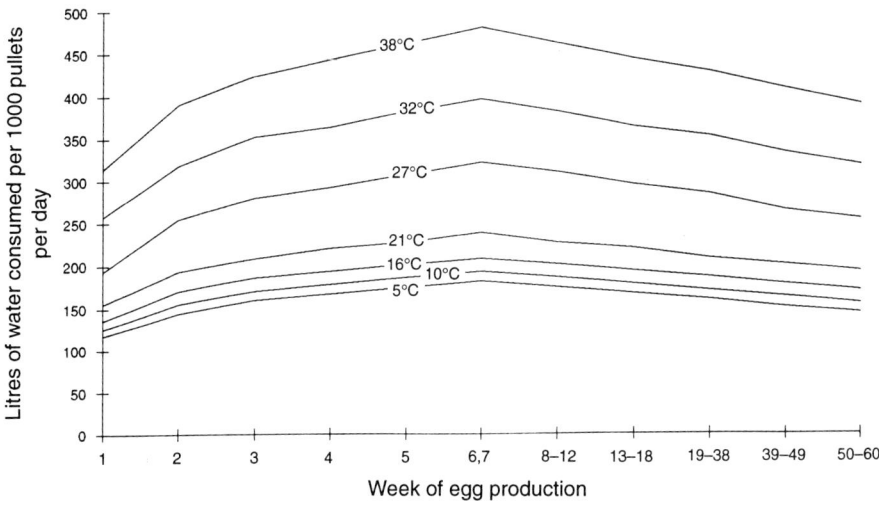

Fig. 9.5. Water consumption of standard laying Leghorn pullets in cages.

tight space to the drinking trough. This is more likely to occur in older and moulted flocks since their combs are larger and they therefore require a greater space above the water troughs. Water troughs should be cleaned more often in hot weather to improve water flow and reduce obstructing feed or algae growth in the troughs.

Acclimatization

Hutchinson and Sykes (1953) were the first to study heat acclimatization in inbred Brown Leghorn hens exposed to a hot, humid climate by measuring changes in body temperatures. Smith and Oliver (1971) reviewed the early literature on laying hen acclimatization to high temperature and suggested that the process is mainly associated with a low basal metabolic rate at high temperatures. Several studies have been published since that time showing that laying hens are able to survive hot lethal conditions if they are previously exposed to a daily intermittent heat stress situation (Hillerman and Wilson, 1955; Deaton *et al.*, 1982; Arad and Marder, 1984; Sykes and Fataftah, 1986b). The increased heat tolerance in those studies was reflected in the lower body temperatures, higher panting rates and decreased evaporative water loss. Strain differences in the response to heat stress were reported by both Arad and Marder (1984) and Sykes and Fataftah (1986b), but it was not clear whether these differences are a reflection of body size and metabolic rate or some other genetically determined character. White laying strains were not always more heat-tolerant than brown strains. Khan (1992) confirmed that a laying hen's acclimatization to heat stress varies with strain of bird and that deep body temperature has a significant influence on egg production. There are variations in laying hens in response to summer stress with respect to deep body temperature. Fataftah (1980) has shown that heat tolerance is somewhat labile and can be increased or decreased considerably by reducing or increasing energy intake. Another factor that has been shown to affect acclimatization is the presence or absence of water. Arad (1983) showed that water depletion reduces heat tolerance. Sykes and Fataftah (1986b), on the other hand, could not demonstrate that hens which remained in positive water balance were any more heat-tolerant than those that did not.

The value of acclimatization should be considered in relation to survival and production during acute heat stress. It may be desirable to allow laying birds to be exposed to temperatures of 29–33 °C before a very hot day is expected even though it may have been possible to keep maximum house temperature below this level.

Effects of temperature on egg quality

There is general agreement among researchers that high ambient temperatures have a negative effect on egg quality. Sauveur and Picard (1987) reviewed the literature on the effects of high ambient temperature on egg quality. Balnave (1988) reviewed factors affecting eggshell calcification and methods of optimizing calcium supply. The adverse effects of high ambient temperature, shell gland lesions, inadequate mineral supplies and bicarbonate ions on eggshell quality were all considered in this review. Many researchers have reported a reduction in egg weight associated with increases in environmental temperature (Payne, 1966; Stockland and Blaylock, 1974; DeAndrade et al., 1977; Vohra et al., 1979). These workers have reported decreases in egg weight ranging from 0.07 to 0.98 g per egg for every 1 °C rise in temperature. Mueller (1961) found that maintaining laying hens in a cycling temperature of 13–32–13 °C resulted in the production of smaller eggs than from hens kept at a constant 13 °C.

Shell quality has also been shown to be reduced as environmental temperatures rise. Harrison and Biellier (1969) found very quick reductions in shell weight as temperatures increase. This was confirmed later by Wilson et al. (1972), DeAndrade et al. (1977) and Wolfenson et al. (1979). Data from the University of California (North and Bell, 1990) show that the decrease in shell thickness during the summer is greater in eggs produced by older birds. The difference in shell thickness between winter and summer was reported to be 10 μm in 50-week-old hens and 17 μm in 60-week-old hens. Shells less than 356 μm in thickness amounted to 47%

in summer versus 30% in winter. This reduced shell thickness at high temperature has been attributed to reduced calcium intake as a result of the reduced feed intake. High temperatures are known to increase respiratory rate, resulting in respiratory alkalosis, which alters the acid–base balance and blood pH. Attempts to improve shell thickness through modification of acid–base balance by adding sodium bicarbonate to feed or giving carbonated water have been found helpful (Odom *et al.*, 1985). Grizzle *et al.* (1992), in a study of the nutritional and environmental factors involved in eggshell quality, suggested that midnight lighting programmes provide a means of supporting eggshell quality in older laying hens during the summer months without a significant reduction in egg production. Eggs in a hot environment should be collected more often and cooled quickly in a properly equipped egg storage room to maintain their internal quality. More care should be taken in handling eggs in hot areas because of the reduced shell quality.

Layer management practices

Several management tips have been recommended by different workers that help to reduce the detrimental effects of high temperature on laying hens. This section will present some of these recommendations.

Ernst (1989) recommends getting the flock up early in the summer to encourage feed consumption before temperatures begin to rise. Feed consumption should be measured weekly and ration compositions adjusted to match intake. Tadtiyanant *et al.* (1991) recommended the use of wet feed to increase feed consumption in laying hens under heat stress. They showed that the use of wet feed gave a 38% increase in dry matter intake when compared with the use of dry feed at 33.3°C.

The removal of comb or wattles from commercial layers is not recommended in high temperature regions because these organs, with their good surface blood supply help in cooling.

It is helpful during hot weather to plan work schedules so that hens are not disturbed during the hot part of the day. Furthermore, the use of low light intensities can help to reduce bird activity and thus heat production (Ernst *et al.*, 1987).

Lighting effects

Nishibri *et al.* (1989) studied diurnal variation in heat production of laying hens at temperatures of 23 and 35°C. Heat production was higher during the light period than during darkness. At 23°C, the differences between the light and dark periods were greater than at 35°C. Body temperature was higher during the light period than during darkness. Li *et al.* (1992) observed that both heat production and abdominal temperature in laying hens declined with decreasing light intensity and this was considered by these authors to result from changes in physical activity. These workers also observed that above 28°C, abdominal temperature increased with both environmental temperature and feed intake, indicating that the heat production associated with feed intake adds to the heat load of high environmental temperature.

Al-Hassani and Al-Naib (1992) tested two lighting regimes (night versus day) and three nutritional treatments (pellet, mash and 4-hour withdrawal of mash) on egg quality in brown layers. Their results indicated that in countries like Iraq, characterized by great diurnal variation in ambient temperature, night lights coupled with pelleted feed and 4-hour feed withdrawal during the day can help to alleviate heat stress in laying hens. Oluyemi and Adebanjo (1979) tested various combinations of lighting and feeding patterns in Nigeria on medium-strain commercial layers. They found that night feeding under a reversed lighting programme (6.0 p.m. to 6.0 a.m.) produced a significantly higher level of egg production than daytime feeding. Nishibri *et al.* (1989), in a study on diurnal variation in heat production of laying hens, reported that body temperature was higher during the light period than during the dark periods of the day. It is therefore recommended that laying hens are not provided with lights during the hot periods of the day, since this will help hold down body temperature. Another practice that helps in holding down body temperature is reducing traffic through the laying house and keeping it at a minimum so that birds will not be unnecessarily excited.

Cage space and shape

Cage space or density significantly affects performance in hot climates. This is because higher densities make ventilation more difficult in those areas. Extensive studies by North Carolina State University have shown an advantage of about 10 eggs per hen when density is reduced from 350 cm^2 to 460 cm^2 per bird in group cages located in fan-ventilated houses. The literature on this subject has been reviewed by Adams and Craig (1985), who found that increasing the density of hens in cages significantly reduced the number of eggs per hen housed, decreased feed intake, increased feed required per dozen eggs and increased mortality. Reducing floor space per hen from an average of 387 cm^2 to 310 cm^2 reduced eggs per hen housed by 16.6, increased mortality by 4.8% and decreased feed consumption by 1.9 g hen^{-1} day^{-1}. Reducing space from 516 cm^2 to 387 cm^2 reduced egg production by 7.8 eggs hen^{-1} housed, increased mortality by 2.8% and decreased feed consumption by 4.3 g hen^{-1} day^{-1}. Teng *et al.* (1990) tested stocking densities of layers in cages with two or three birds per cage in Singapore. Egg production and feed consumption decreased with decreasing floor space per bird. There was a difference of nine eggs per bird during a period of 350 days of production. Their results indicated that birds can be housed at 387 cm^2 per bird in a hot climate, provided sufficient feeding space is available. Gomez Pichardo (1983) evaluated performance of laying hens in Mexico when housed three or four in cages measuring 1350 cm^2, thus allowing 450 or 337.5 cm^2 per bird. They observed that, at the higher space allowance, the weight of eggs produced per m^2 was 28% greater than at the lower space allowance. Rojas Olaiz (1988), also in Mexico, looked at the effect of housing semi-heavy and light breeds housed three or four birds to a cage of 1800 cm^2 during the second production cycle. Birds housed three to a cage had lower mortality, higher egg production and body-weight and higher Haugh unit scores than those housed four to a cage, but the latter produced a higher total egg weight per cage. There were no significant differences between the two groups in egg weight or feed consumption. Egg production was significantly affected by cage shape. Hens in shallow cages produced 5.8 more eggs hen^{-1} housed than those in deep cages. Research has demonstrated that performance of hens in shallow cages (reverse-type cage) is better than in the deep cage. The greater eating space in these shallow cages may be helpful in maintaining feed consumption in hot regions. Ramos *et al.* (1990) did not find a difference in performance between birds housed in deep versus shallow cages. Ulusan and Yildiz (1986) studied the effects of climatic factors in Turkey on hens caged singly versus those caged in groups of four with the same space per hen. For group-caged birds, egg production was significantly correlated with ambient temperature and with relative humidity. For singly caged birds, egg production was significantly correlated with relative humidity. For both groups, egg production was significantly correlated with atmospheric pressure.

External and internal parasites

High ambient temperatures usually increase the population of insects responsible for the transmission of disease. Houseflies (*Musca domestica*) and related species are very active in hot climates and are involved in the transmission of several poultry diseases (Shane, 1988a). Droppings under cages should be allowed to cone and dry out, or be cleaned up completely at intervals of less than 7 days. This is because the fly's life cycle in the heat is about 7 days. Chemical fly treatments are an aid, but do not replace good management. Therefore premises should be kept clean, dry and tidy.

In hot climates, both endo- and ectoparasites can be a problem all year round, particularly in extensive management systems. Warm and humid conditions favour the propagation of endoparasites, including round worms (*Ascaridia* spp. and *Hetarakis* spp.) and hair worms (*Capillaria* spp.) as reported by Shane (1988b). Commercial layer operations located in hot climates which use cages and apply good management and hygiene are less prone to parasitism. A sample of each flock should be inspected for mites and lice each month. Chemical sprays for flocks in cages and powders for floor-housed flocks are usually recommended. Worms can also present a problem and should be brought under control as soon as recognized.

Wet droppings

Layers drink more water when they are on wire than when kept on a litter floor. Furthermore, they drink more when temperatures rise and therefore they eliminate more water through the droppings. Water consumption also increases with increase in production. At 70% production, 1000 pullets in cages drink 201 l day^{-1} at 21°C while at 90% production they drink 239 l day^{-1} at the same temperature.

Wet droppings are affected by several factors, such as relative humidity and temperature of the outside air, relative humidity and temperature inside the house and the amount of air moving through the house. High levels of protein and salt in the ration have been shown to increase moisture in the droppings. Wet droppings rarely occur because of the amount of salt in the feed unless a mixing error has occurred at the feed mill. Sodium levels in the diet have to be very high before one begins seeing significant increases in manure moisture. Sodium levels should be kept in the range of 0.15–0.20% of the diet. The early introduction of the high calcium layer diet has in many cases caused an increase in manure moisture. The effects have been transitory, but in some flocks have persisted for several weeks past peak. The use of a prelay ration has been helpful in these situations. The use of high levels of barley in the ration has been shown to increase water consumption and wet droppings. Using crumbled feeds has also been shown to increase wet droppings.

Some feed additives have been shown to increase moisture in the droppings. Keshavarz and McCormick (1991) reported that the use of sodium aluminium silicate at 0.75% of the diet increased the dropping moisture with or without sodium correction and when tested in both summer and winter. Leenstra *et al.* (1992), in a study on the inheritance of water content and drying characteristics of droppings of laying hens, showed that both of these characteristics can be improved by selection without negative consequences on production traits. Leaky watering devices are a major contributor to excessive water in the droppings collection area. Wet droppings also increase obnoxious odours coming from ammonia and bacterial action in the droppings. Figure 9.6 shows normal manure accumulation underneath laying cages compared to watery manure accumulation.

Fig. 9.6. Left: Normal manure accumulation underneath laying hens in cages. Right: Watery manure underneath laying hens in cages.

Conclusions and Recommendations

1. A major problem of rearing pullets in hot climates is obtaining acceptable body-weights at sexual maturity.
2. Pullets should be weighed frequently, starting as early as 4 weeks of age, and weight and uniformity should be watched closely. When uniformity is a problem, small birds should be sorted out and penned separately at about 5 weeks of age.
3. During the hot months of the year, about 10% fewer pullets should be started for a given space as compared with the cool months. This will mean increased floor space per pullet, along with more water and feeder space.
4. Pullets at housing should be above the breeder's recommended target weight, because heavier hens will consume more feed and this will result in higher peaks and better persistency in production.
5. High environmental temperatures depress feed intake in growing pullets and the percentage decrease in feed consumption varies from 1.3% per 1°C rise in temperature at 21°C to over 3% decrease at 38°C.
6. Although results of studies on the influence of temperature on nutrient requirements of replacement pullets are conflicting, pullet growth can be improved at high temperature by increasing nutrient density of the diet.
7. A feeding programme for growing pullets that would allow the

protein level to be gradually lowered and the energy level to be raised in the diet is needed under most hot weather situations. Extending the feeding period of a starter beyond 6 weeks of age may be necessary to attain the desired body-weights.

8. Several methods have been suggested for increasing feed consumption in growing pullets. Feeding crumbled feeds has been shown to help. Feed consumption can also be encouraged by increasing frequency of feeding and stirring feed between feedings.

9. Ambient temperature is by far the most important factor affecting water intake. Growing Leghorn pullets drink about twice as much water per day at 38°C as they do at 21°.

10. Acclimatization of replacement pullets to high temperature is an area that has not bean researched adequately. The questions to be raised are how long replacement pullets will retain their ability to withstand heat stress and what effect this acclimatization will have on subsequent performance.

11. Beak trimming of growing pullets at an early age (7–10 days) is recommended for hot regions because it is less stressful and not detrimental to body-weight at housing.

12. Intermittent and biomittent lighting for growing pullets has been recommended because pullets tend to perform better in hot regions on this programme than on standard lighting.

13. The average decrease in feed intake of laying hens is about 1.6% per degree centigrade as environmental temperatures increase from 20 to 30°C. Food intake, however, declines an average of 5% per degree centigrade between 32 and 38°C. Therefore, feed consumption should be monitored daily in hot weather to ensure adequate intake of nutrients, i.e. those normally consumed at 21°C. Dietary adjustments can overcome most of the detrimental effects of high temperature on percentage egg production.

14. The use of high-energy layer rations is recommended in hot regions during the early production period (20–30 weeks), because feed consumption during this period is still low and hens could easily become energy-deficient.

15. Every means of increasing energy intake under high temperature conditions should be used, particularly early in the production cycle. Feeding the calcium source separately, along with the use of low-calcium diets, helps to improve feed intake.

16. Several studies have shown that vitamin C supplements in the diet or water help laying hens maintain adequate performance under high temperature conditions.

17. Ration specifications have been presented for starter, grower and prelay rations, as well as for four stages of production for laying hens.

18. Water available for use on poultry farms in many hot regions is high in total dissolved solids. Some studies have shown adverse effects of such

water on eggshell quality. Two treatments have been suggested for this problem, besides desalination of the drinking-water. One is the use of ascorbic acid supplements in the diet or in the drinking-water and the other is the use of zinc–methionine supplements in the diet.

19. Many tests have shown that cooling the drinking-water improves the performance of layers in hot weather.

20. Laying hens are able to survive hot lethal conditions if they have been previously exposed to a daily intermittent heat stress situation.

21. High ambient temperatures have a negative effect on egg quality. Decreases in egg weight range from 0.07 to 0.98 g per egg for every 1°C rise in temperature.

22. Several management practices, such as lighting adjustments, cage space and shape modifications, egg handling and good hygienic practices, have been described to improve performance in hot climates.

References

Adams, A.W. (1973) Consequences of depriving laying hens of water a short time. *Poultry Science* 52, 1221–1225.

Adams, A.W. and Craig, J.V. (1985) Effect of crowding and cage shape on productivity and profitability of caged layers: a survey. *Poultry Science* 64, 238–242.

Al-Hassani, D.H. and Al-Naib, A.Y. (1992) Egg quality as influenced by lighting and feeding regimes of laying hens during hot summer in Iraq. *Proceedings 19th World's Poultry Congress*, Vol. 2, p. 106.

Arad, Z. (1983) Thermoregulation and acid–base status in the panting dehydrated fowl. *Journal of Applied Physiology* 54, 234–243.

Arad, Z. and Marder, J. (1984) Strain differences in heat resistance to acute heat stress, between the Bedouin desert fowl, the white Leghorn and their crossbreeds. *Comparative Biochemistry and Physiology* 72A, 191–193.

Austic, R.E. (1985) Feeding poultry in hot and cold climates. In: M. Youssef (ed.) *Stress Physiology in Livestock*, Vol. 3, Poultry. CRC Press, Boca Raton, Florida pp. 123–136.

Balnave, D. (1988) Egg shell calcification in the domestic hen. *Proceedings of the Nutrition Society of Australia* 13, 41–48.

Balnave, D. (1993) Infuence of saline drinking water on egg shell quality and formation. *World's Poultry Science Journal* 49, 109–119.

Balnave, D. and Yoselewitz, I. (1987) The relation between sodium chloride concentration in drinking water and egg shell damage. *British Journal of Nutrition* 58, 503–509.

Balnave, D. and Zhang, D. (1992a) Responses in egg shell quality from dietary ascorbic acid supplementation of hens receiving saline drinking water. *Australian Journal of Agricultural Research* 43, 1259–1264.

Balnave, D. and Zhang, D. (1992b) Dietary ascorbic acid supplementation improves egg shell quality of hens receiving saline drinking water. *Proceedings*

19th World's Poultry Congress, Vol. 1, pp. 590–593.

Balnave, D., Yoselewitz, I. and Dixon, R.J. (1989) Physiological changes associated with the production of defective egg shells by hens receiving sodium chloride in the drinking water. *British Journal of Nutrition* 61, 35–43.

Balnave, D., Zhang, D. and Moreng, R.E. (1991) Use of ascorbic acid to prevent the decline in egg shell quality observed with saline drinking water. *Poultry Science* 70, 848–852.

Bell, D. (1987a) Flock management – quality pullets. *California Poultry Letter*, March, p. 4.

Bell, D. (1987b) Age of beak trimming and high fibre diets. *Proceedings of the University of California Symposium for Success*, pp. 1–5.

Bohren, B.B., Rogler, J.C. and Carson, J.R. (1982) Performance at two rearing temperatures of White Leghorn lines selected for increased and decreased survival under heat stress. *Poultry Science* 61, 1939–1943.

Chawla, J.S., Lodhi, G.N. and Ichponani, J.S. (1976) The protein requirement of laying pullets with changing season in the tropics. *British Poultry Science* 17, 275–283.

Cheng, T.K., Cook, C. and Hamre, M.L. (1990) Effect of environmental stress on the ascorbic acid requirement of laying hens. *Poultry Science* 69, 774–780.

Daghir, N.J. (1973) Energy requirements of laying hens in a semi-arid continental climate. *British Poultry Science* 14, 451–459.

Daghir, N.J. (1987) Nutrient requirements of laying hens under high temperature conditions. *Zootecnica International*, May, 36–39.

DeAndrade, A.N., Rogler, J.C., Featherston, V.R. and Alliston, C.W. (1977) Interrelationships between diet and elevated temperature on egg production and shell quality. *Poultry Science* 56, 1178–1183.

Deaton, J.W., McNaughton, J.L. and Lott, B.D. (1982) Effect of heat stress on laying hens acclimated to cyclic versus constant temperatures. *Poultry Science* 61, 875–878.

Devegowda, G. (1992) Feeding and feed formulation in hot climates for layers. *Proceedings 19th World's Poultry Congress*, Vol. 2, pp. 77–80.

Douglas, C.R. and Harms, R.H. (1982) The influence of low protein grower diets on spring-housed pullets. *Poultry Science* 61, 1885–1890.

Douglas, C.R. and Harms, R.H. (1990) An evaluation of a step-down amino acid feeding programme for commercial pullets to 20 weeks of age. *Poultry Science* 69, 763–767.

Douglas, C.R., Welch, D.M. and Harms, R.H. (1985) A step-down protein programme for commercial pullets. *Poultry Science* 64, 1137–1142.

El-Zubeir, E.A. and Mohammed, O.A. (1993) Dietary protein and energy effects on reproductive characteristics of commercial egg type pullets reared in arid hot climate. *Animal Feed Science and Technology* 41, 161–165.

Ernst, R.A. (1989) *Hot Weather Management Techniques*. California Extension Leaflet, University of California, Davis, California.

Ernst, R.A., Millan, J.R. and Mather, F.B. (1987) Review of life-history lighting programmes for commercial laying fowls. *World's Poultry Science Journal* 43, 43–55.

Escalante, R., Chernova, I., Herrera, J.A. and Exposito, A. (1988) Effect of body weight at 18 weeks of age on the lifetime performance of White Leghorn pullets. *Revista Cubana de Ciencia Avicola* 16, 53–59.

Fataftah, A.R.A. (1980) Physiological acclimatization of the fowl to high temperatures. PhD thesis, University of London.

Gomez Pichardo, G. (1983) Production of laying hens when housed three or four in cages measuring 1350 cm^2. *Veterinaria, Mexico* 14, 268–270.

Grizzle, J., Iheanacho, M., Saxton, A. and Broaden, J. (1992) Nutritional and environmental factors involved in egg shell quality of laying hens. *British Poultry Science* 33, 781–794.

Harrison, P.C. and Biellier, H.V. (1969) Physiological response of domestic fowl to abrupt change of ambient air temperature. *Poultry Science* 48, 1034–1045.

Henken, A.M., Groote, A.M.J. and Vanderhel, W. (1983) The effect of environmental temperature on immune response and metabolism of the young chicken. 4. Effect of environmental temperature on some aspects of energy and protein metabolism. *Poultry Science* 62, 59–67.

Heywang, B.W. (1947) Diets for laying chickens during hot weather. *Poultry Science* 27, 38–43.

Heywang, B.W. (1952) The level of vitamin A in the diet of laying and breeding chickens during hot weather. *Poultry Science* 31, 294–300.

Hillerman, J.P. and Wilson, W.O. (1955) Acclimatization of adult chickens to environmental temperature changes. *American Journal of Physiology* 180, 591–595.

Hutchinson, J.C.D. and Sykes, A.H. (1953) Physiological acclimatization of fowls to a hot, humid environment. *Journal of Agricultural Science, Cambridge* 43, 294–322.

Ingram, D.R. and Kling, C.E. (1987) Influence of Ethacal feed component on performance of heat-stressed White Leghorn hens. *Poultry Science* 66, 22 (Abstract).

Jensen, L.S. (1977) The effect of pullet nutrition and management on subsequent layer performance. *Proceedings 37th Semiannual Meeting, American Feed Manufacturers Association Nutrition Council*, pp. 36–39.

Kechik, I.T. and Sykes, A.H. (1974) Effect of dietary ascorbic acid on the performance of laying hens under warm environmental conditions. *British Poultry Science* 15, 449–457.

Keshavarz, K. and McCormick, C.C. (1991) Effect of sodium aluminosilicate, oyster shell and their combinations on acid–base balance and egg shell quality. *Poultry Science* 70, 313–325.

Khan, A.G. (1992) Influence of deep body temperature on hen's egg production under cyclic summer temperatures from 72 to 114°F. *Proceedings 19th World's Poultry Congress*, Vol. 2, pp. 128–131.

Leenstra, F.R., Flock, D.K., Van den Berge, A.J. and Pit, R. (1992) Inheritance of water content and drying characteristics of droppings of laying hens. *Proceedings 19th World's Poultry Congress*, Vol. 2, pp. 201–204.

Leeson, S. and Summers, J.D. (1981) Effect of rearing diet on performance of early maturing pullets. *Canadian Journal of Animal Science* 61, 743–749.

Leeson, S. and Summers, J.D. (1989) Response of Leghorn pullets to protein and energy in the diet when reared in regular or hot cyclic environments. *Poultry Science* 68, 546–557.

Li, Y., Ito, T., Nishibori, M. and Yamamoto, S. (1992) Effects of environmental temperature on heat production associated with food intake and on abdominal temperature in laying hens. *British Poultry Science* 33, 113–122.

McGinnis, C.H., Jr. (1989) Vitamins in pullet nutrition. *Multi-State Poultry Meeting* 16–17 May.

McNaughton, J.L., Kubena, L.F., Deaton, J.W. and Reece, F.N. (1977) Influence of dietary protein and energy on the performance of commercial egg-type pullets reared under summer conditions. *Poultry Science* 56, 1391–1398.

Manner, K., Singh, R.A. and Kamphues, J. (1991) Influence of varying vitamin C sources on performance and egg shell quality of layers at varying environmental temperature. *Proceedings of a Symposium: Vitamine und Weitere Zusatzstaffe bei Mensch und Tier*, pp. 266–269.

Marsden, A. and Morris, T.R. (1987) Quantitative review of the effects of environmental temperature on food intake, egg output and energy balance in laying pullets. *British Poultry Science* 28, 693–704.

Marsden, A., Morris, T.R. and Cronarty, A.S. (1987) Effects of constant environmental temperature on the performance of laying pullets. *British Poultry Science* 28, 361–380.

May, J.D., Deaton, J.W. and Branton, S.L. (1987) Body temperature of acclimated broilers during exposure to high temperature. *Poultry Science* 66, 378–380.

Mohammed, T.A. and Mohammed, S.A. (1991) Effect of dietary calcium level on performance and egg quality of commercial layers reared under tropical environment. *World Review of Animal Production* 26, 17–20.

Moreng, R.E., Balnave, D. and Zhang, D. (1992) Dietary zinc methionine effect on egg shell quality of hens drinking saline water. *Poultry Science* 71, 1163–1167.

Morrison, W.D., Braithwaite, L.A. and Leeson, S. (1988) Report of a survey of poultry heat stress losses during the summer of 1988. Unpublished Report from Department of Animal and Poultry Science, University of Guelph, Ontario, Canada.

Mueller, W.J. (1961) The effect of constant and fluctuating temperature on the biological performance of laying pullets. *Poultry Science* 40, 1562–1571.

National Research Council (NRC) (1984) *Nutrient Requirements of Domestic Animals – Nutrient Requirements of Poultry*, 8th edn. National Academy of Science, Washington, DC.

Nir, I. (1992) Optimization of poultry diets in hot climates. *Proceedings 19th World's Poultry Congress*, Vol. 2, pp. 71–76.

Nishibri, M., Li, Y., Fujita, M., Ito, T. and Yamamoto, S. (1989) Diurnal variation in heat production, heart rate, respiration rate and body temperature of laying hens at constant environmental temperature of 23 and 35°C. *Japanese Journal of Zootechnical Science* 60, 529–533.

Nockels, C.F. (1988) *Proceedings of the Georgia Nutrition Conference for the Feed Industry*, Atlanta, GA, 16–18 November, p. 9.

North, M.D. and Bell, D. (1990) *Commercial Chicken Production Manual*, 4th edn. Van Nostrand Reinhold, New York, 643 pp.

Odom, T.W., Harrison, P.C. and Darre, M.J. (1985) The effects of drinking carbonated water on the egg shell quality of single comb white Leghorn hens exposed to high environmental temperature. *Poultry Science* 64, 594–596.

Oluyemi, J.A. and Adebanjo, A. (1979) Measures applied to combat thermal stress in poultry under practical tropical environment. *Poultry Science* 58, 767–770.

Owoade, A.A. and Oduye, O.O. (1992) Use of slatted floor house for pullet rearing – a trial in Nigeria. *Proceedings 19th World's Poultry Congress*, Vol. 2, p. 136.

Payne, C.G. (1966) Environmental temperature and the performance of light breed pullets. *Proceedings 13th World's Poultry Congress*, pp. 480–484.

Peguri, A. and Coon, C. (1991) Effect of temperature and dietary energy on layer performance. *Poultry Science* 70, 126–138.

Perek, M. and Kendler, J. (1963) Ascorbic acid as a supplement for White Leghorn hens under conditions of climatic stress. *British Poultry Science* 4, 191–200.

Petersen, J., Liepert, B.M. and Horst, P. (1988) Sudden laying stop as adaptation reaction to heat stress. *Deutsche Tierarztliche Wochenschrift* 95, 312–317.

Picard, M. (1985) Heat effects on the laying hen, protein nutrition and food intake. *Proceedings, 5th European Symposium on Poultry Nutrition*, pp. 65–72.

Picard, M., Angulo, I., Antoine, H., Bouchot, C. and Sauveur, B. (1987) Some feeding strategies for poultry in hot and humid environments. *Proceedings 10th Annual Conference of Malaysian Society of Animal Production*, pp. 110–116.

Purina Mills, Inc. (1987) Bio-mittent lighting saves money, increases egg income. *Acculine Leaflet P243F-87A5*, pp. 1–3.

Ramlah, H. and Sarinah, A.H. (1992) Performance of layers in the tropics offered diets with and without supplemental fat. *Proceedings 19th World's Poultry Congress*, Vol. 2, pp. 107–108.

Ramos, N.C., Germat, A.G. and Adams, A.W. (1990) Effects of cage shape, age at housing and types of rearing and layer waterers on the productivity of layers. *Poultry Science* 69, 217–223.

Reid, B.L. (1979) Nutrition of laying hens. *Proceedings Georgia Nutrition Conference*, University of Georgia, Athens, p. 15–18.

Reid, B.L. and Weber, C.W. (1973) Dietary protein and sulphur amino acids for laying hens during heat stress. *Poultry Science* 52, 1335–1340.

Rojas Olaiz, L.A. (1988) Performance of laying hens of semi-heavy and light breeds housed three or four birds to a cage measuring 1800 cm^2 during the second production cycle. *Veterinaria, Mexico* 19, 64–66.

Rose, P. and Michie, W. (1986) Effect of temperature and diet during rearing of layer strain pullets. In: Fisher, C. and Boorman, K.N. (eds) *Nutrient Requirements of Poultry and Nutritional Research*. Butterworths, London, pp. 214–216.

Sauveur, B. and Picard, M. (1987) Environmental effects on egg quality. In: Wells, R.G., and Belyavin, C.G. (eds) *Egg Quality: Current Problems and Recent Advances*. Butterworths, London, pp. 219–234.

Savage, S.I. (1983) *Water in Hot Weather*. Poultry Tips, Cooperative Extension Service, University of Georgia, P.S. 2-1, Athens, Georgia, USA.

Scott, T.A. and Balnave, D. (1989) Self-selection feeding for pullets – Part 2. *Poultry International*, December, 22–26.

Shane, S.M. (1988a) Factors influencing health and performance of poultry in hot climates. *Critical Reviews in Poultry Biology* 1, 247–269.
Shane, S.M. (1988b) Managing poultry in hot climates. *Zootechnica International*, April, 37–40.
Smith, A.J. and Oliver, L. (1971) Some physiological effects of high temperature on the laying hen. *Poultry Science* 50, 912–916.
Smith, A.J. and Oliver, L. (1972) Some nutritional problems associated with egg production at high environmental temperatures. I. The effect of environmental temperature and rationing treatment on the productivity of pullets fed on diets of differing energy content. *Rhodes Journal of Agricultural Science* 10, 3–8.
Stockland, W.L. and Blaylock, L.G. (1974) The influence of temperature on the protein requirement of cage-reared replacement pullets. *Poultry Science* 53, 1174–1187.
Sykes, A.H. (1976) Nutrition–environment interactions in poultry. *Proceedings Nutrition Conference for Feed Manufacturers*, University of Nottingham, UK, p. 17–21.
Sykes, A.H. and Fataftah, A.R.A. (1980) Heat acclimatization in laying hens. *Proceedings 6th European Poultry Congress* 4, 115–119.
Sykes, A.H. and Fataftah, A.R.A. (1986a) Acclimatization of the fowl to intermittent acute heat stress. *British Poultry Science* 27, 289–300.
Sykes, A.H. and Fataftah, A.R.A. (1986b) Effect of a change in environmental temperature on heat tolerance in laying fowl. *British Poultry Science* 27, 307–316.
Tadtiyanant, C., Lyons, J.J. and Vandepopuliere, J.M. (1991) Influence of wet and dry feed on laying hens under heat stress. *Poultry Science* 70, 44–52.
Teng, M.F., Soh, G.L. and Chew, S.H. (1990) Effects of stocking densities on the productivity of commercial layers in the tropics. *Singapore Journal of Primary Industries* 18, 123–128.
Thornton, P.A. and Moreng, R.E. (1959) Further evidence on the value of ascorbic acid for maintenance of shell quality in warm environmental temperature. *Poultry Science* 38, 594–599.
Ulusan, H.O.K. and Yildiz, N. (1986) Effects of climatic factors in poultry houses on egg production, egg weight and food consumption of commercial hybrid white Leghorn. *Ankara Universitsi Vetriner Fakultesi Dergisi* 33, 122–133.
Uzu, G. (1989) Some aspects of feeding laying hens in hot climates. *Rhône-Poulenc Animal Nutrition Report*, 03600 Commentery, France.
Valencia, M.E., Maiorino, P.M. and Keyd, B.L. (1980) Energy utilization in laying hens. III. Effect of dietary protein level at 21 and 32°C. *Poultry Science* 59, 2508–2511.
Vo, K.V., Boone, M.A. and Johnston, W.E. (1978) Effect of three lifetime temperatures on growth, feed and water consumption and various blood components in male and female Leghorn chickens. *Poultry Science* 57, 798–801.
Vo, K.V., Boone, M.A., Hughes, B.L. and Knetshges, K.F. (1980) Effects of ambient temperature on sexual maturity in chickens. *Poultry Science* 59, 2532–2536.
Vohra, P., Wilson, W.O. and Siopes, T.D. (1979) Egg production, feed consump-

tion, and maintenance energy requirement of Leghorn hens at temperatures of 15.6 and 26.7°C. *Poultry Science* 58, 674–680.

Waldroup, P.W., Mitchell, R.J., Payne, J.R. and Hazen, K.R. (1976) Performance of chicks fed diets formulated to minimize excess levels of essential amino acids. *Poultry Science* 55, 243–253.

Wilson, W.O., Siopes, T.D., Ingkasuwan, P.H. and Mather, F.B. (1972) The interaction of temperature of 21 and 32°C and photoperiod of eight and 14 hours on white Leghorn hen production. *Archives fur Geflugelkunde* 2, 41–44.

Wolfenson, D., Frei, F.E., Snapir, N. and Nerman, A. (1979) Effect of diurnal or nocturnal heat stress on egg formation. *Poultry Science* 58, 167–174.

Yoselewitz, I. and Balnave, D. (1989a) Effect of egg weight on the incidence of egg shell defects resulting from the use of saline drinking water. *Proceedings of Australian Poultry Science Symposium* p. 98.

Yoselewitz, I. and Balnave, D. (1989b) The influence of saline drinking water on the activity of carbonic anhydrase in the shell gland of laying hens. *Australian Journal of Agricultural Research* 40, 1111–1115.

Yoselewitz, I. and Balnave, D. (1990) Strain responses in egg shell quality to saline drinking water. *Proceedings of Australian Poultry Science Symposium*, p. 102.

10

Breeder and Hatchery Management in Hot Climates

N.J. Daghir[1] and R. Jones[2]

[1]Faculty of Agricultural Sciences, UAE University, Al-Ain, UAE; [2]Shaver Poultry Breeding Farms, Box 400, 16 Branchton Road, Cambridge, Ontario, Canada N1G 2W1.

Introduction	256
Effects of High Temperature on Reproductive Performance	256
Feeding Breeders in Hot Climates	258
Nutritional experiments	258
Feeding programmes	266
Breeder House Management	269
Egg gathering	269
Nest hygiene	270
Nest type	270
Egg hygiene	271
Water supply	272
Cage vs. floor	272
Water-cooled perches	274
Disease control and prevention	274
Storage of Hatching Eggs	277
Hatchery Design	278
Hatchery Hygiene	278
Incubation in a Tropical Climate	279
Incubator conditions	279
Incubator problems in hot climates	280
Incubation time	281
Prewarming eggs for setting	281
Water Supply and Cleanliness	282
Altitude	283
Chick Processing and Delivery	283
Conclusions and Recommendations	284
References	286

Introduction

Commercial incubation was started around 1400 BC in one or more of the warm regions of the world. The native Egyptian hatcheries are still in operation at the present day, producing a sizeable percentage of the baby chicks in the country. Askar (1927) and El-Ibiary (1946) were among the earliest to describe these hatcheries. Ghany *et al.* (1967) gave a detailed description of the structure and operation of the native Egyptian hatchery and reported at that time that these hatcheries were still producing around 95% of the chick output of the country.

This chapter describes some of the important features of breeder management in hot climates as well as hatchery management and operation problems in such an environment. It covers areas of hatching egg production, such as egg gathering, nest types for hot climates, nest and egg hygiene, storage of hatching eggs, water supply for a hatchery, hatchery design, incubator problems and problems of chick processing and delivery.

Effects of High Temperature on Reproductive Performance

The deleterious effects of high environmental temperature on production have been described in previous chapters. The effects of heat stress on the reproductive hormones have also been covered in Chapter 3. This section will therefore consider the effects of high temperature on selected aspects of reproductive performance, mainly fertility and hatchability.

Nature never intended the modern hen to reproduce itself in a hot environment. In the wild, a hot climate which is always above 23°C does not prevent eggs from continuous incubation. The necessary natural cooling feature normally found in the more moderate climate in the spring is missing in a tropical climate. The retarding of hatching egg temperature, below the embryo development temperature of 23°C, is necessary in order to accumulate sufficient hatching eggs and to have eggs hatch in a predictable time period. In a moderate climate at a certain time of the year, this cooling occurs naturally with night-time temperatures, but in a hot climate and during the summer months in a moderate climate this cooling must be done artificially. The hatching egg when laid is at a temperature of 40–41°C. In the spring of the year, in a moderate climate, the temperature is around 12–20°C. This permits eggs to cool gradually over a 5–6 h period. Therefore, in a tropical environment, the egg handling procedure should try to match this criterion.

Taylor (1949) was among the first to report that the hatchability of eggs decreases with increases in ambient temperature. He observed a drop of 15–20% in hatchability whenever the mean weekly temperature exceeded 27°C. High environmental temperatures have also been associated with lower fertility and poor eggshell quality (Warren and Schnepel, 1940;

Heywang, 1944; Clark and Sarakoon, 1967; Miller and Sunde, 1975). Reductions in eggshell quality have also been related to depressed hatchability and weakening of the embryos before eggs were set (Peebles and Brake, 1987). Moving White Leghorn hens from a 21°C to a 32°C environment decreased their egg weight, egg specific gravity and fertility (Pierre, 1989). The detrimental effects of high temperature on egg weight have been discussed in Chapter 9. Since hatching egg weight influences chick weight and early chick performance, early egg size and maintaining adequate hatching egg size in hot climates is of primary importance. Small chicks are often more difficult to start and early mortality is usually high with undersized chicks. Hess *et al.* (1994) conducted a survey of 17 broiler companies in the southern states of Georgia and North Carolina in the USA, representing placements of approximately 14 million broilers per week. The results of this survey showed that broiler operations routinely setting eggs from young hens (24 weeks old) experienced higher average 7-day mortality than companies setting eggs after 27 weeks of age. Therefore, hatchery operators should avoid setting small eggs by either waiting until a certain percentage production is reached or by waiting for a certain prescribed time after breeder housing. Furthermore, methods of improving early egg size in breeders in hot climates are of great economic importance. Nasser *et al.* (1992) studied the effect of heat stress on egg quality of broiler breeders. Pullets were subjected to cyclic temperatures of 20 to 42°C during a 9-week period starting at 35 weeks of age. There was an inverse relationship between temperature and the specific gravity and shell thickness of eggs. Exposing hens to high temperature induced respiratory alkalosis, as indicated by an elevation in blood pH, accompanied by a lowered haematocrit. These results further illustrate the need for environment-controlled housing for breeders in hot climates.

Seasonal variation in sperm production is well documented and less sperm is produced during the summer months, which reduces fertility. Part of this reduction in sperm production is associated with reduction in feed intake and thus nutrient intake. Jayarajan (1992) studied the effects of season on fertility and hatchability in three different breeds in Madras, India. Fertility was highest in White Leghorns and White Plymouth Rocks during the cold season (December to February) and for Rhode Island Reds during the summer (March to May) in that country. Hatchability was highest for all breeds during the monsoon season (September to November). In a study of large commercial hatcheries in the USA, hatchability of eggs during July, August and September was 5% lower than during the remainder of the year (North and Bell, 1990). These decreases were shown in that country to be due mainly to two reasons; one was decreased feed consumption by the breeders, which caused embryonic nutritional deficiencies, and the other was deterioration in egg quality during the holding period.

Poor hatchability in hot climates may be partially due to thin-shelled eggs. Bennett (1992) compared hatchability of thin- and thick-shelled eggs for 25 broiler breeder flocks. Shell thickness was determined by using specific gravity measurements. Hatchability of thin-shelled eggs was 3–9% less than thick-shelled eggs in 30–60-week old breeder flocks. The shell thickness of hatching eggs dropped below the recommended level (specific gravity = 1.080) by 42 weeks of age. The same author (Bennett, 1992) recommends checking shell quality at least once a month for all broiler breeder flocks. Dipping eggs in salt solutions to measure specific gravity had little or no effect on hatchability. This is important in hot climates because the incidence of thin shells is higher on a year round basis than in temperate regions. Poor hatchability is also closely related to obesity in broiler breeders (Robinson, 1993). Hens that are overweight suffer more from heat stress and lay very erratically. They exhibit low rates of production, oviductal prolapse, poor shell quality and double-yolked eggs.

Chaudhuri and Lake (1988) proposed that there would be an advantage in being able to store semen at high ambient temperature in a simple diluent, particularly for genetic selection programmes and commercial breeding in cages in tropical countries. They were able to store semen up to 17 h at 20 and 40°C in a simple diluent. The diluent was composed of sodium chloride, TES buffer, glucose and antibiotics, and was adjusted to pH 7.4 with sodium hydroxide. For successful storage for 17 h at the high temperature, it was necessary to agitate the diluted semen samples. Samples could be kept in still conditions for 6 h at ambient temperature around 20°C on a bench top. The replacement of chloride with glutamate did not maintain good fertility at high temperature.

Feeding Breeders in Hot Climates

The feeding and feeding programmes of meat-type breeders are very different from those used for egg-type breeders because the former tend to become obese and thus poor performing. Boren (1993) presented some basic rules on broiler breeder nutrition and indicated that nutritional strategies in use today on broiler breeders are to help balance the lower potential for egg production against the economic necessity of maximizing viable hatching egg production and minimizing costs. The management of broiler breeders in the laying period has been reviewed by Dudgeon (1988), who discussed the effects of nutrition, body-weight and environmental temperature on different production parameters.

Nutritional experiments

There are very few studies on the feeding of either Leghorn or broiler breeders in hot climates. In the case of Leghorn breeders, it is probably assumed that the feeding of these birds is not too different from that of regular Leghorn layers in hot climates, bearing in mind the extra fortification of the diet in vitamins and trace minerals, particularly those that are known to be critical for fertility and hatchability. Morrison *et al.* (1988) reported the results of a survey conducted on 49 broiler breeder flocks that either received or did not receive a vitamin and electrolyte water additive during heat stress. Results of this survey showed that males were more susceptible to heat stress than females and younger females experienced a more severe drop in egg production. Both percentage mortality and percentage drop in egg production were reduced in flocks receiving the water additives.

Broiler pullet feeding

Broiler breeder pullets are placed under feed restriction starting at about 14 days of age. An early moderate restriction of growth rate is less stressful than a severe restriction later. This allows for better uniformity and proper fleshing, both of which contribute towards good hatching egg production. There are two methods of restricting growth rate in broiler breeders. There is a qualitative restriction, which utilizes reduced level of nutrients in the feed, and there is a quantitative restriction, which utilizes reduced feed allowances. Although qualitative restriction is simpler and has been shown to maintain average body-weight in accordance with breeder recommendations, uniformity using this procedure is poor. This method of restriction has not been used by breeders to any extent. Two methods of quantitative

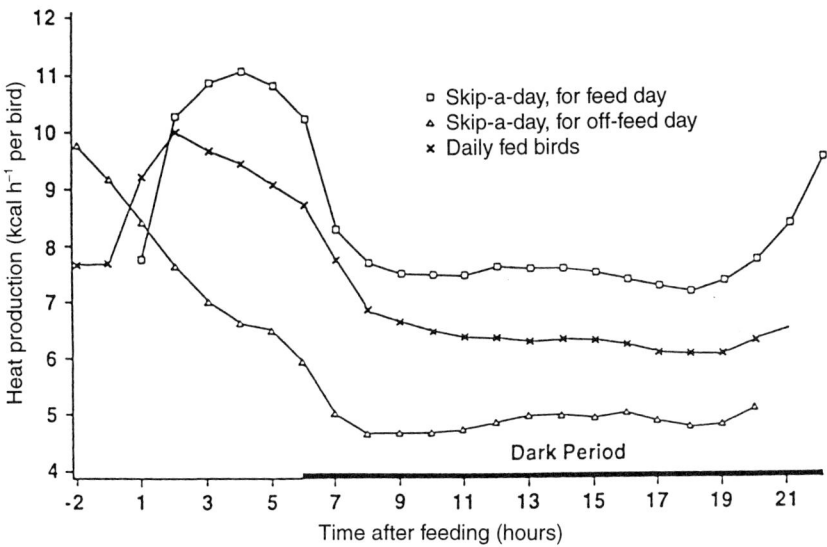

Fig. 10.1. Heat production of broiler pullets (from Leeson and Summers, 1991).

restriction can be used. Birds can be fed either restricted amounts daily or on a skip-a-day programme. The advantage of every-day feeding versus skip-a-day feeding is a feed saving of approximately 1.4 kg per bird to 20 weeks of age (McDaniel, 1991). Leeson and Summers (1991) estimated a saving of up to 10% in feed required to produce a pullet of 2.3 kg bodyweight. For hot climates, it is important to point out that the heat production of birds on every-day feeding was around 10% lower than those on skip-a-day feeding (Fig. 10.1). Feed allowances for growing broiler breeders should be determined on the basis of kcal metabolizable energy (ME) per bird per day and these allowances should be adjusted in relation to body-weight and the condition and uniformity of the pullets. Feed allowances should be calculated on a flock-by-flock basis and more feed is required and probably more feeder space when uniformity is low or when feed quality is poor and intestinal infections are a problem.

To produce the first egg, a pullet should attain a minimum body weight and age. Body protein content has been more closely linked to onset of production than body fat, and properly fleshed birds can mature and begin laying on time. Several workers have studied the effect of high-protein and in many cases high-lysine prebreeder rations on reproductive performance of broiler breeder hens in a temperate climate. Although many reported improved performance (Cave, 1984; Brake *et al.*, 1985; Lilburn and Myers-Miller, 1990), some doubt that it is necessary (Van Wambeke, 1992). This practice may be useful in certain hot-climate areas where poor-quality protein is used and where breeder pullets may be under weight at onset of production. Furthermore, high-protein prebreeder diets do not appear to have any positive effect on well-fleshed and adequately developed pullets. Lilburn *et al.* (1987) studied the relationships between dietary protein, dietary energy, rearing environment and nutrient utilization by broiler breeder pullets. They found that birds exposed to natural daylight and fed 15.5% protein diets had similar caloric efficiencies (kcal g^{-1}) at 15 weeks of age despite dietary density differences of 220 kcal kg^{-1}. This, in their opinion, supported the hypothesis that, above some minimal level of protein intake, caloric intake has the greatest control over body-weight gain in restricted pullets. Hagos *et al.* (1988) studied the effect of different levels of feed restriction on performance of broiler breeders during the growing period in Bangalore, India. They concluded that 30% restriction of a control feed containing 16.6% protein and 2800 kcal ME kg^{-1} and 15% restriction of a low-protein, low-energy diet (15% protein and 2500 kcal ME kg^{-1}) were both adequate to achieve optimum performance. A survey of 49 breeder flock records from Florida in 1984 concluded that one of the major contributors to poor flock performance to peak was inadequate growth. Many of the flocks exhibited little or no body-weight increase around the time of peak production. Peak weekly egg production was reduced by an average of 5%. Leeson and Spratt (1985) calculated the energy requirements of broiler breeders from 20 to 28 weeks of age and compared these requirements with the feed allowances recommended by the breeder companies. Their results showed that the calculated energy requirements between 20 and 25 weeks of age are much higher than provided for in the recommended feed allowance. They suggested that the broiler breeder is in a very precarious situation with regard to energy at the time of sexual maturity. Energy deficiency at this time may delay sexual maturity and influence ovum development of eggs to be laid in the coming 2–3 weeks. The same workers reported that body-weight from 19 to 40 weeks of age was directly related to energy intake. Therefore, weight gain in the pullet until it reaches peak production is critical in hot areas and the use of a high-protein prebreeder ration may be useful.

The onset of lay in broiler breeders can be advanced if greater body-weight is allowed during rearing (Robbins et al., 1988; Yu et al., 1992). Allowing increased body-weight during rearing has been suggested as a means of increasing egg weights in pullets stimulated to lay at an early age (Leeson and Summers, 1980; Kling et al., 1985). Therefore, subjecting breeder pullets to photostimulation at an early age after allowing them to attain greater than recommended body-weights during rearing may reduce rearing time and maximize hatching egg production. Yuan et al. (1994) studied effects of rearing-period body-weights and early photostimulation on broiler breeder egg production. They observed that the onset of lay by broiler breeders can be advanced by early photostimulation and that increased body-weight enhances this. Greater body-weight during rearing, however, does not compensate for reduced early egg weights and results in decreased total egg production and mean egg weights when feed is provided at recommended levels during lay.

Broiler breeder hen feeding

The energy requirement of the breeder hen is considered by most nutritionists as the most limiting nutrient. The breeder hen requires energy for body maintenance and activity, growth and egg production. About 50–75% of its energy needs are for maintenance and activity, while requirements for egg production vary from zero to 35%, depending on rate of production. Growth requirements can be as high as 30% early in the production cycle (18–24 weeks of age) and as low as 5% later in production. Spratt et al. (1990b) in studies on energy metabolism of broiler breeder hens reported that fasting metabolism amounted to 75% of the maintenance energy requirement, while the liver, gut and reproductive tract amounted to 26 and 30% of the total energy expenditure in fed and fasted hens, respectively. The same workers (Spratt et al., 1990a) in studies on the energy requirements of the broiler breeder hen indicated that individually caged hens between 28 and 36 weeks of age in a thermoneutral environment (21°C) require approximately 1.6 MJ (382 kcal) of apparent metabolizable energy (AME) per bird per day for normal growth (3 g per day) and egg production (85%).

Couto *et al.* (1990) studied protein requirements of broiler breeders aged 36–52 weeks and fed 18, 20, 22 or 24 g protein per hen per day. Protein intake had no significant effect on fertility or hatchability, but egg weight was significantly lower in hens fed 18 g protein and body-weight was significantly higher in hens fed 24 g. The same authors suggested that breeder hens should be fed 20.5 g daily. Brake *et al.* (1992) fed breeders from 1 day to 18 weeks of age diets containing 2926 kcal ME kg^{-1} and either 11%, 14%, 17% or 20% crude protein. All treatments received 17% crude protein from 19 to 25 weeks of age. A high correlation was observed between the level of protein in the rearing ration and skeletal growth. The sternum length varied from 119 mm in the 11% crude protein group to 129 mm in the 17% protein group. There was also more feather retention after onset of production in the high-protein groups. This study further illustrates the need for adequate nutrition during the rearing period and the importance of using good-quality feed when feed intake is restricted for body-weight control purposes.

Feeding standards for broiler breeders in cages are practically non-existent. Pankov and Dogadayeva (1988) conducted a study to determine the protein : energy ratio and develop optimum feeding regimes for caged meat-type breeders. Caged broiler breeders, according to these workers, may be fed a diet containing 14.5% protein, provided it is supplemented with lysine and methionine to bring it up to the 16% protein level, with an ME of 11.3 MJ kg^{-1} at a level of 140 g per bird per day up to 42 weeks of age and 135 g therafter. Hazan and Yalcin (1988) investigated the effect of different levels of feed intake on egg production and hatchability of caged Hybro broiler breeder hens in Turkey. The authors concluded that the best feed conversion and optimum hatchability and number of chicks can be achieved by feeding 90% of feed consumed by floor breeders. The diet used by these workers had 16.78% protein and 2748 kcal ME kg^{-1}. This was confirmed by Glatz (1988) in south Australia, who concluded that breeders in cages fed 10% below the standard levels produced more eggs per bird than hens fed at recommended levels. Spratt and Leeson (1987) studied the effects of dietary protein and energy on individually caged broiler breeders. Feed intake was maintained constant across all treatments and thus nutrient intake was varied by diet formulation. Their results show that 19 g protein and 385 kcal ME per day were sufficient to maintain normal reproductive performance through peak egg production.

Rostagno and Sakamoura (1992) studied the effects of environmental temperature on feed and ME intake of broiler breeder hens. They housed broiler breeders in environmental chambers at 15.5, 21.9 and 26.3°C. Water and feed were provided *ad libitum* for 14 days. Daily feed intake, nitrogen-corrected ME (MEn) and nitrogen-corrected true ME (TMEn) decreased linearly as environmental temperature increased. Hen body-weight and rectal temperature were not affected. A 1°C rise in temperature resulted in a decreased intake of 2.43 g feed per hen, 2.10 kcal MEn kg^{-1} body-weight and 2.20 kcal TMEn kg^{-1} body-weight. Egg production, egg weight, egg mass and feed conversion were not affected by environmental temperature.

Vitamin A deficiency decreases the number of sperm and increases the number of non-motile and abnormal sperm (Oluyemi and Roberts, 1979). Babu et al. (1989) reported that Leghorn breeders require 40 p.p.m. vitamin E and 0.67 p.p.m. selenium for maximum hatchability in two different strains in Madras, India. Surai (1992) reported that increasing the level of vitamin E in the breeder diet raised the concentration of α-tocopherol in both the chicken and turkey spermatozoa and thus increased their membrane stability. It is fairly well established that fertility in hot climates can be improved in breeder flocks by the addition of extra amounts of vitamin E if breeder rations contain the usual level of 15–20 mg kg^{-1} of diet. Flores-Garcia and Scholtyssek (1992) studied the riboflavin requirement of the breeding layer and concluded that to obtain optimum hatchability, the breeder diet should contain a minimum of 4.4 mg riboflavin kg^{-1} of feed. Biotin requirements of broiler breeders fed diets of different protein content were studied by Whitehead et al. (1985). Production of eggs or normal chicks was depressed when practical diets containing 16.8% or 13.7% crude protein were not supplemented with synthetic biotin. Biotin requirement was higher with the diet containing 16.8% protein and was estimated to be about 100 μg of available biotin kg^{-1}. The minimum yolk biotin concentration indicative of adequate maternal status was about 550 μg g^{-1}.

In recent years, there has been increased interest in the use of ascorbic acid to overcome some of the deleterious effects of heat stress in breeders. The addition of ascorbic acid to breeder rations of chickens and turkeys has in many cases yielded positive responses. Peebles and Brake (1985) demonstrated increased egg production, hatch of fertile eggs and number of chicks per hen with supplementation of 50 or 100 p.p.m. of broiler breeder rations during the mild summertime stress typical of the southern USA. Supplementation of broiler breeder feeds with 300 p.p.m. ascorbic acid during hot summers in Israel improved offspring performance in both weight gain and feed conversion (Cier et al., 1992a). Mousi and Onitchi (1991) studied the effects of ascorbic acid supplementation on ejaculated semen characteristics of broiler breeder chickens under hot and humid tropical conditions. They concluded that dietary supplementation of 250 mg ascorbic acid kg^{-1} diet is desirable to maintain semen quality during the hot periods (Table 10.1). Dobrescu (1987) reported a 28% increase in semen volume and 31% increase in sperm concentration in turkey toms by the addition of 150 p.p.m. ascorbic acid to breeding tom rations (Table 10.2). Whenever vitamin C is to be added to a pelleted ration, such as in the control of *Salmonella*, it is important to use a stabilized form of the vitamin.

Table 10.1. Effect of dietary ascorbic acid (AA) supplements on semen characteristics of broiler breeder chickens (from Mousi and Onitchi, 1991).

Dietary AA (mg/kg)	Semen vol. (ml)	Sperm concent. ($\times 10^9$ ml^{-1})	Motile sperm/ ejaculate ($\times 10^9$)	Sperm/ ejaculate ($\times 10^9$)	Motility (%)
0	0.47	2.94	0.89	1.42	59.5
125	0.51	2.54	0.79	1.32	58.5
250	0.62	2.86	1.20	1.84	62.6
500	0.77	3.11	1.64	2.4	68.3
SEM	0.07*	0.11	0.19*	0.25*	2.21

*Linear effect of ascorbic acid (significance, $P < 0.05$).

Table 10.2. Effect of ascorbic acid on semen parameters in breeding turkey toms (from Dobrescu, 1987).

	Ascorbic acid (p.p.m.)	
	0	150
Semen vol. (ml)	0.32	0.41
Sperm/ejaculate (10^{12})	2.97	3.88

The requirement of calcium of the breeder hen increases with age. Harms (1987) recommended a daily intake of 4.07 g calcium as an average for the whole production period. Therefore, diets should contain about 3.2% calcium if daily feed intake averages 135 g per hen per day. Birds require slightly more phosphorus at high than at moderate or low temperatures. Breeders need a minimum daily intake of about 700 mg of total phosphorus. Harms (1987) recommended a daily intake of 683 mg. Harms *et al.* (1984) suggested that, for breeder hens maintained in cages, the requirements of both calcium and phosphorus are significantly higher than for those on litter floors. The sodium requirement of the broiler breeder hen was estimated by Damron *et al.* (1983) to be not more than 154 mg per hen daily, while Harms (1987) suggested a requirement of 170 mg per hen daily (as shown in Table 10.5).

Gonzales *et al.* (1991) found that supplementing a maize–soyabean meal diet for 50 days with 20% saccharina (dried fermented sugarcane) for broiler breeder hens and cocks significantly increased the percentage of fertile eggs and the hatchability of total eggs incubated.

The effects of pellets, mash, high protein and antibiotics on the performance of broiler breeders in a hot climate were reported by Cier *et al.* (1992b). No significant differences were observed between crumbles and mash, or from raising protein from 15 to 16.5%. They concluded that, under Israel's climatic conditions, the best biological and economical results are achieved by broiler breeder hens fed daily restricted crumbles, containing 15% protein, and 2700 kcal kg^{-1} of ME.

Ubosi and Azubogu (1989) evaluated the effects of terramycin Q and fish meal on heat stress in poultry production. Body-weight, feed intake and egg production were increased significantly in the group receiving terramycin Q and 3% fish meal. Egg weight was decreased in the group receiving terramycin Q only.

McDaniel et al. (1992) fed acetylsalicylic acid (ASA) to White Leghorn breeders for 13 months of production. When fed at 0.4% of the ration, ASA decreased both fertility and hatchability. Chicks from hens given 0.1% ASA weighed more than chicks from hens given no ASA or levels exceeding 0.1%. ASA fed to layer breeders did not improve hatchability of embryos exposed to increased incubation temperature compared with embryos exposed to control incubation temperatures.

Oyawoye and Krueger (1986) reported that the inclusion of 300–400 p.p.m. of monensin in the feed of broiler breeder pullets from 1 to 21 weeks of age will suppress appetite sufficiently to accomplish restriction of body-weight. Pullets fed monensin were less uniform in body-weight compared with their restricted-feed controls. At low protein levels, the high level of monensin increased mortality.

Male broiler breeder feeding

Several studies have been conducted on feeding programmes for male broiler breeders. Buckner et al. (1986) tested five levels of a diet containing 13% protein and 3170 kcal ME kg^{-1} of feed on male adult broiler breeders. The intake of 91 and 102 g feed per day of such a ration reduced the number of males producing semen at 40 weeks of age compared with 136 g. Brown and McCartney (1986) recommended a daily intake of 346 kcal ME per male per day for normal body-weight maintenance and productivity of broiler breeder males grown in individual cages. Buckner and Savage (1986) fed caged broiler breeder males *ad libitum* diets containing 5, 7 or 9% crude protein and 2310 kcal ME kg^{-1} from 20 to 65 weeks of age. At 24 weeks of age, body-weight, semen volume and sperm counts were reduced for the males fed the 5% protein diet. Average daily protein and energy intake was 10.9, 14.7 and 18.7 g per day and 495, 479 and 473 kcal ME per day for the males fed the 5, 7 and 9% protein diets, respectively. Wilson et al. (1987) reported that broiler breeder males can be fed 12–14% crude protein on a restricted basis after 4 weeks of age with no harmful effects on body-weight, sexual maturity or semen quality. More males fed 12% protein continued production of semen beyond 53 weeks than those fed higher protein levels. Hocking (1994), after a series of experiments conducted at the Roslin Institute in the UK, concluded that low-protein diets (11%) increase semen yields in caged broiler breeder males but have little effect on fertility in floor pens. He

Table 10.3. Percentage fertility of broiler breeder males of different body-weights fed diets differing in protein content (from Hocking, 1994).

Body-weight (kg)	Dietary protein (%)	
	11.0	16.0
3.0-4.0	93.5	16.0
3.0-4.5	94.4	92.4
3.5-4.5	91.5	93.5

further recommended that optimum male body-weight at the start of the breeding period is 3.0 kg and should increase to 4.5 kg at 60 weeks of age. Table 10.3 shows that high-protein diets (16%) improve fertility in heavy birds, while low-protein diets are more beneficial for lightweight birds.

Feeding programmes

Breeder flocks can be fed a wide range of different feeds as long as these feeds meet the minimum requirements suggested in Table 10.6 and are properly balanced. An 18% protein chick starter with 2850 kcal kg^{-1} is usually satisfactory under most conditions. In cases where the 3-week target body weight set by the breeder is not met, then the starter diet may be fed beyond 3 weeks and possibly up to 7 weeks of age. A 15% protein, 2700 kcal kg^{-1} growing ration is suggested. Feed allowances for growers are adjusted weekly in order to maintain target body-weights. This grower diet is also suitable for feeding males separately throughout the breeding period, provided that the breeder vitamin and trace mineral premix shown in Table 10.4 is used rather than the grower premix. A prelay may be used 2–3 weeks prior to the first egg. Such a ration usually contains 15–16% protein and 2750 kcal kg^{-1} with about 2–2.25% calcium. A breeder ration with 16% protein and 2750 kcal kg^{-1} is recommended. Such a ration provides 24 g of protein per breeder per day when the flock is fed 150 g of feed per bird per day. Table 10.4 shows the recommended levels of vitamins and trace minerals per tonne of starter, grower and breeder rations. Waldroup *et al.* (1976) fed diets to broiler breeders in the southern USA that supplied daily protein intakes ranging from 14 to 22 g per day. An intake of 20 g per day supported maximum egg production and maximum egg weight. These workers calculated that, with a feed intake of 146 g per day, a dietary level of 13.7% protein would be adequate. Table 10.5 presents daily nutrient requirements of the broiler breeder female as presented by Harms (1987) and by the National Research Council (NRC, 1994).

Feed allowances for breeder hens are usually determined by egg mass output, body-weight and changes in time to consume feed by the breeder.

Table 10.4. Recommended vitamin-trace mineral levels per tonne of complete feed.

Nutrients	Starter	Grower (restricted)	Breeder
Vitamin A (IU)	10,000,000	10,000,000	12,000,000
Vitamin D_3 (ICU)	1,500,000	1,500,000	2,500,000
Vitamin E (IU)	15,000	15,000	25,000
Vitamin K_3 (g)	2	1.5	2.5
Thiamine (g)	2.5	2.5	2.5
Riboflavin (g)	6.0	5.0	8.0
Pantothenic acid (g)	12.0	10.0	16.0
Niacin (g)	40.0	35.0	40.0
Pyridoxine (g)	4.5	3.5	4.5
Biotin (g)	0.20	0.15	0.30
Folic acid (g)	1.30	1.00	1.30
Vitamin B_{12} (g)	0.015	0.010	0.025
Choline (g)	1300	1000	1200
Iron (g)	96	96	96
Copper (g)	10	10	10
Iodine (g)	0.40	0.40	0.40
Manganese (g)	60	60	80
Zinc (g)	70	70	70
Selenium (g)	0.15	0.15	0.30

Note: Antioxidants should be added at levels recommended by the manufacturer. Antioxidants are especially important in hot climates and where fats are added to the ration. A coccidiostat is to be used for the starter and grower at a level which would allow the development of immunity from day-old to about 12 weeks of age.

Egg mass output usually continues to increase after peak egg production has been reached. Therefore, peak feed, which is usually started from about 40–50% of production, should be maintained for 3–4 weeks after maximum egg production has been reached. Changes in consumption time are good indicators of over- or underfeeding. Changes usually precede alteration in body-weight by 2–3 days and deviation in egg production by 1–2 weeks. Several stressors have been shown to affect the time required to eat the daily allowance, high environmental temperature being one of the most important.

Sex-separate rearing for most broiler breeder strains is recommended and especially for 'high-yielding' males. Males, according to Boren (1993), should reach at least 140% of the female weight before mixing them together. The practice of feeding the males separately from females has become widely used in recent years. This is done to avoid overeating and to keep the body-weight of the male at the right level. Males that become

Table 10.5. Nutrient requirements of meat-type breeders in units per bird per day.

Nutrient	NRC (1994) Females	NRC (1994) Males	Harms (1987) - females
Protein (g)	19.5	12.0	20.6
Arginine (mg)	1110	680	1379
Lysine (mg)	765	475	938
Methionine (mg)	450	340	400
Methionine + cysteine (mg)	700	490	754
Tryptophan (mg)	190	–	256
Calcium (g)	4.00	0.20	4.07
Non-phytate phosphorus (mg)	350	110	300
Sodium (mg)	150	–	170

overweight during the production cycle tend to suffer more from heat stress. Furthermore, the decrease in male fertility can be overcome by this practice of male-separate feeding. Males are usually fed about 120–130 g per bird per day. For maximum benefit of the separate-feeding system, a diet formulated especially for males is essential. McDaniel (1991) recommended 11–12% crude protein and 2800 kcal kg^{-1}, the remaining nutrients being the same as that of a pullet grower diet. Daily nutrient requirements of the broiler breeder male as recommended by NRC (1994) are shown in Table 10.5. When it is difficult to formulate a specific male diet, a grower diet can be used by incorporating in it a breeder vitamin and trace mineral premix. Ration specifications for such a grower are shown in Table 10.6 and can be fed to males from 3 to 64 weeks of age.

Wilson *et al.* (1992) compared restricted feeding with an *ad libitum* programme in males aged 22–58 weeks. Full-fed males showed higher body-weight at 30 weeks of age (5.03 kg compared with 3.85 kg). By 45 weeks, there was no difference in body-weight between the two groups. The full-fed birds produced a high concentration of spermatozoa in their ejaculate during the early production period, but the trend was reversed after 48 weeks of age. The authors concluded, on the basis of the entire cycle, that excess body-weight reduces reproductive capacity. The practice of restricted feeding using a separate male system not only gives better reproductive performance, but also reduces leg abnormalities, especially during the second half of the production cycle.

Table 10.6. Ration specifications for starter, grower and breeder.

Nutrients	Chick starter (0-3 weeks)	Grower: female (3-21 weeks), male (3-64 weeks)	Breeder I (21-40 weeks)	Breeder II (41-64 weeks)
Protein (%)	18	15	16	15
Metabolizable energy				
(kcal kg^{-1})	2850	2700	2750	2700
(kcal lb^{-1})	1295	1225	1250	1225
Megajoules (MJ kg^{-1})	11.9	11.3	11.5	11.3
Calcium (%)	1.00	1.00	3.20	3.40
Available phosphorus (%)	0.45	0.45	0.45	0.40
Sodium (%)	0.17	0.16	0.16	0.16
Chloride (%)	0.15	0.15	0.15	0.15
Potassium (%)	0.60	0.60	0.60	0.60
Lysine (%)	0.90	0.75	0.75	0.70
Methionine (%)	0.40	0.35	0.36	0.34
Methionine + cysteine (%)	0.70	0.60	0.65	0.60
Tryptophan (%)	0.18	0.16	0.17	0.16
Threonine (%)	0.65	0.58	0.60	0.58
Arginine (%)	0.95	0.80	0.80	0.80

Breeder House Management

Proper management procedures can considerably improve performance and reduce mortality in breeders. Whenever ambient temperatures exceed 35°C, mortality can reach 25–30% in a poorly managed house. The greatest losses occur among older and heavier birds, and thus breeders are particularly subject to heat stress.

Egg gathering

When pen temperature approaches 30°C, eggs should be gathered hourly because embryo development is very rapid at that temperature. Eggs should be moved quickly into a room or building which is cool enough to slowly retard egg temperature. A separate building equipped with mechanical refrigeration is preferred. Failure to provide frequent gathering and cooling can lead to embryos developing to a point where they cannot recover. Eggs should be gathered hourly in such situations and placed in a cool clean room for 5–6 hours. This is because it has been known for many years that hatching eggs require 5–6 hours of embryonic development to reach a desired embryo size before they are cooled (Heywang,

1945). More recently, Fasenko *et al.* (1992) conducted a trial on preincubation development of the embryo and subsequent hatchability of eggs held in nests for periods of up to 7 hours under summer conditions in the state of Georgia. Neither embryonic death, fertility, hatchability or hatch-to-fertile ratio were affected by duration of holding time in the nest. Embryos from eggs retained in nests for extended periods showed greater development than eggs removed promptly from nests. Meijerhof *et al.* (1994) studied the effect of differences in nest-box temperature, storage time, storage temperature and presetting temperature on hatchability of eggs produced by broiler breeders of two different ages (37 and 59 weeks of age). A higher temperature in the nest box, longer storage periods and higher storage temperatures, especially at longer storage periods and higher presetting temperatures, significantly reduced the hatchability of fertile eggs from the older birds. For the younger birds, a significant reduction of hatchability was found only for the longest storage period. Table 10.7 shows the effect of nest-box temperature on hatchability of broiler breeder eggs produced at 37 and 59 weeks of age. These authors suggested that the decrease in hatchability of fertile eggs from older birds is related to the increased sensitivity of these eggs to non-optimal preincubation treatments. After cleaning and fumigation, eggs are to be placed in a cold storage room. Failure to cool eggs in this manner can also result in early hatching and a spread in hatching time from first to last chick. To help hatching eggs to cool down, gather eggs on plastic trays. The best tray designs are those with open fabrication. Avoid the use of paper trays until the eggs have cooled. Never pack warm eggs in cartons or cover the eggs until such time as the eggs have lost heat (below 18–10°C).

Nest hygiene

In the vast majority of cases, the egg is clean and free from bacteria when first laid. However, if nests and nest litter are contaminated with micro-organisms, eggs can become infected. In a tropical climate, it is recommended to change nest litter frequently (every 7–14 days) and to add 25 g of paraformaldehyde to each nest after a litter change. Nests should be sprayed twice weekly to suppress bacteria. The best time to carry out this treatment is late in the day. A broad-spectrum disinfectant such as 1.5% formalin plus 1.5% of a 20% quaternary ammonium compound in water is recommended. Another good practice to keep nests clean is closure of the nest at night to prevent birds from roosting and sleeping in them.

Nest type

There are many types of nests in use around the world, some of which are more suitable for warm climates and others for cold climates. A colony-type

Table 10.7. Percentage hatchability of two ages and three simulated nest-box temperature treatments (from Meijerhof et al., 1994).

Age (weeks)	'Nest' temp. (°C)	% Hatchability
37	30	91.6[a]
	20	92.3[a]
	10	92.3[a]
59	30	86.1[a]
	20	88.5[b]
	10	86.5[ab]

[ab]Means within bird age treatment with no common superscript are significantly different ($P < 0.05$).

nest, when fully occupied, would be extremely hot and not suited to a hot climate. An automatic nest, which usually holds more than one bird at a time and is open to front, bottom or rear, is suitable for a warm climate. Most automatic nests have the egg roll-away feature, which is good for a warm climate. The conventional open-front nest, when designed with a wire and open construction with an egg roll-away feature, is very suitable for a hot climate. However, when the base is solid and the egg roll-away feature is absent, this nest is the least desirable for a hot climate. The egg roll-away feature is very desirable for hot climates, because it permits the egg to move away from the hen and thus cause cooling to commence.

Subiharta et al. (1985) studied the effect of two types of nests used in rural Indonesia on hatchability and found that for the bamboo cone nest hatchability was 77.37% and for the wooden box nest it was 66.39%.

Egg hygiene

Eggs that come in contact with faecal material run a high risk of being contaminated with bacteria. There are many ways to clean eggs and disinfect them. Generally speaking, if more than 10% of eggs have faecal contamination, a management problem exists and should be corrected. Birds usually consume most of their feed during the cooler parts of the day. Try to adjust the light on–off times and the feeding times to correspond to the coolest times of the day. Make sure that the first egg collection is finished within 3 hours of the lights coming on. This is to avoid dirty and nest-soiled eggs. Workers should wash and disinfect hands hourly and just prior to egg gathering. Immediately following gathering, eggs should be fumigated with formaldehyde for a 20 min period. One gram of potassium permanganate to 1.5 ml formalin per cubic foot (approximately 0.03 m^3) of fumigation cabinet space is recommended. Proudfoot and Stewart

(1970) evaluated the effects of preincubation fumigation with formaldehyde on hatchability and found that embryo viability was not impaired when the above procedure was followed. Spraying eggs with a disinfectant soon after they are laid is an alternative to fumigation after gathering. The most common disinfectant in use in North America at present is hydrogen peroxide at a concentration of 5% in water.

Water supply

Additional waterers should be placed in the centre of breeder houses during hot weather. This increased supply of water can be provided by installing additional automatic waterers. It is especially important to provide adequate water in places where birds congregate, usually the centre of the house. Water supplied at the right place will save many birds, since they will not move any great distance to get water when they are overheated.

Since meat-type breeders are feed-restricted during growing, the amount of feed restriction greatly influences the quantity of water consumed by these birds. Figure 10.2, produced from data presented by North and Bell (1990), shows the amount of water that such growing pullets should consume at varying ambient temperatures when feed is restricted. Water consumption is optimal when the average daytime house temperature is 21°C. At 10°C, a bird should drink about 81% of the amount consumed at 21°C, 167% at 32°C and 202% at 38°C. When meat-type breeder pullets are placed on a restricted feeding programme, they consume much more water than usual. In fact, they gorge themselves with water in order to feel full and also out of boredom. This often leads to a wet droppings problem. Limiting the time that water is available to broiler breeders will often help to prevent wet litter problems by reducing the total water usage of the birds (Bennett, 1988). Therefore, these pullets are sometimes placed on a programme of water restriction. In hot climates, however, this is not recommended and even in temperate regions, when average daytime house temperatures exceed 27°C, we do not restrict water.

Cage vs. floor

For breeder houses, the industry has for many years moved from the full-litter floor to a two-thirds slatted floor. The slatted floor has given superior results as compared with the full-litter floor. The move to the slatted floor has helped in allowing for about 15% higher density, fewer dirty floor eggs, lower disease incidence and considerable saving in cost of litter. Litter is much easier to keep dry and clean when the feed and water systems are located over the slat area. The same advantages have been found to be true in the hot regions of the world. However, this system is not free of

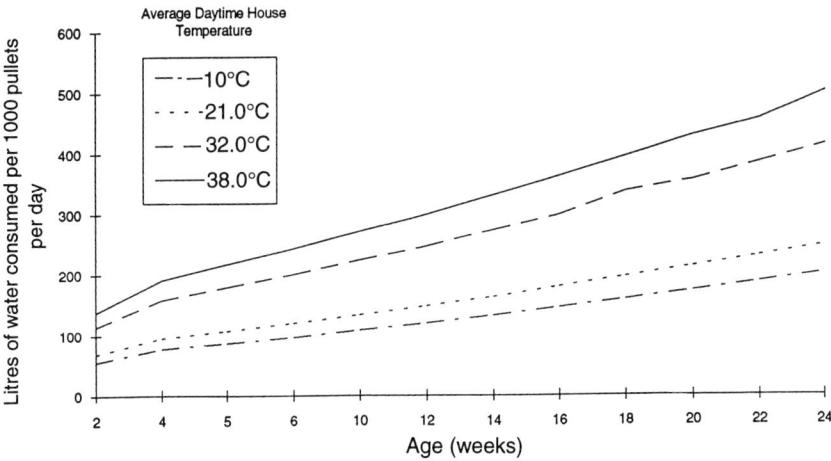

Fig. 10.2. Water consumption by standard meat-type growing pullets for females grown separately, on daily or skip-a-day feed restriction.

problems. High density in breeder houses increases mortality and reduces egg production. Crowding in the centre aisle is a major problem. Some poultryfarmers have reduced mortality by cutting the density so that each hen has 0.225 m² (2.5 square ft) of space.

The use of cages for broiler breeders was started in some European countries and has moved outside Europe in recent years to a limited extent. Its use for breeders in hot climates is still very rare. Renden and Pierson (1982) compared reproductive performance of dwarf broiler breeder hens kept in individual cages and artificially inseminated with dwarf breeders kept in floor pens and naturally mated. Mean body-weights and egg weights were significantly greater for caged hens. There were no differences in feed conversion, hen-day egg production or percentage hatchability between treatments. Fertility of hens housed in floor pens was significantly better than that of hens housed in cages. Leeson and Summers (1985) studied the effect of cage versus floor rearing on performance of dwarf broiler breeders and their offspring. Cage-reared birds were consistently heavier than floor birds. Floor-reared birds had longer shanks. Rearing treatments had little effect on laying performance of birds in cages from 20 to 44 weeks of age. Broiler chicks from the 40-week hatch were reared to 56 days and no differences in weight were seen at 42 or 56 days of age. Yang and Shan (1992) conducted an experiment on 5000 breeder hens at Beijing University in China. They observed that performance in the cage system was superior to that on the floor. Criteria measured were egg production per hen housed, percentage fertility and viability up to 62 weeks of age. They concluded that the cage system will dominate the broiler breeder industry of China very soon. Yalcin and Hazan (1992) studied the effect of different strains and cage density on performance of caged broiler breeders in Turkey. They concluded that medium-weight breeder strains placed at 677 cm^2 per bird achieve the best performance.

In hot climates where environmentally controlled houses with evaporative cooling devices are used, and where labour is cheap for purposes of artificial insemination, cages for breeders appear to be promising.

Water-cooled perches

Simple and economical methods for alleviating the effects of heat stress on breeder performance are lacking. Muiruri and Harrison (1991) studied the effects of roost temperature on performance in hot ambient environments. They found that water-cooled roosts improved hen-day egg production, daily feed intake and hatchability. The water-cooled roosts minimized deleterious effects of heat stress through conductive heat loss from the birds to the roosts. They improved both the productive and reproductive performance of chicken hens.

Disease control and prevention

The 'all-in, all-out' system is the recommended system for broiler breeder projects in hot regions and is the best for disease control and prevention. In hot climates, birds should be handled only during the cool hours of the day. Vaccinations should be performed in drinking-water very early in the morning because the temperature of the water at that time is usually cool. In extreme high temperature, do not water-fast birds more than 1–2 h. Whenever you administer drugs in the drinking-water, remember that, in the case of high environmental temperature, water consumption can be three to four times higher than at a low temperature. Whenever you suspect overdrinking, calculate the dose per bird per day and dissolve the total amount needed per day in a quantity of drinking-water which could be consumed in 3–4 h.

The basic principles of disease control, such as cleaning and disinfection of the premises between flocks, control of potential viruses, isolation of brooding and rearing of young stock away from potential sources of infection, etc. are extremely important in hot climates because the environmental conditions of high temperature and humidity are ideal for microbial growth.

One of the major disease problems that have received serious attention in recent years is the problem of salmonellosis. The reason for this increased interest in the disease was the *Salmonella enteritidis* outbreaks that occurred in both the USA and the UK in 1988. These outbreaks focused on the rising worldwide *Salmonella* epidemic and the urgent need for

Salmonella control in poultry. Today, many countries have developed a fully integrated programme for *Salmonella* control which includes all sectors of the poultry industry and which has as its ultimate objective the maximum reduction of *Salmonella* transmission through poultry products.

The largest number of *Salmonella* species belong to the paratyphoid group. *Salmonella pullorum* and *Salmonella gallinarum*, which are specific to poultry, have been largely eradicated in many countries, but are still causing problems in several developing countries and need to be controlled. *Salmonella enteritidis* is one of several hundred species in this paratyphoid group, but, because of its involvement in food poisoning, it has received most attention. It is believed that the main carrier of *Salmonella* into poultry farms is contaminated feed. However, other natural sources, such as wild birds, insects, rodents and humans can spread this infection easily within any poultry operation. Once the organism is ingested by the bird, it multiplies very rapidly, creating large colonies in the intestinal tract. These organisms may lie dormant in clinically healthy birds for long periods of time. Transmission of *Salmonella* can take place through faecal contamination of the eggshell during or after the laying process and through subsequent eggshell penetration. Horizontal transmission from numerous live carriers or by other vehicles, such as feed (as pointed out earlier) and water, is a very common route of transmission.

Although *Salmonella* is very difficult to eradicate, it can be reduced by following very strict hygienic practices. A *Salmonella* control programme should include a very strict farm and house sanitation programme to prevent horizontal transmission of the bacteria and infection of the flock, and a sound monitoring programme to keep you aware of your disease status and to alert you of any changes that might be taking place. Such a monitoring programme is the most essential for effective *Salmonella* control. Some basic components of a monitoring programme for breeder and hatchery operations are environmental sampling and regular antigen blood testing.

To carry out environmental sampling for breeder-flocks, take swab samples of fresh faecal material every 60 days from each flock, one sample per pen or per 500 birds. Each swab should contain a pool of several spots from the pen. The samples are then processed and incubated in the laboratory and the results obtained in 72 hours.

A regular plate test (P-T) antigen blood test should be performed every 90 days on 150 birds per house. Tested birds should be leg-banded or individually identified so that you can go back and pick out the reactors for culturing. If *Salmonella* is isolated from these, a sensitivity test should be carried out on isolated bacterial cultures and the proper antibiotic used for treatment for 5–7 days' duration. At the end of the treatment period, nests should be fully cleaned and disinfected and nest litter changed. Floor litter should be removed completely and the floor treated with an application of 5% formalin. During this time, eggs should not be set from the infected flock until the lapse of 10 days from the end of treatment.

Monitoring programmes for hatchery operations should include once-a-month testing of dead-in-shell embryos at transfer time. About six to ten specimens should be collected per source flock (up to 30 embryos depending on number of donor flocks). Samples are sent to the nearest laboratory and results can be known within 72 hours. If *Salmonella* shows up, then the follow-up procedure is the same as that for breeder flock sampling. The sensitivity test can be done on the same culture that isolated the *Salmonella* and therefore the idle period in this case can be shorter.

Since feed is considered an important carrier of *Salmonella* into poultry farms, the following steps are recommended.

1. Insist that your feed mill or the mill that you purchase feed from has a *Salmonella* control programme. Although animal by-products are more susceptible to *Salmonella*, plant products can be infected. Pelleted feeds are preferred to mash feeds since pelleting involves some degree of cooking and will expose less surface area to *Salmonella* contamination.

2. Acid treatment of feeds has been found to be effective in reducing *Salmonella*. Propionic acid is most effective against moulds, while formic acid has been found to be effective against *Salmonella*. The use of 0.9% formic acid in feed leads to complete decontamination after 3 days of treatment.

3. Dust has been found to be an important vehicle for *Samonella*. Dust control contributes to production of uncontaminated feed. All equipment must be cleaned and disinfected with a 25 p.p.m. iodine solution.

4. Samples of feed from each load should be taken and identified by delivery date, flock certification number and type of feed and retained for at least 3 months.

5. Store and transport feed in such a manner so as to prevent possible contamination.

Biosecurity for a breeder and hatchery operation is far more critical than for a regular layer or broiler operation. The progeny of a single parent flock populate many farms, with each breeder hen producing over 150 broiler chicks during her lifetime. As Frazier (1990) points out:

> Biosecurity in a breeder operation must consist of a lot more than a pair of throw-away plastic boots. Dust accumulated on outer clothing and one's hair is just as dangerous as dirty footwear. Unsanitized vehicles accumulate contamination day after day. They don't have to appear grossly dirty to be biologically dirty. These are only a few of the many aspects of a minimum biosecurity programme.

Therefore, it is absolutely essential that breeder and hatchery operators develop their own disease monitoring and overall biosecurity programmes based on geographical location and information on disease outbreaks, prevention and control methods known for that area. Biosecurity programmes are often costly, but, to attain the greatest possible success in disease prevention, they have to be instituted and constantly followed.

Storage of Hatching Eggs

As the hatching egg is a highly perishable commodity, it must be stored under specific conditions if its hatching ability and qualities are to be retained. The average egg has between 8000 and 10,000 pores. The pores are the means by which the egg breathes, losing water by evaporation and taking up oxygen, which is transported to the embryo. The egg starts losing water from the time it is laid; the drier the atmosphere, the greater the water loss. With the loss of its moisture, egg vitality is also lost and this affects not only hatchability but also chick quality. A hatching egg storage room requires a uniform temperature and relative humidity level. The longer the storage period, the more critical these two factors become. For storage up to 7 days, a temperature of 16–17°C and 80% relative humidity are recommended. For a longer period of storage, a lower temperature of around 12°C and 85% relative humidity would be more desirable. For an even longer storage period, enclosing eggs in plastic bags and placing the small end of the egg upwards would be beneficial. The beneficial effects of plastic packaging on hatching eggs have been known for many years. Proudfoot (1964, 1965) reported on several preincubation treatments for hatchability and found that hatchability was maintained at a higher level when eggs were enclosed in plastic film during the preincubation period. Gowe (1965) further illustrated that flushing eggs stored in plastic bags with nitrogen gas improved their hatchability. He showed that hatchability of fertile eggs stored 19–24 and 13–18 days in cryovac bags flushed with nitrogen was about 7–8% higher than that of comparable eggs stored for the same periods in cryovac bags and in an atmosphere of air. Obioha et al. (1986) studied the effect of sealing in polythene bags on hatchability of broiler eggs kept at two storage temperatures of 15 and 22°C and held for seven storage periods. Sealing eggs with polythene bags maintained significantly higher hatchability than leaving them untreated under both temperature regimes.

Eggs stored on open incubator trays, even in ideal conditions, are prone to rapid water loss. For storage over 3 days, covering eggs with plastic will extend their shelf-life to about 6 days by reducing water loss. A longer storage time on incubator trays leads to rapid deterioration of egg hatching quality. Baker (1987) studied the effects of storage on weight loss in eggs and found that, irrespective of size, eggs lose an average of 2% in weight when stored at 10°C (50°F) and 5% at 21°C (70°F) after 20 days of storage.

Furuta *et al.* (1992) studied the effects of shipping of broiler hatching eggs from a temperate to a subtropical region in Japan on their hatchability. They concluded that, if hatching eggs are properly shipped, there should be no decrease in hatchability. The most critical factor is the time between laying and setting, which should be within 7 days.

Hatchery Design

An incubator is designed to operate within certain environmental limits. If the environmental conditions are not present year-round naturally, they must be provided by artificial means. For incubators to function correctly, the incubator room must provide three elements.

1. A minimum temperature of 20°C and a maximum of 28°C with an ideal temperature of 25°C.
2. A minimum relative humidity of 50% and maximum of 70% with an ideal humidity of 60%.
3. A minimum of 14 m^3 of fresh outside air per minute per 100,000 eggs in setters, and a minimum of 14 m^3 of fresh air per minute for each 14,000 eggs in hatchers.

Conditions outside these temperature and humidity parameters will cause an interaction in the incubators which produces a poor environment for incubation. Where the environment is hot and humid, large airy rooms are preferable. The following conditions are recommended.

1. A roof height up to 8 m.
2. A side-wall height of 6 m.
3. A ridge opening at roof of 1.4 m and at side-wall of 2.0 m.

For a hot dry climate, evaporative cooling could be advantageous, with a roof height of 6 m and a wall height of 4.5 m.

Hatchery Hygiene

A hatchery should be isolated and positioned on land at a distance of at least 1 km from other poultry facilities. Prevailing wind direction is a major consideration. When locating the hatchery on the site, the wind should blow from the egg receiving and setting end to the hatching rooms and chick room and never vice versa. The hatchery building must have completely separate rooms, with each room having an independent air supply and temperature and humidity requirement and control. The following room separations are suggested: egg rooms, setting rooms, hatching rooms, chick removal, chick tray wash, box storage, staff lunch, and showers and toilets. Where hatching is to take place twice weekly, one hatch room is satisfactory. Should a hatchery hatch more frequently, two or more hatch rooms are desirable to prevent bacterial contamination from one hatch to the next.

Basic hatchery design should provide a corridor separation of setting and hatching rooms. A separation of hatching and chick rooms is also necessary. Incubator rooms functioning with positive pressure and a strong negative pressure in the corridor prevent air movements from one room to the next.

Incubation in a Tropical Climate

Incubation in a tropical climate is not very different from other climates. The key to good incubation is the hatchery environment and the three elements mentioned above required for incubation. A high level of hygiene is needed inside and outside the incubator. The incubator depends on five elements in its design. Management must ensure that each of these elements is functioning correctly at all times. They are heating, humidity, cooling, ventilation and turning. When any one element malfunctions, problems will result.

Incubator conditions

The chicken embryo absorbs heat from its surroundings up until approximately 288 h of incubation. After this, it commences metabolizing its own energy and producing heat. At 288 h of incubation, embryo temperature is approximately 37°C, while by 480 h its temperature is up to 39.5°C. Embryo body temperature must be held close to 37°C if successful hatching is to occur. Controlling this excess heat has been a major concern in artificial incubation for over 2000 years. Today, it is still the most important factor in an incubator.

Modern incubators cool by two methods. Some have incorporated both. When purchasing an incubator for a hot climate, the cooling factors must be fully understood before purchase is made. All modern incubators are designed to use one method of cooling or another, but none of them provides the coolant. This is a separate and usually an expensive item. Two different cooling systems are used.

1. Water cooling. This requires water to be at a temperature of 12°C. In most hot climates, ground water is above 26°C and is of little cooling value.

It is therefore necessary to use water-chilling equipment to reduce water temperature and to circulate it around the hatchery to each incubator.
2. Air cooling. When incubator room air exceeds 28°C, some form of air cooling is required. If the outside air is dry, evaporative cooling can reduce hatchery air temperature below the critical level. However, if the environment is humid, it is necessary to use thermostatically controlled mechanical air conditioning to cool the air around the incubator air intake.

It is impossible to hatch chicks properly in a hot climate without providing the cooling element. The ideal temperature for incubator and chick rooms is 25°C, while for the egg work room it is 23–25°C. Relative humidity in all live rooms in a hatchery should be about 60%.

Incubator problems in hot climates

One of the problems occurring in incubators in a hot climate is an increase in temperature during midday periods causing room temperature to increase with outside temperature because of insufficient cooling capacity. If incubators are water-cooled, cold water supply temperature at incubator entry should be 12°C. The diameter of the cold water supply line is critical in a hot climate. The pipe must be large enough (minimum 2.5 cm diameter) to supply adequate cold water and insulated to avoid water temperature loss and condensation drip. Incubator cooling coils are usually not large enough for hot climates and therefore doubling the number of coils is normal procedure in this type of environment. If incubators are air-cooled, then air-conditioning is necessary to provide sufficient cold air at point of use. The air-conditioner duct should be close to air intake and usually the cool area diameter around intake should not need to exceed 1 m.

Hatcheries located in very dry climates require mechanical means in order to maintain the desired 60% relative humidity in live rooms, i.e. incubator rooms and chick processing rooms. Failure to maintain this desired humidity level will require humidifiers to work harder. Humidifier systems in incubators are designed mainly to correct small differences in humidity. When they operate for long periods of time and cold water is used, there is a cooling effect which prompts incubator heaters to operate more frequently. Under such conditions, hatchery managers restrict ventilation, thus limiting fresh air, which produces poor incubation conditions.

Some chicks hatch much earlier than others in the same incubator. If some chicks are hatching much sooner than others, it is most probably because they have received more heat somewhere along the incubation process. In this case, the following steps are suggested.

1. Ensure all eggs receive the same treatment the day they are laid and during storage.

2. Throughout the storage and prewarming period, eggs in the top and bottom trays should be of the same temperature.
3. Follow the incubator manufacturer's recommended set pattern.
4. It is possible to find a different temperature in the incubators with top or middle being somewhat warmer than the bottom. In this case, at the time of transfer, place eggs from the coolest part of the setters to the warmest of the hatchers.
5. When transferring eggs, transfer the eggs nearest the floor first as this is the coolest location.

Incubation time

Normal incubation time for chicken eggs is 21 days or 504 h from the time the eggs are set to chick removal. There are, however, many reasons for chicks to hatch earlier or later.

1. Breed type: differences between breeds are usually no more than 3 h.
2. Broiler breeds in warm climates can hatch up to 12 h early. The important factor here is to remove the chicks when ready. They are considered ready when 5% are still damp on rear of head. Chicks confined to a hatcher can overheat and dehydrate and be exposed to large volumes of bacteria.
3. As flocks advance in age, eggs produced become larger. Large eggs take longer to lose moisture in the same environment as small eggs and thus they need more time. Therefore set older flocks 3–4 h sooner than younger flocks. If you can incubate eggs from older flocks separately, then operate the incubator at a 5% lower humidity level.
4. When breeders first come into production, egg size is very small and eggshells tend to be thick. Eggs laid during the first 4–6 weeks, though quite fertile, usually have lower hatchability than they do when the pullets are fully mature and laying larger eggs consistently. It has been said that eggs from young pullets with thick shells do not lose sufficient water during incubation. Such eggs can benefit from lower setter humidity levels. However, this is not always easy to provide. It is therefore suggested that a minimum egg size of 52 g be used and that eggs from immature breeders be set 2–3 h ahead of what is considered normal.

Prewarming eggs for setting

Some incubators will benefit from several hours of prewarming eggs prior to placing them in the incubator. Others do not benefit from this practice and prewarming can be detrimental. One should check with the incubator manufacturer about the need for prewarming.

For single-stage incubators, prewarming is not necessary. All eggs are set at one time with good air circulation. All eggs warm up at the same time. Jamesway Big Jay and Super Jay incubators, because of their design, do not require egg prewarming. Air passes from warm to cold eggs and is heated prior to recirculation. However, if the eggs could be prewarmed all to the same warm temperature, hatch time would be more punctual and predictable. Placing a thermometer in an egg from the top tray in a buggy and again in an egg from the bottom tray will reveal temperature spread from top to bottom. Usually, top tray eggs are warmer than bottom tray eggs and therefore hatch slightly sooner. The objective is to have hatch time from first to last chick as close as possible.

Water Supply and Cleanliness

A major concern for a hatchery is the cleanliness of its water supply. Conditions in a hot climate are prone to bacterial contamination of water. As incubators are the ideal environment for bacteria to reproduce, a hatchery must take steps to ensure a clean water supply. Chlorine is usually used in the water supply to provide 5 p.p.m. at point of use. Failure to do this will produce bacterial-related diseases in the baby chick.

Much of the water supply contains minerals and different sediments. These elements tend to clog the small orifices of the incubator humidification system, necessitating very frequent cleaning of these orifices. It is therefore recommended that hatcheries use purified water to supply humidity. This is best accomplished with the use of a reverse osmosis filtration system, which guarantees water purity from virus, bacteria, minerals and sediment. Such systems are not overly expensive and have a short payback period.

A hot climate very often means contaminated water. A high microbial count in the water supply creates problems for the breeders. Contaminated eggshells produce hatchery contamination, which again contaminates eggs and chicks. Egg and incubator rooms should be washed daily, along with all utensils used, with a chlorine-type disinfectant. Floors, the inside of setters and setting rooms should be washed following each and every transfer, using a phenolic disinfectant.

Moulds are a serious problem when they infect chicks. Aspergillosis is a serious problem when it appears in the system. This is a common condition in a hot environment. This mould can infect the litter in the breeder house and be transported to the hatchery in and on eggs and egg carriers. The mould can invade incubators, multiply and infect eggs by penetrating the shell and infecting the lungs of chicks after hatching. In the hatchery, moulds are commonly found in ventilation systems, on evaporative cooling pads and in humidifiers. Moulds should be destroyed from their source, starting with the feed, by the use of a mould inhibitor and in the litter. In hot climates, refuse left following a hatch must be removed from a hatchery site as soon as possible.

Altitude

In a study conducted at an altitude of 5500 feet, Arif *et al.* (1992) reported that hatchability was significantly higher for eggs stored with the small end up (85.8%) than in those with the large end up (74%). Early embryonic mortality did not differ between groups, but late embryonic mortality was lower in eggs stored with the small end up. Christensen and Bagley (1988) examined turkey egg hatchability at high altitudes at two different oxygen tensions and two incubation temperatures. Incubating in a 149 Torr oxygen environment at 37.7°C gave significantly better hatchability than a 109 Torr oxygen environment at 37.5°C. Embryonic mortality data indicated that the higher incubation temperature in combination with increased oxygen tension decreased embryonic mortality during the third and fourth weeks of incubation and resulted in higher hatchability. The data suggest that hatchery managers at high altitudes should supplement with oxygen and incubate turkey eggs at higher temperatures than at lower altitudes.

Chick Processing and Delivery

As mentioned earlier, chicks should be removed from the hatchers when only 5% are left with dampness at the rear of their heads. They should be moved into a well-ventilated room with a temperature of approximately 26°C. At this point, the temperature surrounding the chicks should be lowered from the incubator temperature of 37°C to 33°C for best comfort. A careful monitoring of temperature inside the hatch tray or chick box at this time is important. If the temperature surrounding the chick is too high, quality will be quickly lost. Most people place 100 chicks in the confines of one box. In a hot climate, the number of chicks in a box should be reduced to about 80 (20 per compartment) thus lowering the temperature in the box. Chick boxes need to be well placed while standing in the chick room. If room temperature exceeds 32°C, fans similar in size and power to hatcher fans are needed to blow air through boxes in order to prevent overheating. Chick delivery in a hot climate can be very hazardous, especially if the delivery is for long distances. Chick delivery vans need to be equipped with thermo-king air-conditioners, especially for daytime delivery. In a hot climate, deliveries before sunrise or after sunset are recommended. High environmental temperatures during transportation of neonatal chicks cause dehydration and death. Neonatal chicks cope with heat by evaporation. Therefore, the initial water content of neonatal chicks may be important. Chicks hatching early lose more body-weight in the hatcher than those hatching later. This loss is primarily due to water loss (Thaxton and Parkhurst, 1976). Hamdy *et al.* (1991a) studied the effects of incubation at 45 versus 55% relative humidity and early versus late hatching time on heat tolerance of neonatal male and female chicks. They concluded that chicks that hatched late (i.e. with a short holding period in the hatcher) and coming from eggs incubated at 45% relative humidity had increased heat tolerance in comparison with the other chicks.

Van Der Hel *et al.* (1991) reported that exposing chicks to suboptimal conditions during transportation may lead to impaired production. In order to reduce the negative effects of ambient temperature during and after transport, it is important to assess the thermal limits between which animals can survive without long-lasting negative effect. These workers conducted trials to estimate the upper limit of ambient temperature of neonatal broiler chicks by measuring heat production, dry matter and water loss in the body and yolk sac during 24 h exposure to constant temperature from 30.8 to 38.8°C. Their results showed that this critical temperature was between 36 and 37°C.

Body-weight loss and mortality weights of neonatal chicks are increased when they are transported at high temperatures. After arrival at a farm and placement at normal thermal conditions, chicks that have survived exposure to high temperature will eat and grow less and may have a higher risk of dying than those exposed to lower temperatures (Ernst *et al.*, 1982). A high heat tolerance is therefore an important attribute. Hamdy *et al.* (1991b) reported that a 10% lower than normal incubation relative humidity did not negatively affect performance of chicks after heat exposure. Chicks hatching early had a higher risk of dying after exposure to heat (39°C for 48 h) than late-hatching ones. Chicks hatching late from eggs incubated at low relative humidity were most heat-tolerant.

Conclusions and Recommendations

1. High environmental temperatures have been shown to depress fertility and hatchability. Decreases in fertility are due to depressions in sperm production, while decreases in hatchability have been shown to be due to reduced feed consumption, which causes embryonic nutritional deficiencies and deterioration in egg quality.
2. The feeding of Leghorn-type breeders in hot climates is no different from that of the table egg layer, except for the extra fortification of the diet in vitamins and trace minerals known to be critical for optimum reproductive performance.
3. The feeding and feeding programmes of meat-type breeders are critical in hot climates because of the feed restriction used and differences in

maintenance requirements as well as quality of feeds available in many hot regions. The use of a high-protein prebreeder ration may be useful in those areas where breeder pullets are under weight at onset of production.

4. Certain studies in hot regions have shown that dietary vitamin A, vitamin E, vitamin C and riboflavin are important for maximizing fertility and hatchability in those areas.

5. Broiler breeder pullets should not be severely restricted early in their life cycle in hot climates since a moderate restriction at that age is less stressful. This allows for better uniformity and proper fleshing, both of which contribute to good hatching egg production.

6. Feed allowances for growing meat breeders should be determined on the basis of ME requirements per bird per day and these should be adjusted in relation to body-weight and condition and uniformity of the pullets. More feed and feeder space is required when uniformity is low and when feed quality is poor.

7. The energy requirements of the meat breeder hen, especially early in the production cycle, are considered to be very critical and increase from about 300 kcal ME per bird per day at 20 weeks of age to about 400 kcal ME per bird per day at 28 weeks of age. Therefore, feed allowances during this period should exceed those requirements. The daily protein requirement of the meat breeder has been estimated to be about 20 g per bird per day and feed allowances providing less protein can reduce egg weight and body-weight.

8. A feeding programme has been recommended which consists of an 18% protein chick starter with 2850 kcal kg^{-1}, a 15% protein grower with 2700 kcal kg^{-1}, a 16% protein female breeder with 2750 kcal kg^{-1} and a 12% protein male breeder with 2750 kcal kg^{-1}. When it is difficult to use a specific male diet, the grower diet can be used for feeding males separately throughout the breeding period, provided that it is supplemented with the breeder vitamin and trace mineral premix rather than the grower premix.

9. In the majority of cases, poor hatching is the result of mismanagement at the breeder farm, i.e. poor male and female breeder management. Eggs failing to hatch after a normal incubation period should be examined on a regular basis (every 30 days) and compared with a known standard which would indicate where the problem lies.

10. Many hatcheries are located in areas of convenience rather than areas best suited for a hatchery operation. Some critical factors in locating a hatchery are altitude (not exceeding 1000 m above sea level), proximity to poultry buildings, wind direction, clean and plentiful water supply and design to avoid re-entry of used air.

11. Operation of a hygienic hatchery requires a number of key elements to be in place. Other than a clean location, a sound sanitation programme is needed with a desire and knowledge on the part of management to

maintain a hospital-like environment. The type of disinfectants to be used and how and when they are applied are important considerations. Disinfectants used should be effective against both Gram-positive and Gram-negative organisms. Formaldehyde fumigation (where allowed) should be applied at the appropriate times, i.e. eggs soon after lay, eggs just prior to setting and in the hatchers continuously.

12. The operation of hatcheries in tropical climates is not very different from that in a temperate climate. However, the cooling equipment, clean water and hygienic conditions are more critical factors in a hot climate.

13. Water supply in a hot climate is critical with regard to volume, temperature and purity. A substantial amount of water is required for cleaning and disinfection. This should be a separate supply from that used for cooling and humidification. For cooling, water-chilling equipment is needed while, for humidification, a reverse osmosis system is recommended.

14. During processing and delivery, chicks can be and often are damaged. Temperatures need to be monitored in and around the chick boxes to ensure that chicks are not overheated. In warm environments, horizontal fans should be used to move air through chick boxes while stationary in the hatchery. Thermo-king refrigeration is recommended for use in chick transportation. Chick boxes should be partially filled and not confined to one area during holding, in order to prevent overheating.

References

Arif, M., Joshi, K.L., Shah, P., Kumar, A. and Joshi, M.C. (1992) Hatchability of layer eggs stored and incubated in different positions at 5500 feet altitude. *Journal of Agricultural Science* 118, 133–134.

Askar, M. (1927) Egyptian methods of incubation. *Proceedings of the World's Poultry Congress*, Ottawa, Canada, pp. 151–156.

Babu, M., Mujeer, K.A., Prabakaran, R., Kalatharan, J. and Sudararasu, V. (1989) Effect of vitamin E and selenium supplementation on fertility and hatchability in White Leghorn breeders. *Cheiron* 18, 158–161.

Baker, R. (1987) Effect of storage on weight loss in eggs. *Poultry Digest* 46, 276–278.

Bennett, C.D. (1988) Feed and water intake patterns of broiler breeders. *Canada Poultryman*, February, 22–24.

Bennett, C.D. (1992) The influence of shell thickness on hatchability in commercial broiler breeder flocks. *Journal of Applied Poultry Research* 1, 61–65.

Boren, B. (1993) Basics of broiler breeder nutrition. *Zootechnica International*, December, 54–58.

Brake, J., Garlich, J.D. and Peebles, D. (1985) Effect of protein and energy intake by broiler breeders during the pre-breeder transition period on subsequent

reproductive performance. *Poultry Science* 64, 2335-2340.
Brake, J., Walsh, T.J. and Scheideler, S.E. (1992) Nutritional influences in broiler breeders: frame size, feather drop syndrome and reproduction. *Poultry Science* 71 (Suppl.), 16 (Abstract 48).
Brown, H.B. and McCartney, M.G. (1986) Restricted feeding and reproductive performance of individually caged broiler breeder males. *Poultry Science* 65, 850-855.
Buckner, R.E. and Savage, T.F. (1986) The effects of feeding 5, 7 or 9% crude protein diets to caged broiler breeder males. *Nutrition Reports International* 34, 967-976.
Buckner, R.E., Renden, J.A. and Savage, T.F. (1986) The effect of feeding programmes on reproductive traits and selected blood chemistries of caged broiler breeder males. *Poultry Science* 65, 85-91.
Cave, N.A.G. (1984) Effect of a high protein diet fed prior to the onset of lay on performance of broiler breeder pullets. *Poultry Science* 63, 1823-1827.
Chaudhuri, D. and Lake, P.E. (1988) A new diluent and methods of holding fowl semen for up to 17 hours at high temperature. *Proceedings 18th World's Poultry Congress*, pp. 591-593.
Christensen, V.L. and Bagley, L.G. (1988) Improved hatchability of turkey eggs at high altitudes due to added oxygen and increased incubation temperature. *Poultry Science* 67, 956-960.
Cier, D., Rimsky, I., Rand, N., Polishuk, O., Gur, N., Benshashan, A., Frish, Y. and Ben-Moshe, A. (1992a) The effects of supplementing breeder feeds with ascorbic acid on the performance of their broiler offsprings. *Proceedings 19th World's Poultry Congress*, Vol. 1, p. 620.
Cier, D., Rimsky, I., Rand, N., Polishuk, O. and Frish, Y. (1992b) The effects of pellets, mash, high protein and antibiotics on the performance of broiler breeder hens in a hot climate. *Proceedings 19th World's Poultry Congress*, Vol. 2, pp. 111-112.
Clark, C.E. and Sarakoon, K. (1967) Influence of ambient temperature on reproductive traits of male and female chickens. *Poultry Science* 46, 1093-1098.
Couto, H.P., Soares, P.R., Rostagno, H.S. and Fonseca, J.B. (1990) Nutritional protein requirements of broiler breeding hens. *Revista da Sociedade Brasileira de Zootecnia* 19, 132-139.
Damron, B.L., Wilson, H.R. and Harms, R.H. (1983) Sodium chloride for broiler breeders. *Poultry Science* 62, 480-482.
Dobrescu, O. (1987) Vitamin C addition to breeder diets. *Feedstuffs*, 2 March, 18.
Dudgeon, J. (1988) The management of broiler breeders in the laying period. *Proceedings 4th International Poultry Breeders Conference*, pp. 1-7.
El-Ibiary, H.M. (1946) The old Egyptian method of incubation. *World's Poultry Science Journal* 2, 92-98.
Ernst, R.A., Weathers, W.W. and Smith, J.M. (1982) Effect of heat stress on growth and feed conversion of broiler chicks. *Poultry Science* 61, 1460-1461.
Fasenko, G.M., Wilson, J.L., Robinson, F.E. and Hardin, R.T. (1992) Effects of nest holding time during periods of high environmental temperature on pre-incubation embryo development, hatchability and embryonic viability in broiler breeders. *Poultry Science* 71 (Suppl.), 125 (Abstract 373).

Flores-Garcia, W. and Scholtyssek, S. (1992) Effect of levels of riboflavin in the diet on the reproductivity of layer breeding stocks. *Proceedings 19th World's Poultry Congress*, Vol. 1, p. 622.

Frazier, M.N. (1990) Broiler breeder management. *Vineland Update*, No. 32.

Furuta, F., Shinzato, G., Higa, H. and Shinjo, A. (1992) Effects of shipping of broiler eggs from the temperate to the subtropical regions on their hatchability. *Proceedings 19th World's Poultry Congress*, Vol. 1, p. 682.

Ghany, M.A., Kheir-Eldin, M.A. and Rizk, W.W. (1967) The native Egyptian hatchery: structure and operation. *World's Poultry Science Journal* 23, 336–345.

Glatz, P.C. (1988) The effect of restricted feeding during lay on production performance of broiler breeder hens. *Proceedings 18th World's Poultry Congress*, pp. 959–960.

Gonzales, L.M., Elias, A., Valdivie, M., Berrio, I.I., Fraga, L.M. and Rodriquez, C. (1991) A note on the fertility and hatching rate in heavy hens and roosters fed saccharina. *Cuban Journal of Agricultural Science* 25, 191–193.

Gowe, R.S. (1965) On the hatchability of chicken eggs stored in plastic bags flushed with nitrogen gas. *Poultry Science* 44, 492–495.

Hagos, A., Devegowda, G. and Ramappa, B.S. (1988) Restricted feeding of broiler breeders during the growing period. *Proceedings 18th World's Poultry Congress*, pp. 966–967.

Hamdy, A.M.M., Van Der Hel, W., Henken, A.M., Galal, A.G. and Abid-Elmoty, A.K.I. (1991a) Effects of air humidity during incubation and age after hatch on heat tolerance of neonatal male and female chicks. *Poultry Science* 70, 1499–1506.

Hamdy, A.M.M., Henken, A.M., Van Der Hel, W., Galal, A.G. and Abid-Elmoty, A.K.I. (1991b) Effects of incubation humidity and hatching time on heat tolerance of neonatal chicks: growth performance after heat exposure. *Poultry Science* 70, 1507–1515.

Harms, R.H. (1987) Formulation of broiler and broiler breeder feed based on amino acid composition. *Monsanto Technical Symposium*, pp. 116–127.

Harms, R.H., Bootwalla, S.M. and Wilson, H.R. (1984) Performance of broiler breeder hens on wire and litter floors. *Poultry Science* 63, 1003–1007.

Hazan, A. and Yalcin, S. (1988) The effect of different feeding levels on egg production and hatchability of caged broiler breeders. *Proceedings 18th World's Poultry Congress*, pp. 1099–1101.

Hess, J.B., Wilson, J.L. and Wineland, M.J. (1994) Management influences early chick mortality. *International Hatchery Practice* 8, 27–29.

Heywang, B.W. (1944) Fertility and hatchability when the environment temperature of chickens is high. *Poultry Science* 23, 334–339.

Heywang, B.W. (1945) Gathering and storing hatching eggs in hot climates. *Poultry Science* 24, 434–437.

Hocking, P.M. (1994) Feeding broiler breeder males. *International Hatchery Practice* 8, 17–21.

Jayarajan, S. (1992) Seasonal variation in fertility and hatchability of chicken eggs. *Indian Journal of Poultry Science* 27, 36–39.

Kling, L.J., Howes, R.O., Gerry, R.W. and Halteman, W.A. (1985) Effects of early maturation of brown-egg type pullets, flock uniformity, layer protein level and cage design on egg production, egg size and egg quality. *Poultry Science* 64, 1050–1059.

Leeson, S. and Spratt, R.S. (1985) Nutrient requirements of the broiler breeder. *Proceedings Maryland Nutrition Conference for Feed Manufacturers*, pp. 75–80.

Leeson, S. and Summers, J.D. (1980) Effect of early light treatment and diet selection on laying performance. *Poultry Science* 59, 11–15.

Leeson, S. and Summers, J.D. (1985) Effect of cage versus floor rearing and skip a day versus everyday feed restriction on performance of dwarf broiler breeders and their offspring. *Poultry Science* 64, 1742–1749.

Leeson, S. and Summers, J.D. (1991) *Commercial Poultry Nutrition*. University Books, Guelph, Ontario, Canada, pp. 188–189.

Lilburn, M.S. and Myers-Miller, D.J. (1990) Effect of body weight, feed allowance and dietary protein intake during the prebreeder period on early reproductive performance of broiler breeder hens. *Poultry Science* 69, 1118–1125.

Lilburn, M.S., Ngiam-Rilling, K. and Smith, J.H. (1987) Relationship between dietary protein, dietary energy, rearing environment and nutrient utilization by broiler breeder pullets. *Poultry Science* 66, 1111–1118.

McDaniel, G.R. (1991) Management of breeder replacement stock. *Avian Farms Flock Report* 3, 1–5.

McDaniel, G.R., Balog, J.M., Freed, M., Elkin, R.G., Wellenreiter, R.H., Kuczek, T. and Hester, P.Y. (1992) Response of layer breeders to dietary ASA 3. Effects on fertility and hatchability of embryos exposed to control and elevated incubation temperatures. *Poultry Science* 72, 1100–1108.

Meijerhof, R., Noordhuizen, J.P.M. and Leenstra, F.R. (1994) Influence of pre-incubation treatment on hatching results of broiler breeder eggs produced at 37 and 59 weeks of age. *British Poultry Science* 35, 249–257.

Miller, P.C. and Sunde, M.L. (1975) The effects of precise constant and cyclic environments on shell quality and other lay performance factors with Leghorn pullets. *Poultry Science* 54, 36–40.

Morrison, W.D., Braithwaite, L.A. and Leeson, S. (1988) Report of a survey of poultry heat stress during the summer of 1988. Unpublished report of Department of Animal and Poultry Science, University of Guelph, Ontario, Canada.

Mousi, A. and Onitchi, D.O. (1991) Effects of ascorbic acid supplementation on ejaculated semen characteristics of broiler breeder chickens under hot and humid tropical conditions. *Animal Feed Science and Technology* 34, 141–146.

Muiruri, H.K. and Harrison, P.C. (1991) Effect of roost temperature on performance of chickens in hot ambient environments. *Poultry Science* 70, 2253–2258.

Nasser, A., Wentworth, A. and Wentworth, B.C. (1992) Effect of heat stress on egg quality of broiler breeder hens. *Poultry Science* 71 (Suppl.), 51 (Abstract 151).

National Research Council (NRC) (1994) *Nutrient Requirements of Domestic Animals: Nutrient Requirements of Poultry*, 9th edn. National Academy Press, Washington, DC, pp. 32–34.

North, M.D. and Bell, D. (1990) *Commercial Chicken Production Manual*, 4th edn. Van Nostrand Reinhold, New York, 125 pp.

Obioha, F.C., Okorie, A.U. and Akpa, M.O. (1986) The effect of egg treatment, storage and duration on the hatchability of broiler eggs. *Archiv fur Geflugelkunde* 50, 213–218.

Oluyemi, J.A. and Roberts, F.A. (1979) *Poultry Production in Warm Wet Climates*. Macmillan Press, London, 106 pp.

Oyawoye, E.O. and Krueger, W.F. (1986) The potential of monensin for body weight control and *ad libitum* feeding of broiler breeders from day-old to sexual maturity. *Poultry Science* 65, 884–891.

Pankov, P.N. and Dogadayeva, I. (1988) Standards of feeding battery-caged broiler parent stock. *Proceedings 18th World's Poultry Congress*, pp. 973–975.

Peebles, E.D. and Brake, J. (1985) Relationship of eggshell porosity to stage of embryonic development in broiler breeders. *Poultry Science* 64, 2388–2391.

Peebles, E.D. and Brake, J. (1987) Egg shell quality and hatchability in broiler breeder eggs. *Poultry Science* 66, 596–604.

Pierre, P.E. (1989) Effects of feeding thiouracil on the performance of laying hens under heat stress. *Bulletin of Animal Health in Africa* 37, 379–383.

Proudfoot, F.G. (1964) The effects of plastic packaging and other treatments on hatching eggs. *Canadian Journal of Animal Science* 44, 87–95.

Proudfoot, F.G. (1965) The effect of film permeability and concentration of nitrogen, oxygen and helium gases on hatching eggs stored in polyethelene and cryovac bags. *Poultry Science* 34, 636–644.

Proudfoot, F.G. and Stewart, D.K.R. (1970) Effect of preincubation fumigation with formaldehyde on the hatchability of chicken eggs. *Canadian Journal of Animal Science* 50, 453–465.

Renden, J.A. and Pierson, M.C. (1982) Production of hatching eggs by dwarf broiler breeders maintained in cages or in floor pens. *Poultry Science* 61, 991–993.

Robbins, K.R., Chin, S.F., McGhee, C.G. and Roberson, K.D. (1988) Effects of *ad libitum* versus restricted feeding on body composition and egg production of broiler breeders. *Poultry Science* 67, 1001–1007.

Robinson, F.E. (1993) What really happens when you over-feed broiler breeder hens? *Shaver Focus* 22, 2–5.

Rostagno, H.S. and Sakamoura, N.K. (1992) Environmental temperature effects on feed and ME intake of broiler breeder hens. *Proceedings 19th World's Poultry Congress*, Vol. 2, pp. 113–114.

Spratt, R.S. and Leeson, S. (1987) Broiler breeder performance in response to diet protein and energy. *Poultry Science* 66, 683–693.

Spratt, R.S., Bayley, H.S., McBride, B.W. and Leeson, S. (1990a) Energy metabolism of broiler breeder hens. 1. The partition of dietary energy intake. *Poultry Science* 69, 1339–1347.

Spratt, R.S., McBride, B.W., Bayley, H.S. and Leeson, S. (1990b) Energy metabolism of broiler breeder hens. 2. Contribution of tissues to total heat production in fed and fasted hens. *Poultry Science* 69, 1348–1356.

Subiharta, P.T., Wiloeto, D. and Sabrani, M. (1985) The effect of nest construction on village chicken egg hatchability in rural areas. *Research Report*. Research Institute for Animal Production, Bogor, Indonesia, p. 72.

Surai, P.F. (1992) Vitamin E feeding of poultry males. *Proceedings 19th World's Poultry Congress*, Vol. 1, pp. 575–577.

Taylor, L.W. (1949) *Fertility and Hatchability of Chicken and Turkey Eggs.* John Wiley & Sons, New York.

Thaxton, J.P. and Parkhurst, C.R. (1976) Growth, efficiency and livability of newly hatched broilers as influenced by hydration and intake of sucrose. *Poultry Science* 55, 2275–2279.

Ubosi, C.D. and Azubogu, C.N. (1989) Evaluation of terramycin Q and fishmeal for combating heat stress in poultry production. *Bulletin of Animal Health in Africa* 37, 373–378.

Van Der Hel, W., Verstegen, M.W.A., Henken, A.M. and Brandsma, H.A. (1991) The upper critical ambient temperature in neonatal chicks. *Poultry Science* 70, 1882–1887.

Van Wambeke, F. (1992) The effect of a high protein pre-breeder ration on reproductive performance of broiler breeder hens. *Proceedings 19th World's Poultry Congress,* Vol. 2, pp. 260–261.

Waldroup, P.W., Hazen, K.R., Bussell, W.D. and Johnson, Z.B. (1976) Studies on the daily protein and amino acid needs of broiler breeder hens. *Poultry Science* 55, 2342–2347.

Warren, D.C. and Schnepel, R.L. (1940) The effect of air temperature on egg shell thickness in the fowl. *Poultry Science* 19, 67–72.

Whitehead, C.C., Pearson, A.R. and Herron, K.M. (1985) Biotin requirement of broiler breeders fed diets of different protein content and effect of insufficient biotin on the viability of progeny. *British Poultry Science* 26, 73–82.

Wilson, J.L., McDaniel, G.R. and Sutton, C.D. (1987) Dietary protein levels for breeder males. *Poultry Science* 66, 237–242.

Wilson, J.L., Lupicki, M.E. and Robinson, F.E. (1992) Reproductive performance and carcass characteristics of male broiler breeders fed *ad libitum* or feed restricted from 22 to 30 weeks of age. *Poultry Science* 71 (Suppl.), 50 (Abstract 150).

Yalcin, S. and Hazan, A. (1992) The effect of different strains and cage density on performance of caged broiler breeders. *Proceedings 19th World's Poultry Congress,* Vol. 2, p. 341.

Yang, N. and Shan, C. (1992) Housing broiler breeders in China: cage vs. floor. *Proceedings 19th World's Poultry Congress,* Vol. 2, p. 340.

Yu, M.E., Robinson, F.E., Charles, R.G. and Weingardt, R. (1992) Effect of feed allowance during rearing and breeding on female broiler breeders: 2. Ovarian morphology and production. *Poultry Science* 71, 1750–1761.

Yuan, T., Lien, R.J. and McDaniel, G.R. (1994) Effects of increased rearing period body weight and early photostimulation on broiler breeder egg production. *Poultry Science* 73, 792–800.

Index

abdominal temperature 241-2
acclimatization 36-7, 46, 53, 72
    broilers 207-9
    and indigenous breeds 14
    layers 239-40
    replacement pullets 225-6, 246
N-acetyl-5-methoxytryptamine 47
acetylsalicylic acid 117, 265
acid-base balance 39-40, 73, 117, 118
    broilers 189
    layers 241
ACTH 44-6
activated charcoal 174
adenosine triphosphate 50
adrenal gland 44-6
adrenocorticotrophic hormone 44-6
aflatoxins 158
    in broilers 160-2
    in cottonseed meal 136
    detection and control 173-6
    effect on resistance and
        immunity 163-5
    in eggs and poultry meat 162
    in feedstuffs 159, 160, 166
    in layers and breeders 162
    in peanut meal 137
    susceptibility to 159-60
    and vitamin nutrition 165-6
Africa, poultry industry 3-4

air inlet design 91-6
alarm systems 98
aldosterone 45
Algeria, poultry industry 4
alkalosis 39-40, 73, 112, 114
    in breeders 257
    in broilers 189
    in layers 235, 241
allylisothiocyanate 140
altitude, effect on egg hatchability 283
aluminosilicates 174-5, 244
amino acids
    balance 118, 188
    requirements 106-9
        breeders 268, 269
        broilers 197
        layers 235-6
        replacement pullets 222, 223
    total digestible 109
ammonia 98, 174, 201
ammonium chloride 40, 114, 189-90,
    207, 210
amprolium 199
anaemia 164
anemometer 68, 69, 97
antibiotics
    incorporation into feed 117, 119,
        175, 265
    in salmonellosis 275

anticoccidial drugs, incorporation into feed 117, 198-9
antifungal agents 174
antioxidants in feed 130, 134, 192, 211
Arachis 136-7, 146, 158
arginine 108, 197, 223, 236, 268, 269
arginine vasotocin 42-3
Ascaridia spp. 243
ascorbic acid *see* vitamin C
aspergillosis
    chicks 282
    ducklings 165
Aspergillus spp. 158, 166, 168
    control 175
    peanut meal contamination 137
    screening for 173
aspirin 117, 265
Association of Poultry Instructors and Investigators 2
ATP 50
Australorp, heat tolerance 13
AVT 42-3

β-carotene 111
β-glucans 127, 128
back-crossing 19, 25
bajra 129
bambara groundnut meal 145, 147
Bangladesh, poultry industry 6
barley 126-8, 146, 244
Barred Plymouth Rock, heat tolerance 13
beak trimming
    broilers 200-1
    replacement pullets 226, 246
Bedouin fowl, heat tolerance 53
benne 138-9, 146
Bhutan, poultry industry 6
biological assay, mycotoxins 173
biosecurity
    breeder house 276-7
    broiler house 204, 211
biotin 263
black light test 173
blood pressure, response to heat stress 41-2

body temperature
    abdominal 241-2
    maintenance, and heat stress 32-4
body weight
    broiler breeders 261, 265-6, 268
    broilers 186-7
    replacement pullets 220-1, 245
bone
    effect of ochratoxin on 169
    weakening and breakage 113-14
Brazil, poultry industry 5
breast blister 202
breeder house management 269-77
breeders
    aflatoxicosis in 162
    feeding 258-69
    recommendations 284-6
British thermal unit, definition 68
broiler breeders
    feeding
        hens 261-5
        males 265-6
        pullets 259-61
    overweight 258
    vitamin C supplementation 111
broilers 210-11
    acclimatization 207-9
    aflatoxicosis in 160-2
    beak trimming 200-1
    dewinged 205
    drug administration 198-9
    effect of temperature on body composition 194-6
    effect of temperature on growth and feed consumption 186-7
    energy requirements 187-8
    feed withdrawal 198
    feeding 196-8
    housing 201-5
    industry 185-6
    lighting programmes 209-10
    mineral requirements 189-92
    protein requirements 188-9
    seasonal effects on performance 192-4
    selection for heat tolerance 20-1
    sex-separate feeding 198

Index

vaccination 199–200
vitamin requirements 191–2
water consumption 205–7
bruising
    and aflatoxins 160–1
    and ochratoxins 169
Brunei, poultry industry 6–7
BTU, definition 68
buffalo gourd meal 144, 147
bursa 45, 110
    hypoplasia 164

cages
    breeder house 272–4
    multiple deck system 92, 93
    shape and space 242–3, 247
    single-tier 95
calcium
    in drinking water 206
    plasma concentration 40
    ratio to phosphorus 114, 118, 190
    requirements 112–14, 118
        breeders 264, 268, 279
        broilers 190–1, 197
        layers 233, 236
        replacement pullets 223
Cambodia, poultry industry 6–7
Candida albicans 163
cannibalism 200
Capillaria spp. 243
carbon dioxide 112, 201, 207, 210
carbonic anhydrase 112, 237
cardiac output, response to heat stress 41–2
cardiovascular system 41–2
cassava root meal 141, 147
catecholamines and thermoregulation 46
cereal feedstuffs 126–34
Chaetonium trilaterale 170
chicks
    processing and delivery 283–4, 286
    temperature requirements 68
Chile, poultry industry 5

chloride, plasma concentration 40
chlorine, addition to water 282
chlorogenic acid 138
choline 112
    and feed vitamin stability 192, 193
    requirements 224, 267
    supplementation of sorghum diet 131
citrinin 158, 166–7
climate
    and house design 74
    temperate, poultry breeding in 11–12
coccidiosis 163, 165
coccidiostats, incorporation into feed 117, 198–9
coconut meal 134, 146
Colombia, poultry industry 5
comb
    and heat dissipation 41
    removal 241
    water splashing 35
conduction 34, 70
convection 34, 70, 71
cool-room brooding 68
cooling systems
    artificially ventilated housing 87–9
    broiler houses 202–3
    naturally ventilated housing 83–5
copra 134, 146
corticosteroids 45–6, 191
corticosterone 44–6, 191
corticotrophin-releasing factor 44
cottonseed meal 134–6, 146
CRF 44
critical thermal maximum 33
Cucurbita foetidisima 144, 147
cucurbitacins 144
curtains 82–3
Cyamopsis tetragonaloba 144–5, 147

Dahlem Red
    breeding birds 17
    naked neck strain 18, 19
Dahlem White frizzle strain 18

darkling beetle 79
deoxynivalenol 171-2
dermal melanin gene and heat
    tolerance 17
dihydroergotamine 45
disease control and prevention, breeder
    houses 274-7
disinfectants 204, 211, 270, 272, 282,
    286
domestication of poultry 1
DON 171-2
dried poultry waste (DPW) 143, 147
drinkers 205
droppings, wet 84, 190, 202, 206,
    244, 245
dry-bulb temperature 68
duck farming, Southeast Asia 7
dwarf (*dw*) gene, and heat
    tolerance 16-17, 18, 19

EDTA 130
eggs
    aflatoxin residues in 162
    consumption 2, 5
    discoloration 135
    effect of altitude on
        hatchability 283
    effect of temperature on 73, 112,
        113, 240-1, 256-8, 270, 271
    fumigation/disinfection 271-2
    gathering 269-70, 271
    hatching 277-8
    hygiene 271-2
    prewarming for setting 281-2
    production 2, 3, 4, 106, 107
        effect of temperature on 73
        effect of vitamin C on 110-11
        see also layers
    as protein source 3
    size 236, 257, 281
        effect of temperature on 73
    see also shell; yolk
Egypt, poultry industry 4
emergency systems 98
energy
    balance 31-2

requirements 103-6
    broiler breeder hens 261-2, 269,
        285
    broiler breeder males 265, 269
    broiler breeder pullets 260-1,
        269
    broilers 187-8, 197
    layers 229-34, 236
    replacement pullets 223
environment, operant control of 35-6
epinephrine and thermoregulation 46
Escherichia coli 53
Ethacal 234-5
ethoxyquin 130
ethylenediamine tetra-acetic acid 130
evaporation 71
evaporative cooling 68
evaporative cooling pads
    calculating area required 89-90
    design 90, 91
    evaluating performance 97-8
    maintenance 96-7
    placement 91-6

F gene and heat tolerance 16, 18, 19
fans
    belt-drive 96
    direct-drive 96
    evaluating performance 97-8
    location 91-6
    louvers 90
    maintenance 96
    in naturally ventilated housing 84
    safety guards 91
fat, dietary 73-4, 105-6, 118, 176,
    232-3, 236
fatty acids
    cyclopropene 135
    free 43
    polyunsaturated 195, 211
feathers
    cover 38
    pecking 200
feed
    consumption
        effect of temperature on 37, 73,
            102-3

increasing 105, 241, 246
crumbled 196, 224, 244, 246, 264-5
ingredients 126-47
and metabolic heat production 73-4
mycotoxins in 157-76
non-nutrient additives 116-18, 119, 198-9, 265
pelleted 128, 196, 233-4, 242, 276
'stress' 117
vitamin stability in 192, 193
wet/dry mash 203, 241
feeding
    breeders 258-69, 284-5
    broilers 196-8
    every-day/skip-a-day 260
    intermittent programmes 196-7
    layers 228-36
    phase 235
    replacement pullets 221-4
    sex-separate 198, 267-8
fertility 73, 256-8, 265
fish meal 265
floor
    breeder house 272-4
    structure 36, 76
flunixin 117, 118, 119
fogging systems 84, 85
folic acid 112
formaldehyde 271-2, 286
formalin 270
formic acid 276
foundation, housing 76
fowl cholera 165
frizzle gene and heat tolerance 16, 18, 19
fumonisins 158, 167-8
Fusarium spp. 158, 167-8, 170-2

GH and thermoregulation 44
gingili 138-9, 146
glycosides 145
Gossypium spp. 134-6, 146
gossypol 134-6

groundnut meal 136-7, 146, 158
growth, effect of temperature on 73
growth hormone and thermoregulation 44
guar meal 144-5, 147
gular flutter 39

h gene and heat tolerance 17
hair worms 243
hatchery 256, 285-6
    altitude 283
    design 278
    hygiene 278-9
    water supply 282-3
heart rate, response to heat stress 41-2
heat
    calculation of load in housing 88-9
    methods of loss and gain 70-1
heat exchange mechanisms 37-8
heat-shock cognates 49
heat-shock element 51
heat-shock factor 51
heat-shock proteins 49-53
heat stress
    and air movement 70, 71
    behavioural responses to 34-6
    and maintenance of body temperature 32-4
    and nutrition 101-19
    physiological responses to 36-42
    resistance to *see* heat tolerance
heat tolerance 12-13, 23-5
    development of commercial stocks with 22-3
    experiments on selection for 19-21
    genes affecting 14-19
    population differences in 13-14, 53
    quantitative genes in development of 21
    selection for, under controlled conditions 23
Helianthus annuus 137-8, 146
Hetarakis spp. 243
heterophilia 164

histidine 134
Hong Kong, poultry industry 6–7
Hordeum sativum 126–8, 146, 244
hormones, role in
 thermoregulation 42–9
housefly 243
housing 68
 attic plena 95–6
 broilers 201–5
 closed, fan-ventilated 86–96
 design
  factors affecting 74–5
  related principles 70–4
 evaluating performance 97–8
 flex 96
 high-rise 92, 93–4
 insulation 76–9, 80, 83, 84, 87
 maintenance 96–8
 naturally ventilated 81–6
 replacement pullets 228
 slot-and-fan 92, 94, 95
 structure 75–6
 tunnel 94–5
 vapour barriers 79–81
HSE 51
HSF 51
HSP70 49, 50, 51
HSP90 49, 50
HSP108 52
humidity 11, 19
 in hatcheries 280
 and performance 72, 193
hydrocortisone 45
hydrocyanic acid 141
hydrogen peroxide 272
hydrogen sulphide 201
hyperthermia 39, 190
hypoproteinaemia 164
hypothalamic–pituitary adrenal axis and
 thermoregulation 44–6

id gene and heat tolerance 17
immune system 45–6, 51, 110, 163–5, 169, 171
immunoassay for mycotoxins 173
incubation 256
 in hot climates 279–82
 time 281
incubator 278
 conditions 279–82
 in hot climates 280–1
 Jamesway 282
India, poultry industry 6
indigenous breeds, heat tolerance 25
Indonesia, poultry industry 6–7
infectious bursal disease 110, 165
insects
 damage to insulation materials 79
 as parasites 243
insulation
 artificially ventilated housing 87
 naturally ventilated housing 83, 84
 poultry houses 76–9, 80
ipil-ipil leaf meal 142–3, 147
Iraq, poultry industry 5

jojoba meal 145, 147
Jordan, poultry industry 5
jungle fowl, heat tolerance 53

K gene and heat tolerance 17, 18

Laos, poultry industry 6–7
Lasalocid 199
latent heat 70
Latin America, poultry industry 5
layers
 acclimatization 239–40
 aflatoxicosis in 162
 cage shape and space 242–3
 effect of temperature on egg quality 240–1
 energy requirements 229–34
 feed consumption 228–9
 feeding programmes 235–6
 lighting requirements 241–2
 management 241
 mineral requirements 234–5
 parasites 243–4

protein requirements 229-34
recommendations 246-7
vitamin requirements 234-5
water quality and
consumption 237-9
Lebanon, poultry industry 6
length, conversion between metric and
imperial units 70
Leucaena leucocephala 142-3, 147
leucine 108
LH 47
LHRH 47-8
lice 244
lighting programmes
biomittent 227, 246
broilers 209-10
intermittent 209, 211, 227, 246
layers 241-2
replacement pullets 227, 246
linamarin 139, 141
linase 139
linoleic acid 129, 195, 236
linolenic acid 195
linseed meal 139, 146
litter management 201-2, 275-6
lower critical temperature 32-3, 72
lower lethal temperature 72
luteinizing hormone 47
luteinizing hormone-releasing
hormone 47-8
lymphocytopaenia 164
lysine 104, 105, 230
in coconut meal 134
requirements 106-8
broiler breeders 262, 268, 269
broilers 189, 197
layers 237
replacement pullets 223
supplements 139

magnesium, plasma concentration 40
maize 126, 158
Malaysia, poultry industry 6-7
malvalic acid 135
manioca/manioc meal 141, 147
Marek's disease 165

mating, effect of temperature on 73
melatonin 47
mesotocin 43-4
metabolizable energy *see* energy
methionine 108, 166, 230, 236
requirements
broiler breeders 262, 268, 269
broilers 189, 197
layers 237
replacement pullets 223
in sesame meal 138
supplements 131, 237, 247
Mexico, poultry industry 5
Middle East, poultry industry 5-6
millet 128-9, 146
mimosine 142
minerals
in drinking water 206
requirements 112-16
broiler breeders 264
broilers 189-92
layers 234-5
replacement pullets 224
mites 244
monensin 117, 265
monoamines 47-8
Morocco, poultry industry 4
MT 43-4
Musca domestica 243
mustard seed meal 140, 146
Myanmar, poultry industry 6-7
mycotoxins 157-72
detection and control 172-6
myrosinase 140

naked neck (*Na*) gene, and heat
tolerance 14-16, 18-19, 38
NE and thermoregulation 46
Nepal, poultry industry 6
nest
hygiene 270
type 270-1
neurohypophyseal hormones and
thermoregulation 42-4
Newcastle disease 163-4, 165
nicarbazine 117, 198-9

nicotinic acid 112
Nigeria, poultry industry 4
nitrate 206
norepinephrine and
    thermoregulation 46
nutrient requirements
    breeders 258-69
    broilers 186-92
    layers 228-36
    replacement pullets 221-4
    under heat stress 101-19

ochratoxins 168-9
oleic acid 195
oosporein 169-70
operant control of environment 35-6
Oryza sativa 129-31, 146
ovulatory hormones 47
oxytetracycline 175
oxytocin 43
oyster shell 112, 113, 118, 233
ozone 175

P gene and heat tolerance 17
Pakistan, poultry industry 6
palm kernel meal 142, 147
palm oil 111
panting 38-9, 43, 72, 73, 203, 239
paraformaldehyde 270
parasites, layers 243-4
peacomb gene and heat tolerance 17
peanut meal 136-7, 146, 158
pearl millet 129
Penicillium 158, 166
Pennisetum typhoides 129
perches, cooled 38, 202-3, 274
performance
    effects of temperature on 72
    and humidity 72, 193
    seasonal effects on 192-4
peripheral resistance, response to heat
    stress 41-2
peritonitis 172
Philippines, poultry industry 6-7

phosphate, plasma concentration 40
phosphorus requirements 114-15, 118
    broiler breeders 264, 268, 269
    broilers 190, 197
    layers 136
    replacement pullets 223
phytase 135
phytates 132
phytic acid 139, 145
pineal gland 47
plasma
    changes in ion concentration in
        response to heat stress 40-1
    pH 39, 45
polyphenols 145
potassium
    plasma concentration 40-1
    requirements 115-16, 118
        broiler breeders 269
        broilers 190, 191, 197
        layers 236
        replacement pullets 223
potassium biphosphate 190
potassium chloride 190, 207, 210
poultry industry
    constraints on 8-9
    future development 7-9
    history 1-2
    present status 2-7
poultry meat
    aflatoxin residues in 162
    consumption 2, 5
    production 2, 3, 4
    as protein source 3
power
    costs 75
    failure 209
    source 74
    standby generator 98
prelay rations 235
progesterone 47
propanolol 45
propionic acid 167, 174, 176, 276
protein
    feed supplements 134-41
    requirements 106-9
        broiler breeder hens 262

broiler breeder males 265-6
broiler breeder pullets 260
broilers 188-9
  layers 229-34
  replacement pullets 221-2
  single-cell 140-1, 146
  total digestible 109
psychrometer 68, 69
pullets
  acclimatization 225-6
  beak trimming 226
  body-weight 220-1
  feeding 221-4
  handling 227-8
  housing 228
  light requirements 227
  management 226
  protein requirements 221-2
  recommendations 246-7
  water consumption 225
pyridoxine 112, 139

radiation 34, 70
rape-seed meal 109
red jungle fowl 1
relative humidity 38, 68, 74, 193
replacement pullets *see* pullets
reproductive hormones and thermoregulation 47-8
reproductive performance 47, 256-8
reserpine, incorporation into feed 45, 117, 119
resistance to disease, effect of aflatoxins on 163-5
resistance values (R) 76, 77, 78, 80
respiration, response to heat stress 38-40
rete ophthalmicum 38
reverse triiodothyronine 48
Rhode Island Red, heat tolerance 13
riboflavin 263, 285
rice
  bran 129-30, 146
  by-products as feedstuffs 129-31, 146
  kani 130-1, 146
  polishings 129-30, 146
rodents, damage to insulation materials 79
roof
  cooling in naturally ventilated housing 83
  insulation 202
  reflective coatings/linings 77-9
  structure 75, 87
  whitewashing 203-4
roof vents 75, 83, 85-6
round worms 243

saccharina 264
salmonellosis 163, 165
  prevention and control 274-6
salpingitis 172
salseed 143, 147
sandstorm, protection of housing from 97
Saudi Arabia, poultry industry 5-6
selenium 263
semen
  and diet 265
  quality 263
  storage 258
sensible heat 70
  loss 33, 34, 38
sesame (*Sesamum indicum*) meal 138-9, 146
Setoris spp. 128-9, 146
shade, naturally ventilated housing 86
shell
  quality
    effect of saline drinking water on 237
    effect of temperature on 112, 113, 240-1
  stains 138
  thickness 257-8
Shorea robusta 143, 147
silky gene and heat tolerance 17
Simmoensia chinensis 145, 147
simondsin 145
Singapore, poultry industry 6-7
sinigrin 140

slow feathering gene and heat tolerance 17, 18
sodium
    plasma concentration 40-1
    requirements 264
        broiler breeders 268, 269
        broilers 190-1, 197, 202
        layers 236
        replacement pullets 223
sodium aluminium silicate 244
sodium bicarbonate 112, 114, 190, 241
sodium chloride 190, 237
sodium zeolite 234-5
sorghum (*Sorghum bicolor*) 131-2, 146
South Africa, poultry industry 3-4
South Asia, poultry industry 6
Southeast Asia, poultry industry 6-7
soyabean meal 109, 126, 158
sperm
    abnormalities 263
    seasonal variation in production 257
sprinkling systems 84, 85
Sri Lanka, poultry industry 6
sterculic acid 135
sulphate, plasma concentration 40
sunflower seeds/meal 137-8, 146

T-2 toxin 170-1
tannins 131-2
tapioca 141, 147
teel 138-9, 146
temperate climates, poultry breeding in 11-12
temperature
    conversion between Celsius and Fahrenheit 70
    effect on body composition, broilers 194-6
    effect on egg quality 240-1
    effect on feed consumption
        broilers 186-7
        layers 228-9
        replacement pullets 221
    effect on growth in broilers 186-7
    effect on reproductive performance 256-8
    effect on water consumption
        broilers 205-6
        replacement pullets 225
    ideal 102
    operative 71-2
    and performance 72
    see also body temperature
terramycin Q 265
Thailand, poultry industry 6-7
thermoneutral zone 32, 33, 40, 72
thermoregulation 33, 34, 36, 39
    role of hormones 42-9
thiamine 112, 166
    deficiency 199
threonine 197, 223, 236, 269
thymus 45, 110
    hypoplasia 164
thyroid hormones 43, 48-9
thyroidectomy 48
thyroxine 43, 48
tibial dyschondroplasia 73, 168
til 138-9, 146
trichothecenes 170, 171
triiodothyronine 43, 48
triticale 132-4, 146
trypsin inhibitors 145
tryptophan 197, 223, 236, 268, 269
Turbo house 92, 94
turkey X-disease 158
turkeys, nutrition at high temperatures 116

ubiquitin 50
upper critical temperature 33, 72
upper lethal temperature 72

vaccination 199-200, 274
vapour barriers, poultry housing 79-81
Venezuela, poultry industry 5
ventilation
    artificial 86-96
    evaluating 97-8
    natural 81-6

negative-pressure systems 86
positive-pressure systems 86
vitamins
   requirements 109-12
      in aflatoxicosis 165-6
      broiler breeder hens 263-4
      broilers 191-2
      layers 234-5
      replacement pullets 224
      stability in feed 192, 193, 227
   vitamin A 111, 118, 165, 227, 234, 285
      deficiency 263
   vitamin B_6 112, 139
   vitamin C 109-11, 118, 227, 285
      supplementation
         broiler breeders 263-4
         broilers 191
         layers 234, 246
   vitamin D_3 111-12, 165
   vitamin E 111-12, 227, 263, 285
   vitamin K 165
Voandizeia subterrenea 145, 147
vomitoxin 171-2

'walking the birds' 207
wall structure 76, 87
water
   carbonated 40, 113-14, 189, 238, 241
   consumption
      broilers 205-7
      effect of heat stress on 37, 40
      layers 237-9
      replacement pullets 225
   desalination 237
   saline 234, 237
   supply
      breeder house 272, 273
      hatchery 282-3
wattle
   and heat dissipation 41
   removal 241
   water splashing 35
wet-bulb temperature 68
White Leghorn, heat tolerance 13, 19-21
White Plymouth Rock, heat tolerance 13
whitewash 203-4
World Poultry Congress 2
World Poultry Science Association 2

yeast single-cell protein (YSCP) 140-1
yolk 132, 133, 135, 136, 138
yucca 141, 147

zearalenone 172
zeolites 175, 234-5
zinc 166, 237, 247
zone of hyperthermia 33
zone of hypothermia 33
zone of least thermoregulatory effort 33-4
zone of minimum metabolism 33, 34
zone of normothermia 32, 33, 40, 72